INNOVATION IN FLIGHT:
RESEARCH OF THE NASA LANGLEY RESEARCH CENTER
ON REVOLUTIONARY ADVANCED CONCEPTS FOR AERONAUTICS

By
Joseph R. Chambers

NASA SP-2005-4539

Library of Congress Cataloging-in-Publication Data

Chambers, Joseph R.
  Innovation in flight : research of the NASA Langley Research Center on
revolutionary advanced concepts for aeronautics / by Joseph R. Chambers.
      p. cm.
  Includes bibliographical references and index.
  "NASA SP-2005-4539."
  1. Aeronautics--Technical innovations--United States. 2. Research
aircraft--United States. 3. Langley Research Center--Research. I.
United States. National Aeronautics and Space Administration. History
Office. II. Title.
  TL521.312.C44 2005
  629.130072'073--dc22

                                        2005026776

ACKNOWLEDGMENTS

I am sincerely indebted to the dozens of current and retired employees of NASA who consented to be interviewed and submitted their personal experiences, recollections, and files from which this documentation of Langley contributions was drawn. The following active and retired Langley personnel contributed vital information to this effort:

Donald D. Baals	Louis J. Glaab	Vivek Mukhopadhyay
Daniel G. Baize	Sue B. Grafton	Daniel G. Murri
E. Ann Bare	Edward A. Haering, Jr. (NASA	Thomas A. Ozoroski
J.F. Barthelemy	Dryden)	Arthur E. Phelps, III
Dennis W. Bartlett	Andrew S. Hahn	Dhanvada M. Rao
Steven X. S. Bauer	Roy V. Harris	Richard J. Re
Bobby L. Berrier	Jennifer Heeg	Wilmer H. Reed, III
Jay Brandon	Jerry N. Hefner	Rodney H. Ricketts
Albert J. Braslow	William P. Henderson	A. Warner Robins
Dennis M. Bushnell	Paul M. Hergenrother	Gautam H. Shah
Richard L. Campbell	Joseph L. Johnson, Jr.	William J. Small
Peter G. Coen	Denise R. Jones	Stephen C. Smith (NASA Ames)
Fayette S. Collier	Donald F. Keller	M. Leroy Spearman
Robert. V. Doggett, Jr.	Lynda J. Kramer	Eric C. Stewart
Samuel M. Dollyhigh	Steven E. Krist	Dan D. Vicroy
Cornelius Driver	John E. Lamar	Richard D. Wagner
Clinton V. Eckstrom	Robert E. McKinley	Richard A. Wahls
James R. Elliott	Domenic J. Maglieri	Richard M. Wood
Michael C. Fischer	Mark D. Moore	Jeffrey A. Yetter
Charles M. Fremaux	Robert W. Moses	Long P. Yip
Neal T. Frink	Thomas M. Moul	Steven D. Young

I would like to express my special gratitude to Noel A. Talcott and Dr. Darrel R. Tenney, who provided the inspiration and mechanism to undertake this activity.

The efforts of the following active and retired industry members in reviewing the material significantly enhanced the accuracy and content of the final product: The Boeing Company: A. L. "Del" Nagel, Rudy N. Yurkovich, Eric Y. Reichenbach, M. Emmett Omar, Michael J. Janisko, Robert A. Woodling, D. L. "Tony" Antani, Norbert F. Smith, Daniel Smith, Arthur G. Powell, Robert H. Liebeck, and William E. Vargo. Lockheed-Martin Corporation: Barry Creighton. Gulfstream Aerospace: Michael P. Mena. Northrop Grumman Corporation: Heinz A. Gerhardt, Dave H. Graham, Dale J. Lorincz and Vince Wisniewski.

Thanks also to Percival J. Tesoro for the book design, Alicia V. Tarrant and Jeffrey B. Caplan for outstanding assistance in photographic research in the Langley files, Denise M. Stefula and Kay V. Forrest for editing, and Christine R. Williams for printing coordination.

Special thanks and gratitude go to Dennis M. Bushnell, Senior Scientist of the NASA Langley Research Center, for his support and comments in the preparation of this document. His personal example and dedication to the pursuit of revolutionary technology have served as the stimulus and encouragement sought by thousands of his peers during his outstanding and remarkable NASA career.

Joseph R. Chambers
Yorktown, VA
August 22, 2005

CONTENTS

PREFACE

This document is intended to be a companion to previous books by the author: NASA SP-2000-4519, Partners in Freedom: Contributions of the Langley Research Center to U.S. Military Aircraft of the 1990's, and NASA SP-2003-4529, Concept to Reality: Contributions of the Langley Research Center to U.S. Civil Aircraft of the 1990's. Material included in the previous volumes provides informative and significant examples of the impact of applications of aeronautics research conducted by the NASA Langley Research Center on important U.S. civil and military aircraft of the 1990s. These contributions occurred because of the investment of the Nation in the innovation, expertise, and dedication of a staff of researchers and their unique facilities at Langley. Within that research environment, literally thousands of revolutionary concepts and advanced technologies for aeronautics have emerged, directed at challenges and barriers that impede the advancement of the state of the art in aircraft design and mission capabilities. Unfortunately, in the world of technology only a handful of advanced concepts are ever applied, due to a number of reasons. Factors that inhibit the application of advanced research technology are numerous and varied in nature, including cost, environmental impact, safety, complexity, reduced or inadequate funding and human resources, world events, perceived or actual risk, technical barriers, and others.

The objective of this publication is to discuss the importance of innovation and the role of revolutionary advanced concepts within the aeronautics research community, and to provide information on typical advanced research projects conducted by Langley and its partners on topics that have not yet been applied by the military or civil aviation industry to production aircraft. Detailed information is first provided to describe each advanced concept, the projected benefits that could be provided if the technology is applied, and the challenges faced by the NASA research team to reduce the risk of application. Next, descriptions of specific research activities on the concepts identify the key projects, accomplishments, personnel, and facilities involved in the development of each concept. Finally, perspectives are provided on the current status of the subject concepts, including discussions of factors or future events that might intensify interest in their use for future applications. Many of the concepts described herein are subjects of ongoing NASA research thrusts, for which significant technical challenges are in the process of being addressed. Some of the research activities discussed were conducted and completed in past NASA projects; however, evolving requirements for military or civil aircraft systems demand a reexamination of the potential and current feasibility of the principles involved.

This document is intended to serve several purposes. As a source of collated information on revolutionary concepts, it will serve as a key reference for readers wishing to grasp the underlying principles and challenges related to specific revolutionary concepts. Hopefully, such information will provide valuable background that can serve as starting knowledge bases for future research efforts and minimize the so-called "reinvention of the wheel" syndrome. More importantly, the information identifies major obstacles to advanced aeronautics technology, thereby providing a sensitivity for multi-faceted research projects to ensure a higher likelihood of application. A definition of current barriers to application is extremely valuable for use in the future, when new breakthroughs in various technical disciplines may eliminate or minimize some of the critical barriers that have traditionally blocked the application of some of these specific revolutionary concepts. Finally, a review of the material will hopefully inspire the nontechnical (as well as technical) communities that aeronautics is not a "mature science" and that considerable opportunities exist to revolutionize the future.

The written material has been prepared for a broad audience and does not presume any significant technical expertise. Hopefully, it will provide informative and interesting overviews for researchers engaged in aeronautics activities, internal NASA policy makers, national policy makers, NASA stakeholders, the media and the general public. A bibliography is provided for technical specialists and others who desire a more in-depth discussion of the concepts.

INNOVATION IN FLIGHT:
RESEARCH OF THE NASA LANGLEY RESEARCH CENTER
ON REVOLUTIONARY ADVANCED CONCEPTS FOR AERONAUTICS

By
Joseph R. Chambers

Innovation: The Seed Corn of Tomorrow

*"The pointy end of NASA's technological spear must stretch beyond "good work and evolutionary ideas"
to revolutionary, high-payoff concepts—with a sensitivity to the barriers that limit application."*

The foregoing statement by Dennis M. Bushnell, Senior Scientist of the NASA Langley Research Center, captures what many believe to be the critical strategy in maintaining a superior and vibrant aeronautical research capability for the United States. In his 42-year career at Langley, Bushnell has authored more than 240 technical papers on an impressive variety of technical subjects and presented over 260 invited lectures and seminars at numerous international meetings. Widely regarded as an international leader for his personal contributions to innovative research and his futuristic perspectives in science and technology, he is an outspoken advocate for creativity and the pursuit of "stretch goals" that challenge NASA's aeronautics program. Bushnell's perseverance and actions have nurtured one of NASA's key missions in aeronautics—the conception and maturation of breakthrough, revolutionary technologies. By definition, revolutionary technologies—such as the swept wing and jet propulsion—can radically change the very nature of aeronautical technology, resulting in unprecedented levels of capability, rather than incremental benefits.

*Dennis M. Bushnell, Senior Scientist of the
NASA Langley Research Center.*

In today's world, where the public is generally unaware of NASA's activities and contributions in aeronautics, the subject of innovative, revolutionary NASA research is even more unknown. Previous NASA publications, such as NASA SP-2000-4519, Partners in Freedom: Contributions of the Langley Research Center to U.S. Military Aircraft of the 1990's, and NASA SP-2003-4529, Concept to Reality: Contributions of the Langley Research Center to U.S. Civil Aircraft of the 1990's, document some of the more important contributions of NASA that have been applied to modern military and civil aircraft. However, thousands of revolutionary, high-risk, high-payoff research projects have not yet resulted in applications for various reasons, including technical risk, economics, and environmental impact.

The goal of this publication is to provide an overview of the topic of revolutionary research in aeronautics at Langley, including many examples of research efforts that offer significant potential benefits, but have not yet been applied. The discussion also includes an overview of how innovation and creativity is stimulated within the Center, and a perspective on the future of innovation. The documentation of this topic, especially the scope and experiences of the example research activities covered, is intended to provide background information for future researchers. By nature, the technical interests of the aircraft industry—and by necessity those of NASA—are highly cyclical and often shaped by external factors. For instance, laminar flow control and powered lift tend to reemerge in priority every two decades or so. Being able to go back and review what has been done in the past is quite valuable. Hopefully, reviewing the extent of specific past projects will identify appropriate points of departure to advance these concepts to future applications. With this background, repetitive or unproductive research options might be filtered at an early stage in order to avoid so-called "reinvention of the wheel". In addition, discussions of specific technical challenges that blocked the application of the example concepts will prove to be valuable in the future, if breakthroughs in enabling technologies result in removal of the past barriers. Examples of such occurrences include advances in composite fabrication technology permitting the previously unacceptable application of forward-swept wing configurations, and the introduction of turbofan engines (with relatively cool exhaust flow) permitting the use of externally-blown flaps for high-lift configurations.

Within an aeronautical "research and development" community, such as the Langley Research Center, the knowledgeable observer can distinguish two distinct thrusts of activity, both of which provide critical advances in the leadership of this Nation in aerospace technology and its end products. In a simplistic view, the "development" efforts tend to address evolutionary opportunities, with an emphasis on providing incremental improvements in capabilities, solving known or unexpected multidiscipline systems-level problems that arise, ensuring that the technology readiness level is sufficiently high, and that risk has been lowered to an acceptable level for applications. At the other end of the spectrum, "research" efforts address revolutionary, breakthrough concepts within disciplines or at the integrated configuration level. These activities are typically high risk, involve radical departures from conventional technology, and are capable of providing revolutionary benefits that can change the conventional paradigm. Such research is typically viewed with skepticism by many, and numerous challenges and barriers must be addressed before the development stage can be reached. What is appreciated, however, is that innovative research is the "seed corn" that provides tomorrow's advances in aeronautics.

NASA research centers are populated with inquisitive, highly capable research staffs that can, with appropriate resources, facilities and stimulation, lead the world in innovative research. Virtually every research professional enters his or her first work duties at these locations with innate curiosity and interest in advancing technology in selected technical disciplines. The attributes of an innovative person include not only the technical prowess and expertise to accomplish revolutionary breakthroughs, but also an appreciation of the current state of the art, past research efforts, and barriers to successful applications.

The environment available to the researcher within NASA has provided resource opportunities to fund the cost of conducting research studies, including the use of unique wind tunnels, laboratories, computational facilities, and flight testing. With the encouragement and approval of management

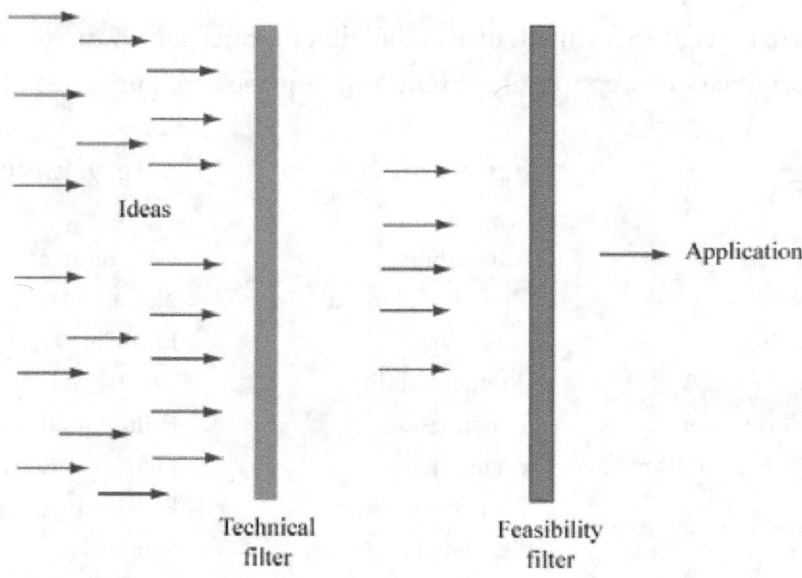

Filters that limit the ultimate application of advanced technology.

and technical peers, a continuous stream of revolutionary concepts has come from the imaginative minds of Langley's staff for potential transition to aeronautical applications. Unfortunately, the path from idea or concept to application is mined with severe challenges and barriers, and only a handful of thousands of advanced concepts survive to emerge as applications, as depicted in the accompanying sketch.

Initially, the new idea must meet and satisfy technical requirements for its intended usage. To pass through this first filter, the researcher is challenged to prove that the concept will work as envisioned, and that the technical benefit can be realized. Some research ideas at Langley have passed this level of maturity in relatively quick order, but others have required decades of frustrating attempts to reach technical closure, requiring extraordinary dedication and efforts by researchers.

When the technical/scientific filter has been successfully met by new technology, a second, even more daunting filter must be faced—a determination of the "real-world" feasibility of the concept. Embedded in the feasibility issue are assessments of the technical readiness level, cost/benefit trades, environmental impact, safety, market demand, risk, and numerous economic issues. Such issues typically pose a significant challenge to NASA researchers, who have limited experience in the profit-making stimuli of the business world, and the highly proprietary methods used by industry to evaluate and calibrate the worthiness of new technology. Nonetheless, appropriate systems-level studies or teaming arrangements with industry or others with sufficient qualifications to predict these factors must be undertaken to determine the application value of the new concept.

Bushnell refers to the challenges in this filter as the "ilities", which for advanced aeronautical concepts (individual disciplinary concepts or unconventional airplane configurations) can include:

Engineering	Economics	Safety/Environmental
Producability	Affordability	Environmental (icing, etc.)
Maintainability	Operational profit	Noise
Reliability	Fuel usage	Emissions
Inspectability	Product liability	Robustness
Capability (performance)	Timeliness	Failure modes
Flexibility (growth capability)	Exclusivity	Flying qualities
Repairability	Regulatory issues	Ride quality (turbulence)
Operability	Certification	Contrails
Durability	Risk	Sonic boom
Compatibility (airport)	Competitive status	Structural integrity
	Resource availability	Emergency egress

Obviously, the tasks facing the researcher and his/her team for applications of new concepts involve many, many potential "show stoppers" that demand attention and solution. Despite the scope of assessments required, NASA has successfully contributed critical technology to the Nation's aircraft military and civil aircraft fleets.

The remainder of this document is designed to provide an overview of some of the revolutionary technology concepts that have emerged from Langley studies, but have not yet been applied because of certain factors referred to in the previous discussion. Following these examples, a final section describes the efforts that have taken place at Langley to preserve and stimulate its reputation for the conception and development of innovative aeronautical technologies.

Supersonic Civil Aircraft: The Need for Speed

Concept and Benefits

Since the first days of commercial airliners, passengers have always placed high priorities on speed, cost, safety, reliability, and comfort in air transportation. Travel speed, and its beneficial impact on personal mobility and business interactions, has been one of the more dominant factors for the traveler. In the late 1950s, as faster swept-wing jet transports replaced their propeller-driven ancestors for public transportation, and the military began to pursue large supersonic bomber designs such as the XB-70, the aviation industry and the government began serious efforts to develop the next logical progression in the quest for speed: a supersonic commercial transport. If an economically feasible, environmentally friendly supersonic transport (SST) could be introduced into the air transportation system, the benefits of significantly increased cruise speed would attract a large segment of the passenger market (especially in the business sector), possibly driving conventional subsonic transports from a large portion of the international air transport marketplace.

The subsequent international rush to SST applications resulted in abortive programs. An ill-fated national effort within the United States for a U.S. SST was terminated in 1971 without a viable aircraft, a brief and unsuccessful introduction of the Russian Tu-144 SST occurred, and the commercial introduction of the French/British Concorde proved to be an impressive technical success but a hopeless economic failure. Nonetheless, the quest for supersonic civil

Supersonic transports offer revolutionary benefits in travel time and personal productivity.

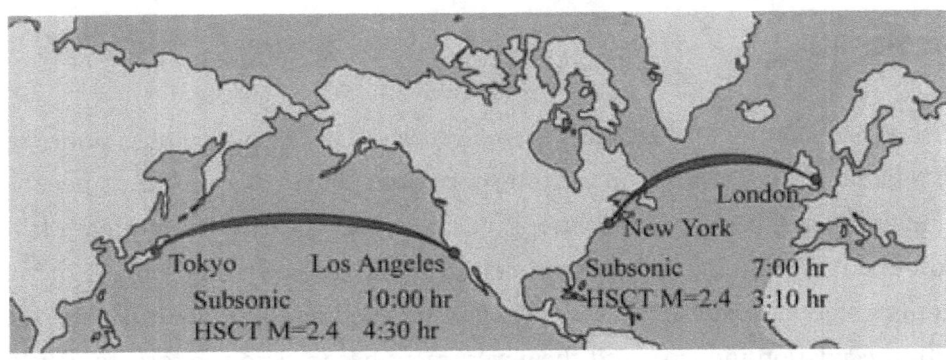

Comparison of international travel times for a conventional subsonic transport and an advanced Mach 2.4 high-speed civil transport.

aircraft has sporadically persisted. In the 1990s, visions of second-generation supersonic transports that could fly 300 passengers at more than 1,600 mph (Mach 2.4) were the focus of extensive research efforts involving U.S. industry and government, as well as Europe and Japan. Such an airplane could speed across the Pacific from Los Angeles to Tokyo in a mere 4.5 hours, revolutionizing air travel and, perhaps, conventional business paradigms. Unfortunately, a multitude of technical and nontechnical obstacles has prevented supersonic air travel and its potential benefits.

Challenges and Barriers

Obstacles to the introduction of supersonic commercial transports are arguably the most demanding of any aircraft type. In addition to a formidable array of technical challenges within virtually all critical aircraft engineering disciplines, a multitude of issues involving environmental protection and international politics must be addressed and resolved. Although research agencies such as NASA are uniquely qualified to help provide the technology necessary to meet mission requirements for SST operations, the resolution of many barriers to implementation, such as political and emotional issues, are far beyond the capabilities of researchers. The enormous difficulty in introducing viable supersonic civil aircraft into the commercial fleet is highlighted by the fact that, although the world has seen almost continuous, yearly introduction of new subsonic jet transport aircraft into service over the last 45 years, no new supersonic civil aircraft have entered operations in the last 30 years.

The following discussion of technical, environmental, economic, and political issues provides a background of some of the more critical factors preventing operational application of supersonic civil aircraft. Many of these issues have been identified and resolved by NASA and its partners, but many important issues continue to remain unresolved, preventing the deployment of a U.S. SST.

Technical Issues

The most fundamental technical challenges to an SST configuration involve the ability to meet the mission requirements in an efficient, economically viable, environmentally acceptable fashion. At the forefront of these requirements is the need for efficient aerodynamic performance at supersonic speeds. Geometrical shaping of the airframe, efficient performance of the propulsion system at subsonic, transonic, and supersonic speeds, and airframe/propulsion integration must be accomplished in a manner to provide sufficient lift/drag (L/D) ratio for efficient supersonic and subsonic cruise, as well as during takeoff and landing. At supersonic speeds, the dominant aerodynamic challenge is drag reduction. Whereas current subsonic jet transports can exhibit L/D ratios approaching 19 at cruise conditions, the most efficient supersonic transport designs today exhibit L/D ratios of less than 10, indicative of the tremendous drag increase associated with additional wave drag created in supersonic flight. With relatively low L/D values, the supersonic transport will require significantly more fuel per minute at cruise conditions. Obviously, supersonic aerodynamic design methodology is a mandatory part of the designer's tool kit for this class of aircraft.

Stability and control requirements for supersonic transports employing highly swept wings can result in configuration compromises and trade-offs that markedly decrease supersonic aerodynamic performance. For example, the low-speed longitudinal stability and control characteristics of supersonically efficient, highly swept arrow wings are usually unacceptable. In particular, aerodynamic flow phenomena over such wings at low-speed conditions result in flow separation on the outer wing at moderate angles of attack. This separation leads to the intolerable longitudinal instability known as "pitch up" and also to marginal ability to control lateral motions, particularly in sideslip conditions during approach to landing. As a result, the nearly optimum arrow-wing planform may have to be modified to a less efficient supersonic shape (approaching a delta wing) to ensure satisfactory flying characteristics at low speeds.

The aerodynamic performance of an SST configuration at off-design conditions, such as subsonic cruise or during extended air traffic landing delays or diversions to other airports, is a key factor in the fuel and weight required for mission capability. Because highly swept configurations exhibit aerodynamic and propulsive efficiencies that are much less than conventional transport designs at subsonic speeds, a large fuel reserve requirement must be met for contingencies during normal operations. In fact, for typical supersonic transport configurations the weight of fuel reserves can be equal to or greater than the useful payload. As an example of the significance of this issue, the Concorde exhibited an L/D ratio of only about 4 at subsonic loiter and approach conditions, resulting in heavy fuel usage, poor engine efficiency, and high noise levels. On flights from Europe, the airplane arrived at New York with 15 tons of spare fuel (1.5 times the payload), and an additional 30 tons of mission fuel was needed to carry the 15 tons of spare fuel.

Integrating airframe and propulsion components into the SST configuration is especially challenging. In addition to maximizing aerodynamic and propulsive performance, the designer must also minimize weight to achieve adequate payload capability. Representative supersonic transports designed for Mach 2.0 carry a payload of only about 6 percent of takeoff gross weight compared with subsonic transports that typically accommodate about a 25-percent payload fraction. Thus, aircraft empty weight (and weight-saving concepts) becomes especially critical for supersonic vehicles.

In the area of aircraft materials, the supersonic transport may have to incorporate exotic and expensive fabrication techniques and materials. In this regard, selection of design cruise Mach number may dictate a departure from conventional structures. Conventional aluminum materials are not able to withstand the withering temperatures at cruise speeds much above Mach 2.5, requiring the use of stainless steel or titanium for the airframe. Resulting cost associated with these advanced materials has been a major factor in limiting design cruise speeds, even though some supersonic airplane productivity studies show maximum benefits near Mach 3.0.

Along with strenuous aerodynamic and structural design challenges, the supersonic transport must meet demanding propulsion requirements. Efficient engine operations over the subsonic and supersonic flight envelopes must be attained, engine inlet and nozzle configurations must be efficient and robust, and engine components and subsystems must be capable of extended high-temperature operations with minimal maintenance and maximum reliability. Efficient engine cycles for supersonic cruise typically lead to low bypass ratio engine configurations in contrast to the high bypass ratio designs used by subsonic transports. The low bypass ratio engine produces a high velocity jet efflux that creates extremely high noise levels if not attenuated by noise suppressors.

Operational field length requirements impose severe constraints on engine sizing and attendant issues, such as takeoff noise, for supersonic cruise vehicles. The inherent low-lift, high-drag characteristics of highly swept supersonic wings may result in long takeoff runs, driving designers to higher thrust engines that are mismatched for optimum cruise applications. In addition, higher engine settings may exacerbate takeoff noise levels and affect the ability to comply with noise regulations. Extremely large, heavy engine noise suppressors may be required, further aggravating the inherent weight issues previously noted. Takeoff and airport community noise is one of the most difficult barriers to implementation of future supersonic civil transports, especially with advanced subsonic transport aircraft having successfully met dramatic reductions in noise levels required by new regulations.

An aircraft traveling through the atmosphere at supersonic speeds (above about 660 mph) continuously produces air-pressure waves similar to waves created by a ship's bow. When supersonic flight became a reality around 1950, the impact of these waves and the accompanying sonic boom was unexpected. Aerodynamicists were aware of the shock waves associated with supersonic motion, but they did not expect these shock waves to reach the ground for high flying aircraft. However, when supersonic fighters were introduced in the 1950s, people were startled by the booms and some buildings and structures were damaged by low-flying supersonic aircraft. As military aircraft increased their supersonic missions over populated areas, complaint and damage claim numbers grew.

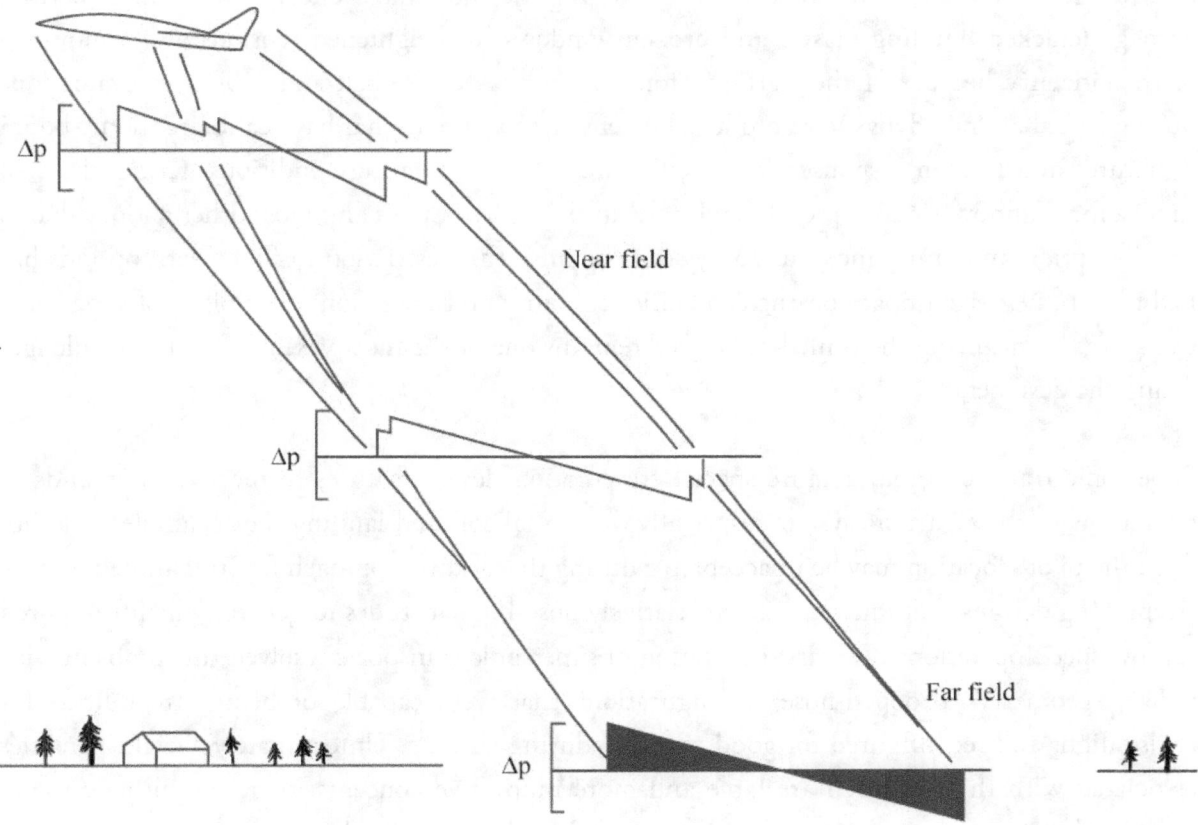

Pressure fields created by supersonic shock waves result in sonic booms.

A typical airplane generates two main shock waves, one at the nose (bow shock) and one off the tail (tail shock). Local shock waves coming off the canopy, wing leading edges, and engine nacelles tend to merge with the main shocks at some distance from the airplane. The resulting pressure pulse is a characteristic "N" shape. To an observer on the ground, this N pulse is felt as an abrupt compression above atmospheric pressure, followed by a rapid decompression below atmospheric pressure and a final recompression to atmospheric pressure. The total change takes place in less than half of a second (about 0.23 seconds for Concorde) and is felt and heard as a double jolt or boom. The relative strength of the boom overpressure is dependent on several factors, including the altitude, speed, length, shape, and weight of the generating aircraft as well as atmospheric conditions.

The strongest sonic boom is usually felt directly beneath the airplane and decreases on either side of the flight path. A turning, or accelerating, supersonic airplane may concentrate the set of shock waves locally where they intersect the ground and produce a focused "super boom." For example, the SR-71 airplane, maneuvering at Mach 3.0 and an altitude of 80,000 ft, creates window-rattling, double-crack sonic boom signatures on the ground. Human subjects have strongly opposed unexpected encounters with sonic booms. The highly undesirable effects run from structural damage (cracked building plaster and broken windows) to heightened tensions and annoyance of the citizenry because of the startle factor. Ground-based research and subjective evaluations during actual aircraft flybys have produced extensive sets of data that have calibrated sonic boom signatures and human responses for specific aircraft and operating conditions. Currently, civil supersonic flight over land is prohibited by law in the United States and most other nations due to the disruption and annoyance caused by sonic booms. This restriction in flight path options has had a deep, negative impact on the economic feasibility and operational flexibility of supersonic transports, and solving the sonic boom issue remains one of the most vexing technical challenges facing the designer.

Supersonic transports may require special crew-station design features to meet the demands of operations in the air traffic system, especially during takeoff and landing. For example, visibility from the pilot's location may be unacceptable during the landing approach for streamlined, highly swept wing designs that must operate at relatively nose-high attitudes to generate the lift required for low-speed operations. The first generation of supersonic transports resolved this problem with variable-geometry "drooped-nose" configurations that were capable of being streamlined for cruise flight and reconfigured for good visibility during landing. Unfortunately, weight penalties associated with this approach are large and more innovative concepts, such as synthetic vision (discussed in a separate section of this document), have been explored to minimize this penalty.

The Tu-144 supersonic transport used a variable-geometry drooped nose for improved visibility during low-speed operations.

The unique, slender shaping of the supersonic transport creates flying quality issues and control system considerations radically different from conventional subsonic transports. For example, the slender shape of such airplanes may result in roll response and lateral-directional handling qualities that differ substantially from those of conventional subsonic transports. Slender aircraft also require certain unconventional augmentation systems in the control system design. Pilot cockpit displays may require additional information beyond that used for subsonic flight, especially for the sensitive cruise flight conditions. In addition, the efficient and safe integration of SST aircraft in the airport terminal area requires analysis of the impact of the speed differential between supersonic and subsonic configurations during loiter, approach, and landing.

Other technical issues posing special challenges for supersonic commercial aircraft are concerns over potential hazardous effects on aircrew and passengers due to radiation exposure at the high altitudes (over 55,000 ft) appropriate for supersonic cruise and the potential harmful impacts of supersonic transport fleet emissions on the Earth's ozone layers. Depletion of significant amounts of upper atmosphere ozone could result in an increase in skin cancer incidences on Earth. These environmental challenges have been prominent research topics in all supersonic transport studies.

Nontechnical Issues

In addition to the foregoing technical challenges, commercial introduction of supersonic transport configurations must address and resolve certain nontechnical hurdles involving economics, industry market projections and investment strategies, and actual or perceived risk. Foremost among these concerns is the level of surcharge for passengers, which is above that associated with typical subsonic jet travel. The fact that supersonic air travel will cost the passenger more than comparable subsonic fares is accepted as a primary characteristic of this air transportation mode. For example, the cost to fly as a passenger on the Concorde on a typical transatlantic trip was 10 times (1,000 percent) more than fares available on some subsonic transports. However, the passenger base upon which revenues will depend for this mode of travel will probably consist of business people in need of rapid business interactions, or the wealthy in need of exclusive and unique travel experiences. The operational experiences with the Concorde fleet revealed that 80 percent of its passengers were business travelers and close to 80 percent were repeat travelers. In contrast to the business traveler, however, the general public regards air travel as a commodity, readily willing to sacrifice in-flight cruise speed (with surcharges) for lower ticket prices offered by airlines flying large wide-bodied subsonic transports. The level of surcharge and the public's willingness to accept it are, therefore, key factors determining the ultimate economic feasibility of supersonic transportation.

Business strategies and industry market projections play an important role in the potential insertion of supersonic transports in the international air transportation system. Competing sectors of the market, such as extremely large subsonic transports (over 500 passengers), very efficient long-range subsonic transports, and other travel options play key roles in the willingness of industry to invest in supersonic technology. Evolving technologies in other fields, such as telecommunications and virtual, computer-based meetings, also have large impacts on the demand for business travel, the prime market for supersonic transports.

Arguably, one of the most powerful negative influences affecting supersonic transports is the continuing worldwide rise in aviation fuel prices. Airline concerns over fuel costs have consistently resulted in abrupt rejection of near-transonic or supersonic cruise capability in new transport aircraft. Instead, in times of fuel crises airlines seek fuel-efficient aircraft and a willingness to reject high cruise speeds if necessary in order to remain economically viable.

Finally, the unprecedented costs and risks associated with developing and certifying a revolutionary new product line like a supersonic transport could easily result in catastrophic consequences for the aviation industry and business investors. This factor—combined with the disappointing experiences with Concorde and the Tu-144, and an appreciation of the high technical and political risks associated with developing a supersonic transport—has resulted in a pessimistic, disinterested atmosphere within the aviation community except for the business jet sector.

Langley Activities

Background

No other undertaking in the aeronautics research activities of the National Aeronautics and Space Administration (NASA), or its predecessor, the National Advisory Committee for Aeronautics (NACA), approaches the magnitude of human and monetary resources expended on the conception, development, and assessment of supersonic civil airplane configurations. Thousands of researchers—contributing expertise from a wide variety of technical disciplines and representing all NASA (and NACA) research centers, NASA Headquarters, and NASA's industry, military, and university partners—have participated for over 40 years in the quest for economically and environmentally feasible supersonic airplanes. The scope and details of this huge research effort are far beyond the intent of this book. Even a constrained attempt to identify key individuals and events within Langley will undoubtedly result in unintentional omissions of central figures and events. The following discussion is an attempt to collate and provide a high-level summary highlighting some Langley roles in this monumental endeavor. Hopefully, the casual reader will understand and appreciate the challenges and barriers unique to applying the technology and the role that Langley has played in attempting to resolve these issues.

Excellent additional sources suitable for the technical specialist and others interested in the details of domestic and international activities in this area are readily available in the literature. Much of the NASA formal document base has now been declassified, but the results of some programs are not in the open literature at this time. In addition to the extensive technical papers and reports of NASA and its contractors, excellent overviews of the international supersonic transport experiences are available. Two publications are particularly recommended for further information and perspectives. The technical summary *Supersonic Cruise Technology* (NASA SP-472) by F. Edward McLean provides an in-depth review of the advancement in technology and NASA's role in supersonic transport technology through the early 1980s. In *High Speed Dreams*, Erik M. Conway constructs an insightful history that focuses primarily on political and commercial factors responsible for the rise and fall of American supersonic transport research programs. Material from both publications has been liberally used in this document.

Chronology of Langley Involvement

The research efforts of Langley Research Center in supersonic technology involve five distinct phases. The first phase of research, during the NACA era, began in the mid-1930s and lasted until the NACA was absorbed by NASA in 1958. During those years, the fundamental understanding of supersonic flight was developed and refined, the aerodynamics of basic shapes and aircraft configurations were explored, experimental methods and facilities were developed, and manned supersonic flight was demonstrated with several NACA and military research aircraft. Results of this pioneering NACA research from Langley on supersonic technology and design methodology were subsequently used by the U.S. industry and the Department of Defense (DoD) in designing the famous military "Century-Series" fighters and other high-speed vehicles of the 1950s.

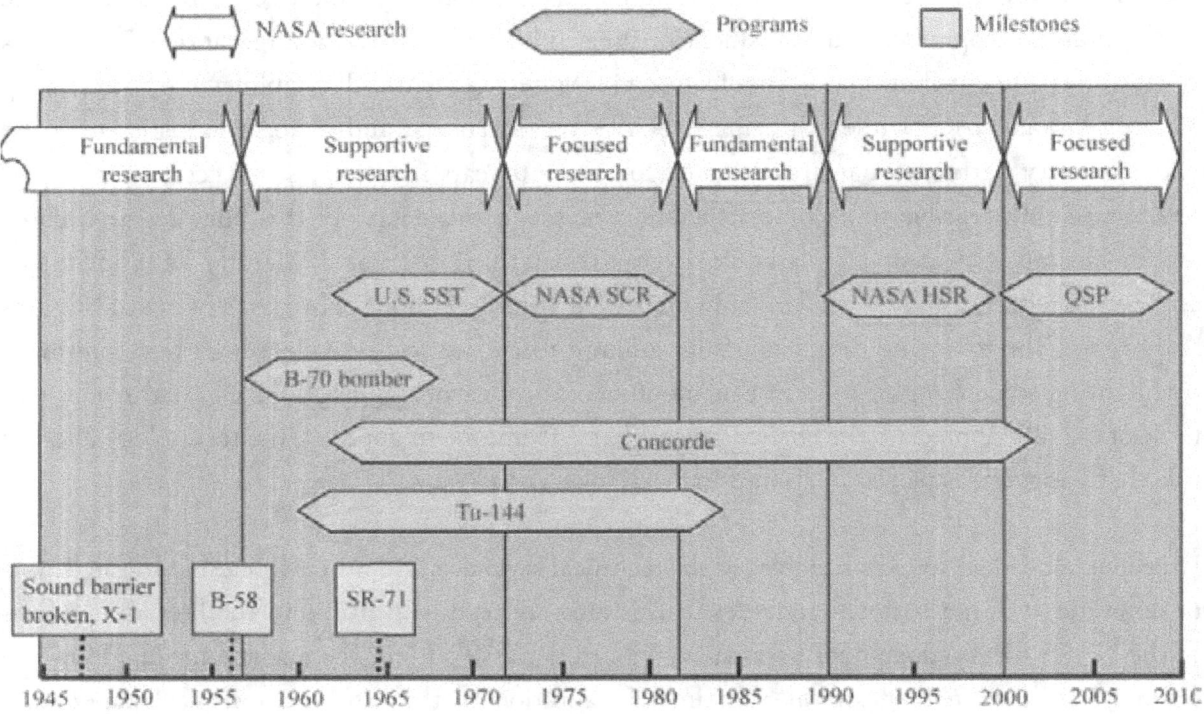

Chronology of supersonic research at NASA Langley Research Center.

In the next phase of U.S. supersonic research, from 1958 to 1971, Langley became a major participant in support of the development of two high-priority national projects: the supersonic cruise XB-70 bomber and the SST. In those years, supersonic aerodynamic design methods were refined, critical phenomena such as the sonic boom became research topics, and efficient supersonic-cruise configurations emerged.

When the SST Program was terminated by Congress in 1971, NASA continued to seek out solutions to the technical issues of supersonic flight in a third phase of supersonic research, known as the NASA Supersonic Cruise Research (SCR) Program, which was led by Langley and conducted between 1971 and 1981. In the SCR Program, aggressive technical advances were made in the areas of propulsion, noise, and takeoff and landing performance. Unprecedented accomplishments in the development of unique materials and fabrication processes for supersonic-cruise applications were also contributed. Following the SCR Program's termination in 1982, NASA's supersonic research efforts declined significantly.

During the Reagan administration of the late 1980s, growing domestic and international enthusiasm over the potential benefits of hypersonic flight and the X-30 (National Aerospace Plane) Program spun off a renewed interest in high-speed flight. This interest resulted in a fourth major phase of NASA's focused supersonic research, known as the NASA High-Speed Research (HSR) Program, which was managed by Langley for the Agency. Extensive NASA/industry research and development activities in the HSR Program began in 1989 and ended in 1999.

The fifth (current) phase of NASA's supersonic research and development program began following the termination of the HSR Program in 1999. After the program's cancellation, interest in and funding for commercial supersonic transports plummeted. However, sporadic interest in economically feasible supersonic business jets began to surface within industry, and collaborative efforts involving systems-level analysis and a few key technical disciplines (such as noise and sonic boom) were established within NASA. Several of these activities continue within NASA's aeronautics program today.

Early NACA and NASA Supersonic Configuration Research

Fundamental research related to flight at supersonic speeds had occurred within the NACA and NASA for at least three decades before the emergence of any serious consideration of supersonic civil aircraft. Early activities included rapid advances in areas such as analysis of compressible flows, development of unique testing techniques and facilities for transonic and supersonic studies, development and validation of analytical and computational methods for aerodynamic analyses, and extensive experimental investigations in wind tunnels and flight. An expansive discussion of Langley's early contributions to high-speed flight and the key staff members involved is given in James R. Hansen's book, *Engineer in Charge* (NASA SP-4305).

Pioneering efforts in compressible aerodynamics and high-speed flight by such Langley legends as Eastman Jacobs and John P. Stack provided the technical and managerial leadership to keep Langley in the forefront of supersonic activities on an international scale. Stack's hard driving, personal interest in advanced high-speed aircraft and his aggressive tactics in international research activities and committees played a key role in the advancement of research maturity and relevance of the contributions from Langley's staff. His dedication and advocacy for a transonic research airplane helped spark the national interest that resulted in the remarkable accomplishments of the X-1 project, the conquest of the sound barrier, and the race for supersonic capability.

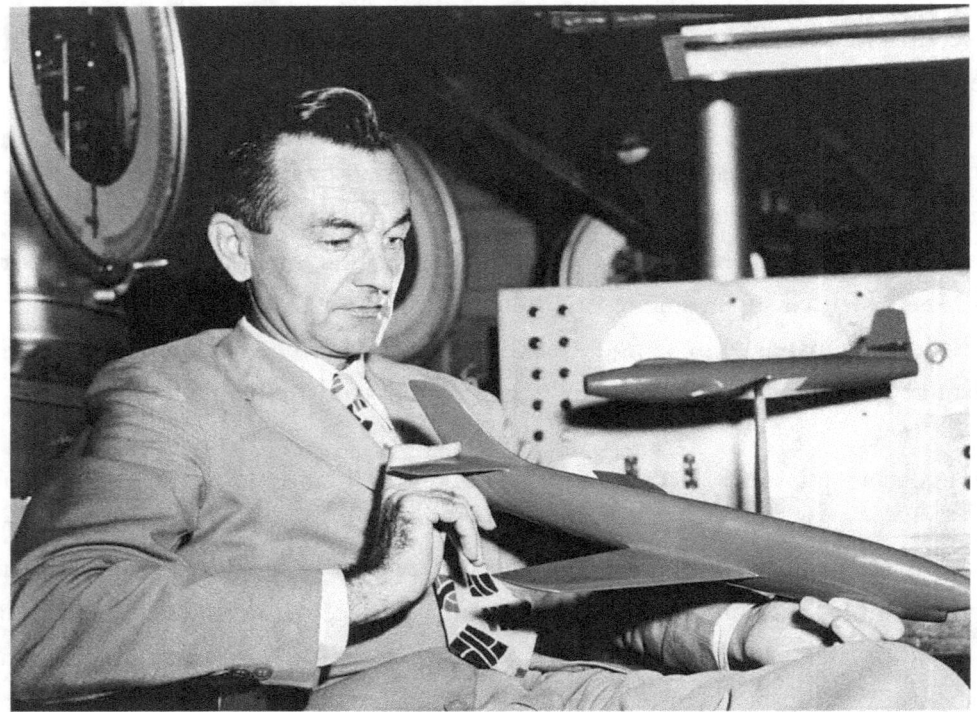

John P. Stack, leader of Langley efforts in high-speed flight.

Langley staff members responded to Stack's leadership and challenges with unprecedented research accomplishments that led to international recognition for leadership in the field. Langley's contributions were remarkable:

- Leadership and participation in the X-1 and subsequent high-speed X-series aircraft programs that demonstrated the feasibility of supersonic and hypersonic flight

- Robert T. Jones' sweepback theories, which led the way toward future high-speed wing configurations

- Richard T. Whitcomb's development of the area rule, which unlocked the puzzle of efficient transonic and supersonic flight

- Conceptual development and implementation of slotted walls for transonic wind tunnels, which permitted valid ground testing at transonic speeds

- Acquisition of extensive and valuable design data on the effects of configuration variables, such as wing sweep, on high-speed characteristics

- Development of a series of unique wind tunnels that became national treasures, including the 4- by 4-Foot Supersonic Pressure Tunnel and the Unitary Plan Wind Tunnel

- Establishment and staffing of a new high-speed flight center that would become known as the Dryden Flight Research Center

- Initiation of pioneering research of operating problems associated with high-speed flight

- Leadership in the assault on hypersonics that led to the X-15 program's historical accomplishments

- Pioneering research by Harvey H. Hubbard, Domenic J. Maglieri, and Harry W. Carlson on the sonic boom phenomenon, which would become the ultimate barrier to civil supersonic overland flight

One notable contribution made by Langley's staff during the NACA years was development of airplane configurations for efficient flight at supersonic and hypersonic speeds. This pioneering research resulted in concepts and design methodology still in use today. Much of this research resulted in innovative "arrow wing" designs with highly swept leading and trailing edges. Such configurations became the cornerstone of Langley's high-speed research activities.

Highly swept arrow wings are nearly optimum shapes for supersonic flight.

The supersonic arrow wing designed by Langley's Clinton E. Brown and F. Edward McLean appears in the NASA insignia.

One particular arrow wing design has endured throughout NASA's history. In the late 1950s, Clinton E. Brown and F. Edward McLean developed an arrow wing for supersonic applications based on analytical studies and subsequent wind-tunnel tests that verified the configuration's high aerodynamic efficiency at high speeds. Following the test program, the wind-tunnel model was put on display and observed by James J. Modarelli, a visitor from the NACA Lewis Laboratory. Modarelli headed the Lewis Research Reports Division when NACA Executive Secretary John Victory requested suggestions for an appropriate seal for the new NASA organization. Impressed by the sleek, futuristic aspects of the Langley model, Modarelli and his graphic artists included an interpretation of it in a seal design that was subsequently adopted and later modified to an insignia that became unofficially known as the NASA "meatball." In the logo, the sphere represents a planet, the stars represent space, the red chevron is the rendering of the Brown/McLean wing representing aeronautics, and an orbiting spacecraft going around the wing represents the space program This insignia was officially used from 1958 to 1975, when it was replaced by a stylized NASA "worm," and was then returned to NASA usage as directed by Administrator Daniel S. Goldin in 1992. Brown and McLean later advanced during their careers to become historic NASA managers in high-speed research.

Perhaps the most descriptive characterization of Langley throughout its history in aeronautical research has been its capability as a "one-stop shopping center" of technology in most of the critical disciplines required for aircraft applications. Certainly this personality emerged in the area of supersonic aircraft design, with world-class staff and facilities associated with aerodynamics,

The XB-70 bomber, designed for supersonic cruise at Mach 3.0.

structures and materials, flight dynamics and control, noise and other environmental issues, advanced instrumentation, and computational methods. Coupled with a high demand for consultation by industry peers and active participation in aircraft development programs (especially leading-edge military activities), Langley's staff of experts was poised and ready when the Nation's interest turned to supersonic cruise vehicles.

The Military Incubator

Langley's intimate partnership with the military community resulted in extensive use of Langley's vast collection of aerodynamic design data, wind-tunnel facilities, and expertise in military supersonic programs during the 1950s, such as the famous Century Series fighters (F-102, F-105, etc.), the B-58 bomber (capable of short supersonic "dash" mission segments), and planning for advanced supersonic vehicles within the Air Force. In the late 1950s, a single Air Force program—the XB-70 bomber—was to stimulate NASA's involvement in what would ultimately become the basis for a civil supersonic transport program.

With projected entry into service for the 1965 to 1975 time period, the XB-70 (known initially as the highly secret WS-110A project) was initiated as a result of Air Force interest in a supersonic strategic bomber replacement for the B-47. North American Aviation was awarded the XB-70 contract in December 1957, but the excitement of potential production was quickly chilled 2 years later when the Eisenhower administration decided that the intercontinental ballistic missile

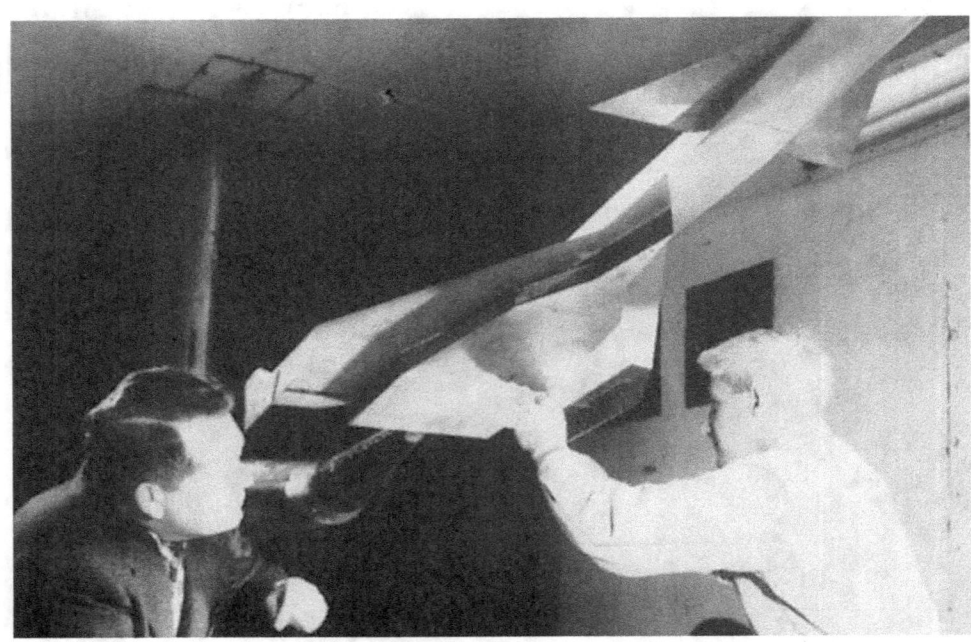

Delwin R. Croom (left) led tests of an early XB-70 design in the Langley High Speed 7- by 10-Foot Tunnel in 1957.

Free-flight model testing of the early XB-70 design was conducted in the 30- by 60-Foot Tunnel in 1957.

The XB-70 was used in NASA research on supersonic drag and sonic booms.

(ICBM) had replaced the manned bomber as primary deterrent for potential enemies of the United States. This directive was eventually amended, then reimposed by Defense Secretary Robert S. McNamara of the Kennedy administration. In 1964, the program was reduced to construction of two XB-70 aircraft for research flights. In addition to the ICBM issue, one factor possibly leading to the cancellation of XB-70 production was that another supersonic cruise aircraft—to be developed as the SR-71—was already underway and had, in fact, been flying as the Lockheed A-12 for 2 years before the XB-70 was rolled out. The decision to eliminate production of the XB-70 signaled the end of acknowledged interest within the Air Force for a long-range supersonic cruise bomber.

The first XB-70 research aircraft flew in September 1964 and attained its design cruise speed of Mach 3.0 (2,000 mph) on October 14, 1965. This remarkable airplane was engineered before high-speed computers or automated procedures would be incorporated into the design process. The XB-70 had two windshields. A moveable outer windshield was raised for high-speed flight to reduce drag and lowered for greater visibility during takeoff and landing. The forward fuselage was constructed of riveted titanium frames and skin. The remainder of the airplane was constructed almost entirely of stainless steel. The skin was a brazed stainless steel honeycomb material. Six General Electric turbojet engines, each in the 30,000-lb-thrust class, powered the XB-70. Flight testing of the two research aircraft provided unprecedented data on aerodynamics, structures and materials, flying qualities, and sonic boom phenomena.

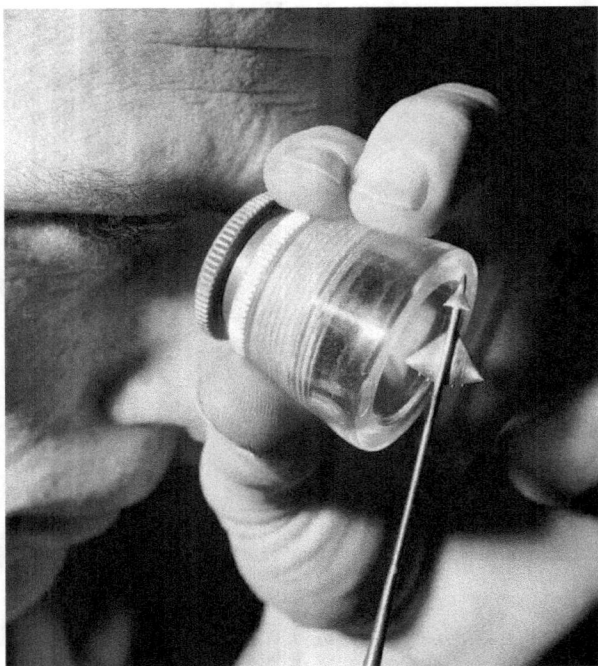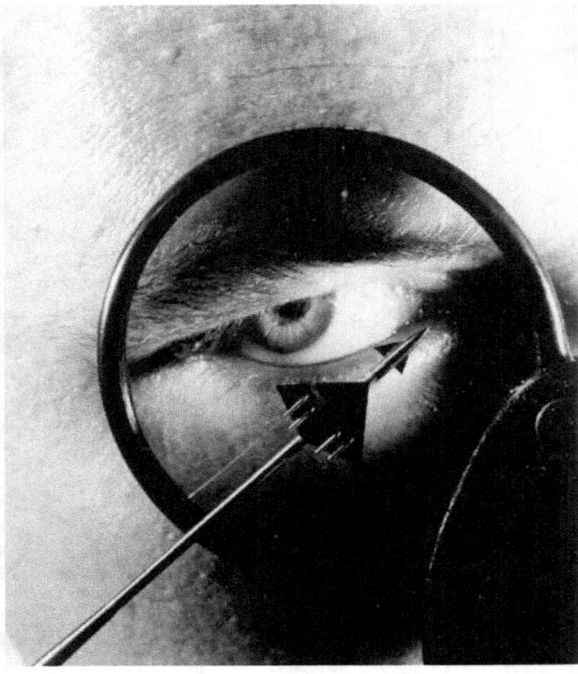

Small wind-tunnel models used in the Langley 4- by 4-Foot Supersonic Tunnel to study pressure fields associated with sonic booms.

Because of its relative size and speed capability, the XB-70 became the technological focal point of an embryonic industrial and governmental interest in supersonic civil transports. Langley had conducted developmental wind-tunnel testing of the airplane configuration in some of its tunnels, and staff members, such as Cornelius (Neil) Driver, had responded to requests from the Air Force for independent assessments of the XB-70 design's supersonic efficiency. NASA began formulating a supersonic transport technology research program around the XB-70 in 1962. In early 1966, NASA allocated funding for elaborate instrumentation for the second XB-70 airplane with plans to collect extensive data for supersonic design purposes. Unfortunately, this particular airplane was destroyed in a crash following a midair collision during a military photographic session at Edwards Air Force Base on June 8, 1966. Following an instrumentation retrofit of the first XB-70 by the Air Force and NASA, initial flights were begun to measure sonic boom characteristics and exploration of SST flight envelopes. The Air Force subsequently withdrew from the program and NASA took over as the sole XB-70 sponsor in March 1967.

During its research missions, the XB-70 was used by NASA for sonic boom characterization and supersonic drag correlations between flight and wind-tunnel predictions. Both areas involved participation of Langley's staff, including Domenic Maglieri and John P. (Jack) Peterson. Boom study results produced unprecedented details on the physical characteristics of boom propagation and coverage from various altitudes and speeds. The drag correlations showed that several improvements in the extrapolation of wind-tunnel data to flight conditions were required. The final flight of the XB-70 was on February 4, 1969, when it was delivered by NASA to the Air Force Museum at Wright-Patterson Air Force Base, Ohio, where it is on display today.

The U.S. National Supersonic Transport Program

The XB-70's demise as a potential production aircraft for the U.S. Air Force in the early 1960s was regarded as a major blow by those interested in seeing the normal extension of military technology to commercial use. The loss of a potential proving ground for supersonic cruise technologies impacted the entire industry outlook for supersonic civil transports. In addition, opposition to supersonic flight had already risen from environmentalists concerned over noise, pollution, radiation, safety, and sonic booms associated with a fleet of such aircraft.

NASA, however, had continued to produce an extensive background of supersonic technology that was rapidly maturing for applications to supersonic cruise flight. Within its research mission, the Agency could logically focus some of its talents toward the potential benefits of this vehicle class. Langley's John P. Stack exerted his powerful leadership in many venues. After Federal Aviation Administration (FAA) representatives visited Langley in November 1959 to review prospects for a U.S. SST, Stack orchestrated a pivotal briefing by Langley staff members to top FAA officials the following month on the status and outlook of supersonic transports. This meeting and Stack's enthusiasm helped begin the advocacy for an SST Program within the FAA. Contributors included: John Stack and Mark R. Nichols—state of the art and performance; Harvey H. Hubbard and Domenic Maglieri—noise and sonic boom; Eldon Mathouser—structures and materials; Thomas L. Coleman—loads; Ralph W. Stone—flying qualities; Joseph W. Wetmore—runway/braking requirements; James B. Whitten—operations; John M. Swihart and Willard E. Foss—propulsion; and Thomas A. Toll—variable geometry.

Stack represented NASA in 1961 during discussions with the FAA and the DoD to formulate a cooperative U.S. SST Program and to define agency responsibilities. President John F. Kennedy charged the FAA with responsibility to provide leadership and fiscal support to the program, with NASA providing basic research and technical support. The FAA published a Commercial Supersonic Transport Report in June 1961 stating that the DoD, NASA, and the FAA were the appropriate government agencies to define the program. Stack had formed an in-house Supersonic Transport Research Committee in June 1961 to coordinate and guide Langley's supersonic research. While the FAA sought and obtained funding from Congress in 1962 for the proposed program, NASA researchers at Langley, Ames, Lewis, and Dryden explored configuration concepts and technologies that might meet the requirements of a commercial SST.

Stack left NASA in 1962, and Chief of the Full-Scale Research Division, Mark R. Nichols and his assistant, Donald D. Baals, became Langley leaders in the formulation of SST research. Nichols and Baals were central figures for NASA SST efforts, and their personal leadership led to the success of NASA's supersonic cruise efforts at that time. More importantly, they set into motion research programs producing results that have endured to the present day in terms of design methodology and technology for future supersonic vehicles. The research team that Nichols and Baals led during the U.S. SST Program consisted of some of the most outstanding high-speed aerodynamic experts in the history of Langley. Aerodynamic theory and performance optimization were contributed by Harry W. Carlson, Roy V. Harris, F. Edward McLean, and Richard T. Whitcomb; and supersonic transport conceptual designs were conceived by A. Warner Robins. M. Leroy Spearman, another key researcher, was head of Langley's supersonic wind tunnels. (Spearman retired from Langley in 2005 after 61 years, the longest career service in Langley's history.) With jurisdiction over the high-speed tunnels and staff, Nichols' organization was poised to lead the Nation in SST research.

A significant initial act of Nichols and Baals was to challenge their staff with designing candidate supersonic transport configurations for assessments of relative merits and research requirements of different approaches. With designs submitted by several staff members and teams, Langley began extensive in-house analyses and wind-tunnel tests in 1959 of 19 different SST designs referred to as supersonic commercial air transport (SCAT) configurations, also known as SCAT-1 through SCAT-19. Over 7 years additional derivative configurations were also evaluated, resulting in about 40 concepts. Subsonic, transonic, and supersonic wind-tunnel tests of selected configurations were conducted in the Langley facilities. In February 1963 four potential candidate configurations were chosen for more detailed analysis. Although Langley's expertise in supersonic aerodynamic technology was ideally suited to define aerodynamically efficient configurations, assessment effort needed the unique experiences and systems-level expertise of industry to fully assess economic viability and technical advantages of the NASA configurations. NASA, therefore, used part of its funding from the FAA to contract industry feasibility studies of NASA candidate configurations in a SCAT feasibility study.

Langley researchers conceived three of the four down-selected NASA SCAT configurations—SCAT-4, SCAT-15, and SCAT-16—while the fourth configuration, SCAT-17, was developed at Ames Research Center. The Langley designs were heavily influenced by Langley's expertise and experiences with arrow wing concepts and with the variable-sweep concept, also made feasible by Langley personnel. Many Langley leaders, including Mark Nichols, believed that variable-sweep configurations might be the only way to meet the contradictory demands of subsonic and supersonic flight. SCAT-4 was an elegant arrow wing design sculpted by Richard T. Whitcomb, and A. Warner Robins conceived the SCAT-15, which was an innovative variable-sweep arrow

SUPERSONIC CIVIL AIRCRAFT: THE NEED FOR SPEED

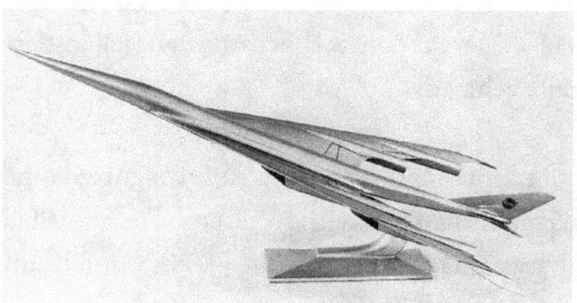

*Langley SCAT designs: SCAT-4 (above left), SCAT-15
(above right), and SCAT-16.*

wing design that used auxiliary variable-sweep wing panels. Robins also designed the SCAT-16, which was a more conventional variable-sweep design. Following contracted industry evaluations by the Boeing Airplane Company and the Lockheed California Company—as well as an in-house assessment by NASA—for a representative SST mission (3,200-nmi mission with 125 passengers at cruise Mach of 2 or 3), it was concluded that a Mach 3 airplane with titanium (rather than aluminum) construction would be required. The SCAT-16 and SCAT-17 designs were identified by industry as the most promising configurations, while the SCAT-15 and SCAT-4 were judged to have major issues. The SCAT-15, with its dual wings, would have excessive weight; and the SCAT-4, with its optimized arrow wing, would have major stability and control issues.

Study results were reviewed at a conference on supersonic transport feasibility studies and supporting research held at Langley in September 1963. At this historically significant meeting, key papers that would influence SST efforts for decades thereafter (to the present day) were presented by Langley researchers, and general perspectives and conclusions were drawn that have proven timeless. Key Langley presenters were Laurence K. (Larry) Loftin, Jr., Mark R. Nichols, Donald D. Baals, Roy V. Harris, Jr., F. Edward (Ed) McLean, Harry W. Carlson, M. Leroy Spearman, A. Warner Robins, Richard T. Whitcomb, Jack F. Runckel, Emanuel Boxer, Richard R. Heldenfels, Thomas A. Toll, Herbert F. Hardrath, Richard A. Pride, Robert W. Boswinkle, Jr., Harvey H.

Innovation in Flight

27

Hubbard, and Domenic J. Maglieri. This elite group of researchers quickly became national leaders in their respective fields of specialization for supersonic vehicles.

Although review results were informative and illuminating, technologists recognized the tremendous amount of work that would have to be accomplished before the SST could be a feasible venture. Considerable research would be necessary to mature required technologies in virtually all areas, especially in propulsion and sonic boom. Despite the cautious conclusions of the 1963 study, which indicated that a commercial supersonic transport was not yet feasible, a few months later the FAA initiated a request for proposals (RFP) that began the U.S. SST Program.

Several factors, including international politics and concern, had precipitated the decision to progress with the SST Program, even with the recognized technical immaturity and environmental challenges. The Nation was still responding with concern and technical embarrassment to the Soviet Union's launching of its Sputnik satellite on October 3, 1957; and an announced agreement of the British and French to collaborate on their own SST in 1962 sent waves of concern over potential technical inferiority through the U.S. technical community and Congress. On June 5, 1963, in a speech before the graduating class of the United States Air Force Academy, President Kennedy committed the Nation to "develop at the earliest practical date the prototype of a commercially successful supersonic transport superior to that being built in any other country in the world . . ." and designated the FAA as the manager of the new program.

During FAA proposal evaluations of submissions from Lockheed, Boeing, North American, General Electric, Pratt and Whitney, and Curtiss-Wright, over 50 NASA personnel participated by defining evaluation criteria and evaluating proposals, with additional wind-tunnel testing and analytic studies wherever required. Most evaluators were from Langley, with William J. (Joe) Alford, Jr., serving as the lead interface with the FAA. Design requirements for the U. S. SST were driven by an attitude that the airplane had to be faster and larger than the competing Concorde design. Thus, the specifications for the competition included a cruise Mach number of 2.7, a titanium structure, and a payload of 250 passengers.

The proposals of Boeing, Lockheed, General Electric, and Pratt and Whitney were accepted, but North American and Curtiss-Wright were eliminated early during evaluations. North American had chosen a delta/canard design (Model NAC-60) similar in some respects to the XB-70; however, the design's aerodynamic drag was excessive. Boeing had been working potential supersonic transport designs since the mid-50s and had narrowed its candidates to either a delta design or a higher risk variable-sweep airplane potentially capable of more efficient and quieter operations during takeoff and landing. Although concern existed over the weight of this configuration (massive 40,000-lb pivots would be required for the variable-sweep wing), it was chosen as the Boeing entry in the competition (Model 733-197). Lockheed chose a fixed-wing double-delta design (Model CL-823), which it felt would be a simpler, lighter airplane. The performances of both designs were judged to be substantially short of the design goals and the contractors conducted extensive design cycles during the development process. Boeing's fears of weight growth for the variable-sweep design had become real, and the configuration had required extensive changes due to other concerns, including moving the engines to the airplane's rear under the horizontal tail (Model 2707-100). The FAA advised Boeing and Lockheed to explore potential use of a derivative of the NASA SCAT-15 design, which was showing substantial performance improvement over industry designs.

Researcher Delma C. Freeman, Jr., inspects a free-flight model of the Lockheed L-2000 Supersonic Transport design (left) used in tests in the Langley 30- by 60-Foot Full-Scale Tunnel.

Time-lapse photography shows the variable sweep wing of early Boeing Model 733-290 Supersonic Transport design. Free-flight testing was conducted in the Langley Full-Scale Tunnel.

Boeing continued to develop its variable-sweep design (Model 733-290), encountering heating problems with the low horizontal tail position and pitch-up problems when the tail was moved to a T-tail position. In the next design cycle, a revised configuration (Model 733-390) retained the low tail and featured an arched fuselage, long landing gear struts, and upturned engine nacelles.

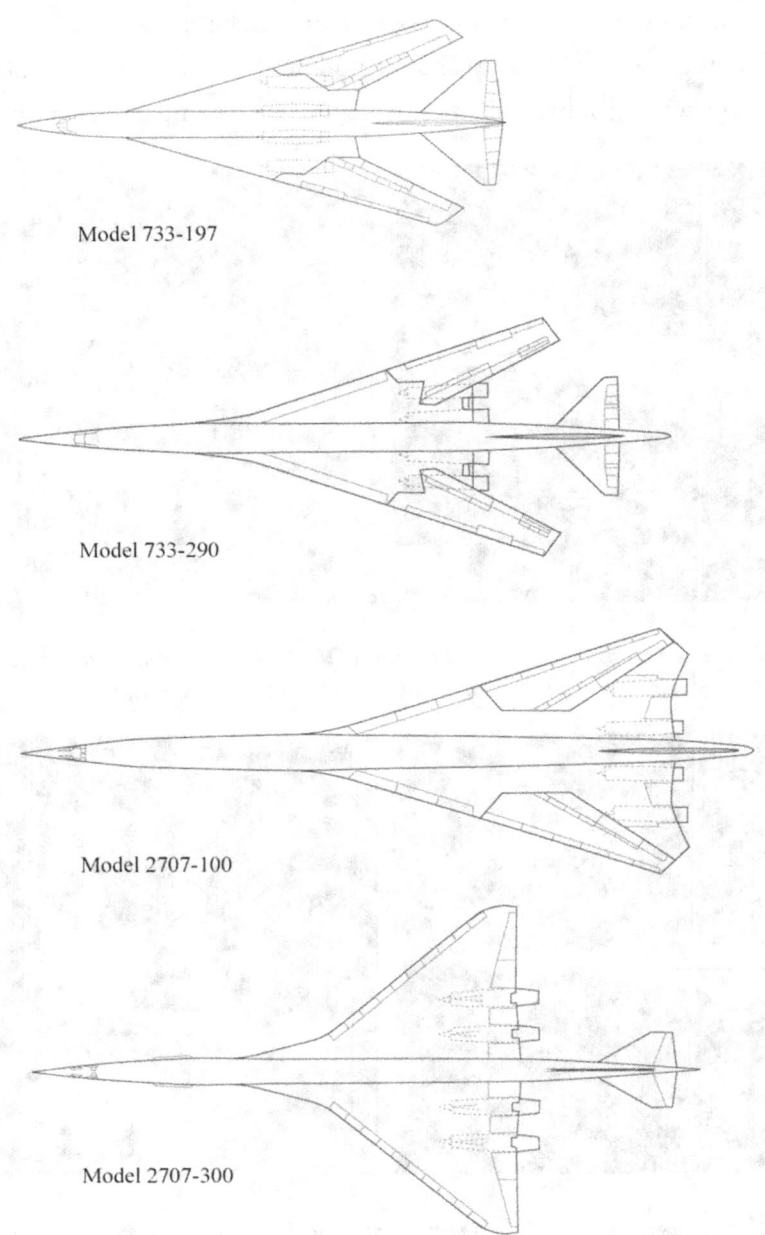

Model 733-197

Model 733-290

Model 2707-100

Model 2707-300

Evolution of the final Boeing Supersonic Transport configuration.

Innovation in Flight

Engineering mock-up of the Boeing 2707-300 Supersonic Transport.

The revisions, however, led to high weight and degraded performance. Then, in a dramatic change, Boeing moved the four engines to beneath the horizontal tail (Model 2707-200). Meanwhile Lockheed had been concurrently developing a refinement of its double delta (Model 2000-7A) as well as a version of the refined NASA SCAT-15 (Model L-15-F). Boeing was ultimately declared the competition winner on December 31, 1966, with its 2707 design. After the award, Boeing finally gave up on its variable-sweep airplane concept in 1968 and adopted a double-delta SST design, known as the Model 2707-300. Boeing complied with an FAA directive to continue analysis of its refined SCAT-15 design (Model 969-336C) and make a final configuration selection by late 1968. Boeing then selected the 2707-300 but continued its studies of the SCAT design.

Research activities at Langley during the SST competition were, of course, intensive efforts involving wind-tunnel tests, controversial assessments, and consultation in several areas including aerodynamics, structures, materials, flight controls, noise, sonic boom, propulsion integration, aeroelasticity, landing loads, stability and control, operating problems, integration of SST and subsonic air traffic, and flight dynamics (including free-flight models). Detailed discussion of the very important contributions of Langley's staff during the SST Program is not feasible in the present document, and the reader is referred to the formal NASA reporting publications listed in the bibliography for details and sources of additional information.

Charles L. Ruhlin inspects a 1/17-scale semispan flutter model of the Boeing 2707-300 during flutter clearance testing in the Langley 16-Foot Transonic Dynamics Tunnel in 1970.

An important contribution by Langley during this period was the vital role played by Harry W. Carlson, Roy V. Harris, Jr., and Wilbur D. Middleton in the development of analytical tools and theory for the design and evaluation of complex supersonic configurations. Until their efforts, supersonic design processes were based on simple isolated wings, and attempts to combine volume and lift effects produced inconsistent results. The breakthrough came when the Langley researchers attacked the problem by treating the effects of volume at zero lift and the effects of lift at zero volume—a much more manageable approach. They then added optimization codes to further embellish the analysis, and the resulting methodology was enthusiastically embraced by industry (especially the Lockheed SST team). Along with the advanced design process, Langley contributed the first application of high-speed computers to the aerodynamic design of supersonic aircraft. This advance in technology, spearheaded by Roy Harris, dramatically shortened the time required for analysis of supersonic aerodynamics, resulting in a quantum leap in engineering analysis capability. First applied by Harris in his adaptation of a Boeing code for the prediction and analysis of zero-lift supersonic wave drag, his computer methods shortened the manual task of analyzing wave drag for a single Mach number for a single configuration from over 3 months to just a few days. His work also represented the first "wire frame" rendering of aircraft configurations for analysis. Computer programmer Charlotte B. Craidon used Langley's massive IBM 704 Data Processing

System to provide Harris valuable assistance in his time-savings breakthrough. When Harris' zero-lift drag program was coupled with analysis methods by Clinton E. Brown and F. Edward McLean for lifting conditions, the combination provided designers with powerful tools for analysis of supersonic aerodynamic performance. Industry has adapted many of these Langley-developed computer-based techniques and still uses them today.

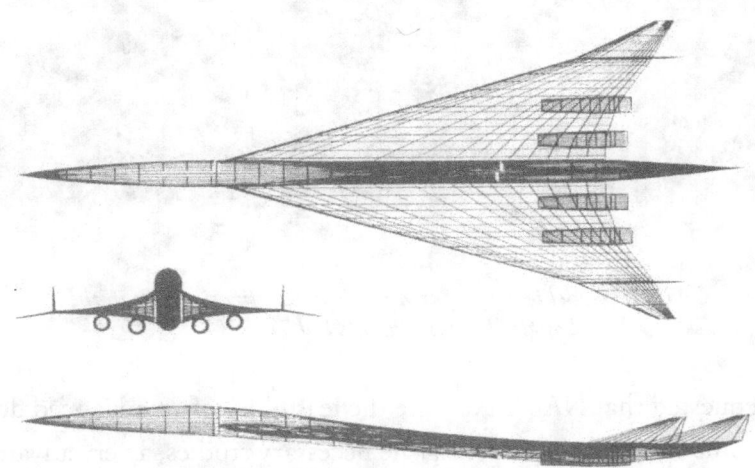

Langley's application of computer use to supersonic aerodynamic design was a major breakthrough in design capability. Shown is the SCAT-15F design.

Throughout the Boeing-Lockheed competition, NASA personnel at Langley, Ames, Lewis, and Dryden provided data and analysis to the individual contractors as well as consultation on how to improve their designs. Langley also pursued improved versions of its original SCAT concepts as potential alternatives to the Boeing and Lockheed designs. In 1964, advances in the sophistication of aerodynamic theory and computer codes at Langley allowed a team of Harry Carlson, Ed McLean, Warner Robins, Roy Harris, and Wilbur Middleton to design an improved SCAT configuration, called the SCAT-15F. This fixed-wing version of the earlier variable-sweep SCAT-15 configuration was designed using the latest, computer-based supersonic design methodology that Langley had developed, resulting in a L/D ratio of 9.3 at Mach 2.6, an amazing 25 to 30 percent better than the previous state of the art at that time. The new Langley theories permitted the researchers to shape the wing with reflex, twist, camber, and other parameters that nearly optimized its supersonic capability. When informed by the design team of the SCAT-15F's projected performance, Larry Loftin (Langley's top manager for aeronautics) did not believe the prediction. Subsequent wind-tunnel results, however, proved the estimates accurate and the SCAT-15F remains to this day one of the most aerodynamically efficient SST designs ever conceived. Carlson, McLean, Robins, Harris, and Middleton were awarded a patent for the SCAT-15F design in 1967.

*SCAT-15F model being prepared for flow visualization tests in the
Langley Unitary Plan Wind Tunnel.*

In 1966, the FAA requested that NASA examine the feasibility of an advanced domestic SST with acceptable levels of sonic boom. In conducting the necessary studies, a very advanced aerodynamic configuration was required. Analysis and further development of the SCAT-15F design quickly became a high priority activity at Langley, and Mark Nichols and Don Baals directed extensive computational analysis and wind-tunnel testing of several models in virtually every subsonic, transonic, and supersonic tunnel at Langley. Key facilities and test leaders included the 16-Foot Transonic Tunnel (Richard J. Re and Lana M. Couch), the 8-Foot Transonic Pressure Tunnel (Ralph P. Bielat and John P. Decker), the 30- by 60-Foot (Full-Scale) Tunnel (Delma C. Freeman), the Unitary Plan Wind Tunnel (Odell A. Morris, Dennis E. Fuller, and Carolyn B. Watson), and the High-Speed 7- by 10-Foot Tunnel (Vernon E. Lockwood and Jarrett K. Huffman). The aerodynamic studies were augmented with an investigation of the predicted sonic boom characteristics of the configuration (Harry Carlson and Domenic Maglieri).

Dynamic stability studies using free-flight models were also undertaken in the 30- by 60-Foot Tunnel, and piloted ground-based simulator studies of the handling qualities of the SCAT-15F during approach and landing were conducted using Langley simulators. Although studied by teams at both Lockheed and Boeing as an alternate configuration, the SCAT-15F was not adopted by either contractor during the SST Program. A major issue with the SCAT-15F design was its low-speed stability and control characteristics. The highly swept arrow-wing planform exhibited the longitudinal instability (pitch up) tendency typically shown by arrow wings at moderate angles of attack, as well as the possibility of a dangerous unrecoverable "deep stall" behavior. The deep stall followed the pitch-up tendency and was characterized by an abrupt increase in airplane angle of attack to extremely high values (on the order of 60 degrees) where the L/D ratio became much less than 1, and longitudinal controls were ineffective in reducing the angle of attack to those values required for conventional flight. This potentially catastrophic characteristic was unacceptable. Several Langley tunnel entries were directed at the pitch-up problem and the development of modifications to alleviate it.

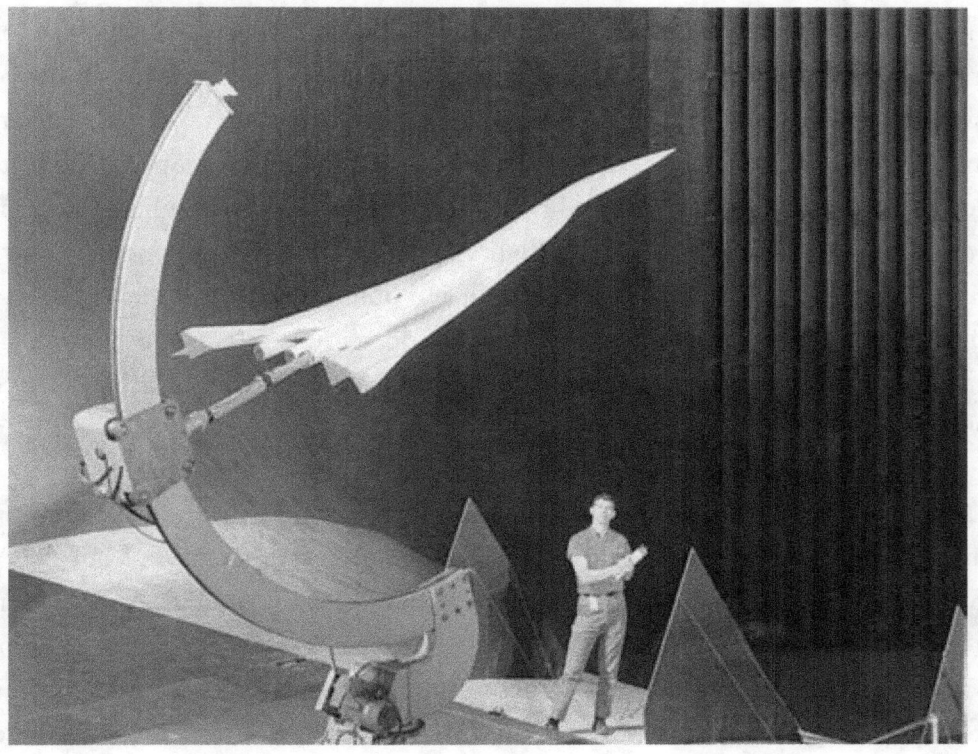

Model of the SCAT-15F design mounted in the Langley Full-Scale Tunnel for tests to alleviate an unacceptable "deep stall" characteristic.

Meanwhile, Marion O. McKinney and Delma C. Freeman, Jr. led exhaustive tests to alleviate the unacceptable SCAT-15F deep-stall characteristic in the Langley 30- by 60-Foot (Full Scale) Tunnel. After extensive testing, a combination of modifications—including a 60-degree deflection of the wing leading-edge flap segments forward of the center of gravity, a "notched" wing apex, Fowler flaps, and a small aft horizontal tail—eliminated the deep-stall trim problem.

All Langley research efforts to make the SCAT-15F a satisfactory configuration were quickly exchanged with Boeing on a virtually real-time basis during 1968, and formal reports were collated and transferred to the industry in 1969. Many lessons learned and design approaches derived from the SCAT-15F experience in solving low-speed problems were critical inputs for later NASA supersonic transport technology programs in the 1980s and 1990s.

Despite promising progress in configuration research, the most critical obstacle to the SST remained the sonic boom issue. In June 1961, DoD, NASA, and the FAA had released the "Commercial Supersonic Aircraft Report," known as the SST Bluebook. This report concluded that the development of a commercial supersonic transport was technically feasible, but that a major research and development program would be required to solve a major problem: the sonic boom. It became essential to know the level of sonic boom exposures that might be accepted by the public. Special supersonic overland flight projects were needed to augment disjointed Air Force data, calibrate theory, assess community reaction, define structural damage due to booms, and assess boom effects on the environment. Domenic J. Maglieri, Harvey H. Hubbard, David A. Hilton, Vera Huckel, and Roy Steiner participated in these studies. Special flight tests were conducted at St. Louis from November 1961 through January 1962 (B-58 flights to confirm theory), at Edwards Air Force Base in 1963 (F-104 flights to determine effects of booms on two general aviation airplanes in flight and ground operations), and public exposure flights at Oklahoma City in 1964 and Chicago in 1965. Langley staff also participated in supersonic overland flights of an F-104 aircraft in Colorado to determine if booms could trigger avalanches on mountains covered with heavy snow.

The Office of Science and Technology established a Coordinating Committee on Sonic Boom Studies in 1965 under the management of the U.S. Air Force with members from the Air Force, NASA, Stanford Research Institute, the Environmental Science Services Administration, and the U.S. Department of Agriculture. This organization sponsored a concerted study of human acceptance and response of typical house structures to booms from 1966 to 1967 using overland flights of XB-70, F-104, F-106, B-58, and SR-71 aircraft at Edwards Air Force Base. NASA contributed data requirements, data scheduling, and test operations and boom research (including equipment installation, pressure signature acquisition, and analysis). The program conducted over 350 supersonic overland flights and extensive analysis of boom characteristics.

In some tests, the public was exposed to sonic boom overpressures of up to 3 lbs/ft^2. SST overland flights were expected to result in levels of 1.5 to 2.0 lbs/ft^2. No one level of overpressure was found below which public acceptance was assured; but the results clearly indicated that levels of 1.5 to 2.0 lbs/ft^2 were unacceptable for human exposure. Further, the results suggested that exposure must be considered in terms of frequency, intermittency, time of day or night, and the particular signature.

The Acoustical Society of America summed up the state of the art of sonic boom during symposia at St. Louis in 1965 and Houston in 1970. Survey papers were given on the nature of the sonic boom, sonic boom estimation techniques, design methods for minimization, atmospheric effects on sonic boom, the impact of airplane operation on sonic boom signature, and the effect of sonic booms on people. The final survey paper at that conference emphasized that the operation of a supersonic transport would probably be constrained to supersonic flight over water only or supersonic in low population corridors. Research and technical meetings on sonic boom continued at a steady pace between 1965 and 1970, with NASA hosting conferences in Washington DC in 1967, 1968, and 1970. Langley's sonic boom experts, including Maglieri, Carlson, McLean, and Hubbard, were major participants in all the foregoing activities.

Langley's overall support of SST research also included several novel studies, one being an in-flight simulation of SST configurations using the Boeing 367-80 transport (prototype for the 707 series). Langley contracted with Boeing to modify the "Dash 80" as a low-speed in-flight simulator for assessments of the approach and landing characteristics of representative double-delta and variable-

Boeing's 367-80 was used at Langley for in-flight simulator studies of handling qualities of supersonic transport configurations.

geometry SST concepts during instrument flight rules (IFR) conditions. The program's objectives were to study the handling qualities of two representative SST configurations, evaluate potential problem areas and stability augmentation requirements, obtain indications of the minimum acceptable handling qualities, determine effects of operation on the unconventional "backside" of the power required curve, and obtain approach and landing data applicable to criteria and certification requirements for SST transports. Flown at Langley by pilots from NASA, Boeing, the FAA, and airlines from May to October 1965, the venerable Dash 80 used computer-based aerodynamic inputs from Langley staff to replicate the responses of SST designs to pilot inputs, producing pioneering information on the handling qualities and control system requirements for SST designs during approach and landing. Key Langley participants in the program included Project Manager Robert O. Schade, Harold L. Crane, Albert W. Hall, William D. Grantham, Robert E. Shanks, Samuel A. Morello, and test pilots Lee H. Person and Jere B. Cobb. Philip M. Condit, a new Boeing engineer, was the lead Boeing participant for the study as an expert in aerodynamics and stability and control. Condit later became President of The Boeing Company in 1992 and then Chief Executive Officer in 1996.

Amid extensive controversy, technical issues, international politics, and environmental concerns, the U.S. Congress reduced the SST Program's funding in December 1970 and canceled the program in March 1971. The House, by a vote of 217 to 203, deleted all SST funds from the Department of Transportation (DoT) appropriation for fiscal year 1972. An amendment to restore SST funds was defeated in the Senate 51 to 46. On May 1, 1971, the Senate approved $156 million in termination costs. Thus, after 8 years of research and development and an expenditure of approximately $1 billion, the United States withdrew from the international supersonic transport competition. The program's cancellation was a severe blow to the enthusiasm of NASA researchers, especially the Langley participants in supersonic cruise research. In addition, the cancellation dampened and obscured the significant impact of advances in the state of the art that had occurred due to Langley's involvement. Supersonic aerodynamic theory and design methodology had been aggressively advanced and validated thanks to the introduction of high-speed digital computer codes that included the effects of extremely complex aerodynamic phenomena. Langley's researchers had also contributed unprecedented advances in the knowledge of sonic boom phenomena. The significance and understanding of sonic booms had clearly not been fully appreciated by the technical community before the SST program. Langley personnel developed and validated sonic boom estimation procedures, including analysis of near-field sonic boom characteristics that led to a new field of sonic boom minimization. Extensive participation by Langley during measurement of sonic booms from various supersonic aircraft, including data gathered on atmospheric effects and possible damage to dwellings, resulted in extremely significant information for future supersonic aircraft programs.

Other Langley Research center efforts and contributions during the SST Program included valuable improvements in understanding the characteristics, design variables, and fabrication problems of titanium material. As previously discussed, conventional aluminum material will not survive in the 500 °F temperature environment encountered at cruise speeds of Mach 2.7 and above, requiring the use of alternate materials. Design methodology, including the use of advanced computer codes for structural design, was developed. Also, Langley began a major effort in stratospheric emission technology, contributing to a much better understanding of atmospheric pollution phenomena. Finally, Langley researchers contributed a substantial database on stability, control, and handling qualities for advanced arrow-wing type configurations that emerged as the most efficient approach for future SST designs.

The Supersonic Cruise Research Program

Program Genesis

In early 1972, the Nixon administration directed NASA to formulate a supersonic research program that would answer difficult technical and environmental problems requiring resolution for a viable commercial supersonic transport. Under the leadership of William S. Aiken, Jr., of NASA Headquarters, an intercenter team formulated a program called the NASA Advanced Supersonic Technology (AST) Program. Agency lead role for the program was assigned to Langley. The new program's goals were to build on the knowledge gained during the SST program and to provide—within 4 years—the supersonic technology base that would permit the United States to keep options open for proceeding with the development of an advanced supersonic transport, if and when it was determined that it was in the national interest.

Several events at Langley influenced the ultimate leadership of the AST Program. At the time of the request, Robert E. Bower had become Director for Aeronautics and an Aeronautical Systems Office (ASO) had been formed under Thomas A. Toll. Within ASO, the Advanced Supersonic Technology Office (ASTO) headed by David G. Stone was given responsibility for the emerging AST activities. In 1971, Langley's Cornelius (Neil) Driver had been assigned to a temporary position for presidential and congressional assistance in assessing emerging military aircraft (U.S. Navy F-14 and U.S. Air Force F-15). Driver spent his early career in supersonic wind-tunnel research, including participation in the XB-70 and early SST studies. During his congressional assignment, he became personal friends with ex-test pilot William M. (Bill) Magruder, the top FAA official in the final, turbulent months of the SST Program. Magruder regarded the possible breakup of industry SST design teams as a potential disaster for national interests in aviation and impressed Driver with devising a program that would keep the national expertise in place. Upon his return to Langley, Driver joined the ASTO organization as Stone's deputy and became active in planning the AST effort, which maintained a sensitivity to including industry teams in the NASA effort. Meanwhile, Mark Nichols' High Speed Research Division had started to ramp up planning for a technical program in support of AST.

In 1972, an AST Working Group was formed to define a technical program plan under the leadership of F. Edward McLean. McLean, who would later become manager of the technical AST effort in 1974, was also a key participant in the activity's startup and operations. His outstanding contributions in the areas of advanced aerodynamics and sonic boom theory were impressive credentials for his role in the program. Like Driver, his technical expertise in supersonic configurations, industry relations, and hard-driving personal dedication and management style were keys to the successful accomplishments of the new NASA activity. Both Driver and McLean operated within the new Aeronautical Systems Division (ASD) of Adelbert L. (Del) Nagel that replaced Toll's office in 1974. Driver headed the Vehicle Integration Branch while McLean led the Advanced Supersonic Technology Project Office. Langley was extremely fortunate to have two such outstanding leaders assigned to the AST Program. In 1978, McLean retired and Nagel departed NASA for Boeing. Driver then headed up the ASD organization, while management of the Advanced Supersonic Technology Office passed from Vincent R. Mascitti to Domenic Maglieri.

Because of concern that potential opponents of future SST programs would misinterpret the program acronym as a program for developing an advanced supersonic transport, the name was changed to Supersonic Cruise Aircraft Research (SCAR) in 1974. Further sensitivity over the use of the word "aircraft" led to yet another modification in the program name, resulting in the final name, Supersonic Cruise Research, or SCR, in 1979. The SCR Program's pace and funding were intentionally cut back by Congress to delay the issue of another SST battle, and the funds available to NASA researchers were extremely limited. This situation led to a fundamental integrated technology approach, which NASA referred to as a "focused" program. The SCR activity involved all NASA aeronautical centers and many aerospace companies, research organizations, and universities. Although the program was managed by NASA Headquarters, day-to-day operations were the responsibility of Langley's ASD with research tasks being accomplished by NASA Center staffs and industry. the program's sponsored disciplinary research included aerodynamic performance (Langley and Ames), propulsion (Lewis), structures and materials (Langley, Dryden, and Ames), stability and control (Ames, Dryden, and Langley), and stratospheric emissions impact (all Centers).

The program also included an important element known as systems integration studies, which attempted to quantitatively measure the potential impact of various disciplinary technology advances on representative supersonic cruise aircraft. The systems integration activity participants included Boeing, Lockheed-California, and McDonnell Douglas, as well as in-house NASA and local Langley contractor (Ling-Temco-Vought) personnel. In addition to the enthusiastic and capable contributions of industry airframe companies, SCR efforts included the participation of propulsion groups from the General Electric Company and the Pratt and Whitney Aircraft Company. In this approach, the Nation was able to maintain the invaluable expertise and talent that had participated in earlier SST studies, as desired by Bill Magruder and Neil Driver.

Initially, a baseline supersonic transport configuration, known as the AST-100, provided a common reference for the integration teams from industry and NASA. As the program evolved, Rockwell International also came onboard, and each team developed a refined design based on a different cruise speed, generating five slightly different configurations with cruise Mach numbers ranging from 2.2 to 2.7. McDonnell Douglas chose a highly swept arrow-wing configuration similar to the Mach 2.7 NASA SCAT-15F but designed for a lower cruise Mach number of 2.2. Lockheed also chose an arrow-wing concept designed for cruise at 2.55, but with two of its four engines located unconventionally above the wing for noise shielding. The NASA concept used an improved version of the SCAT-15F with a proposed cruise Mach number of 2.7.

Aerodynamics

Within the aerodynamic performance element of the SCR Program, major tasks involved developing and testing advanced aerodynamic supersonic transport concepts, as well as developing and validating aerodynamic design and analysis tools. Work areas included concept development, theory development, and sonic boom.

As previously discussed, the state of the art in supersonic aerodynamic design technology had been brought to a high level of maturity by the SST program's end in 1971. The unprecedented aerodynamic efficiency of the NASA SCAT-15F stood as an example of what could be achieved, but the highly efficient supersonic arrow-wing concept faced deficiencies in subsonic performance, resulting in major issues in off-design performance and poor noise characteristics. Thus, a major effort was directed to improve the low-speed behavior of arrow-wing concepts during the SCR Program, whereas relatively limited effort was expended on improving supersonic cruise efficiency. This low-speed research was especially critical from the McDonnell Douglas perspective, and numerous models of various arrow-wing configurations were explored cooperatively in the Langley 30- by 60-Foot (Full Scale) Tunnel, the Langley 7- by 10-Foot High Speed Tunnel, and the Langley V/STOL Tunnel (now the Langley 14- by 22-Foot Tunnel), as well as tunnels at Ames.

An exhaustive low-speed aerodynamic program led by Langley's Paul L. Coe, Jr., and Joseph L. Johnson, Jr., explored methods to improve low-speed performance (lift and L/D) while retaining satisfactory stability and control. The scope of research included extensive studies of wing planform effects and leading-edge devices as well as innovative technical concepts, such as thrust vectoring to augment lift. Additional studies—for example, an evaluation of the low-speed characteristics of an advanced blended arrow-wing configuration designed by A. Warner Robbins—were also conducted. The collaborative efforts with industry teams led to solutions for many of the stability and control problems, and the low-speed aerodynamic efficiency of arrow-wing concepts was dramatically increased.

In addition to wind-tunnel research to improve low-speed aerodynamic characteristics, Langley's staff also conducted ground-based and in-flight piloted simulator studies of the low-speed handling characteristics of advanced supersonic transports. In 1977, William D. Grantham and Luat T. Nguyen led simulator studies of two advanced configurations: a canard version of the SCAT-15F and a powered-lift arrow-wing airplane with engines located over and under the wing. Conducted in a fixed-base simulator at Langley and in flight using the Calspan Total In-Flight Simulator (TIFS) airplane, the study defined details of the longitudinal and lateral directional stability augmentation systems required for satisfactory characteristics. Another study highlight was the identification of a critical roll-control power deficiency for the airplanes during crosswind operations. This problem is inherent for the highly swept supersonic transport configuration, and these results provided guidance for meeting handling quality requirements.

Much of the data, concepts, and design methodology derived from aerodynamic research in the SCR Program proved invaluable in other NASA and DoD programs. One outstanding example of the value of Langley's supersonic aerodynamic design expertise and methods was the cooperative venture initiated by a request from General Dynamics (now Lockheed Martin) for a joint design effort to develop a supersonic cruise wing for the F-16 fighter. General Dynamics was in close communication with the Langley staff during the SCR period and viewed the vast experience and mentoring of the researchers in supersonic wing design methodology as an extremely important ingredient of the new fighter design. As discussed in NASA SP-2000-4519, *Partners in Freedom*, this joint activity produced the highly successful F-16XL version of the fighter.

Sonic Boom

In 1972, Christine M. Darden and Robert J. Mack continued to advance sonic boom theory. Computer codes were developed to advance boom predictions for various atmospheric conditions,

*The F-16XL benefited from a cooperative supersonic wing design study between
Langley and General Dynamics.*

nose-bluntness effects were analyzed, and models were fabricated for wind-tunnel tests to verify theories. Six-inch models of three wing-body concepts for cruise at Mach 1.5 and Mach 2.7 were prepared for testing and were at that time the largest sonic boom models tested in the Langley 4- by 4-Foot Unitary Plan Wind Tunnel (models tested in the tunnel during the 1960s were from 0.25 in. up to 1 in.). Model size was driven by the need to measure far-field signatures to ensure linear theory was valid—about 30 body lengths away. As confidence in the extrapolation methods grew, signatures could be measured closer to the body and model size could become larger.

The SCR Program continued to explore airplane shaping for minimizing sonic booms. Various design concepts of vehicles having minimum boom design-shaped signatures were derived, and other studies indicated that by altering the boom signature shock rise time and waveform from that of a normal far-field N-wave, the perception of loudness was reduced. Darden and Mack were in the process of additional planning when the SCR Program was canceled, and NASA funding for sonic boom research was then dropped for nearly 6 years.

Stability and Control

The stability and control work area in SCR developed methods for accurately assessing the stability and control characteristics and requirements for large, flexible supersonic cruise aircraft, including the design of active control systems. This area was led by Ames and Dryden, with some participation by Langley in the active controls area.

Propulsion

The propulsion element of SCR was directed at the development of a propulsion system concept that would efficiently meet the conflicting requirements of subsonic and supersonic operations. Specific problems addressed by this critical area at the Lewis Research Center included noise and pollution issues.

The accomplishments and contributions of the superb propulsion research conducted at Lewis under sponsorship of the SCR Program covered engine concepts, noise, and emissions. Many of these activities resulted in design methodology, concepts, and technologies that were rapidly matured for supersonic cruise vehicles. The first 3 years of propulsion-related activity by Lewis and its industry partners resulted in the most important propulsion development in the SCR Program: the variable-cycle engine. In the variable-cycle engine concept, a special engine flow-through design permits operation like a turbojet during supersonic cruise and a turbofan during subsonic flight. Both General Electric and Pratt & Whitney conducted engine test-stand assessments of proposed variable-cycle configurations. The engine concepts did not represent significant advances in propulsion cycle efficiencies over previous supersonic turbofan or turbojet engines. Instead, the new engine concepts provided airflow control appropriate for each phase of flight, thereby enabling low specific fuel consumption, less weight, and reduced noise. NASA then spun off a Variable Cycle Engine (VCE) Program from the SCR Program in 1976, maintaining a close relationship and communications between the two efforts at Langley and Lewis.

Noise technology was a major part of the SCR Program efforts, including exploratory studies of inverted velocity profile nozzles, mechanical noise suppressor technology, acoustic shielding (such as placing engines in an over/under arrangement on the wing), and the use of "minimum noise" flight profiles during terminal area operations. The substantial noise reduction potential of these concepts gave rise to optimism about the potential noise compliance for future supersonic cruise aircraft. The situation was a vast improvement over the state of the art at the end of the SST Program in 1971, when the only noise reduction technology was a vastly oversized, heavy noise suppressor system that severely degraded airplane mission capability.

Structures and Materials

SCR-sponsored structures and materials research attempted to develop structural concepts and materials that would withstand the large load and temperature variations experienced during representative supersonic transport flight operations. Work areas included structural concepts, development of design data and tools, material applications, and fuel-tank sealants. Richard R. Heldenfelds from the Structures Directorate of Langley provided leadership for most of these activities (tank sealant work was led by Ames).

Highlights of the Langley contributions in structures and materials included improved computational methods for structural design and analysis, with an emphasis on rapid design methods to produce structures that met the requirements for strength, divergence, and flutter, including the use of active controls. Other SCR-sponsored research efforts included the development of low-weight, low-cost structural concepts with low-cost fabrication techniques. Many consider the most important output of the entire SCR Program to be a demonstration of the potential use of superplastic forming and diffusion bonding (SPF/DB) of titanium. The process involves heating a sheet of titanium in a mold until the titanium reaches a malleable temperature. Gas is then injected into the mold, and the titanium is either blown into a shape or bonded to another titanium sheet. Using this revolutionary process, potential applications to various parts seem limitless. During the course of the SCR activities, Rockwell (which had accrued experience with the technique as part of its XB-70 activities) and McDonnell Douglas fabricated SPF/DB panels and structural components. Using the SPF/DB process in the design of supersonic transports showed significant advantages over the titanium honeycomb that was used in the Boeing SST design of the 1960s. For example, projected weight savings of changing to a titanium SPF/DB sandwich construction for a fuselage, compared with titanium skin stringer concepts, indicated a reduction of fuselage weight by about 22 percent.

Emissions

Research in stratospheric emissions was a critical area directed at assessing the impact of upper atmosphere pollution by high-flying aircraft. Issues addressed included the chemistry, propagation, and dissipation of jet wakes as well as the natural causes of pollution. The NASA Office of Space Science funded and managed research in this area after October 1976. A number of emission research programs had emerged after 1972, including the Climatic Impact Assessment Program (CIAP), the High Altitude Pollution Program (HAPP), and the NASA Emissions Reduction Research and Technology Program. A major Lewis undertaking, the Emissions Reduction Research and Technology Program identified means for reducing the pollution of jet engines. By far the most important outcome of the 1970s research was the accidental discovery of the ozone problem. By 1980, DoT and NASA-funded research had demonstrated that SSTs were less threatening to the ozone layer than initially thought, leading to a relatively well-accepted belief that future combustor technology could lead to acceptable transport configurations. By-products of this research activity included the fact that the major threat to the Earth's ultraviolet shield was chlorofluoromethanes (CFM), which subsequently led to a ban on CFM-powered aerosol sprays in the United States in 1978, a worldwide ban on CFM manufacture in 1987, the formalization of stratospheric research at NASA, and a permanent NASA program to monitor the stratosphere's composition.

Other Activities

The program's mission performance integration element assessed the impact of individual disciplinary advances, including a measurement of the technological advances of the integrated disciplines. The work was critical to determine whether progress was being made toward an economically viable, environmentally acceptable commercial supersonic transport. Along with providing systems-level progress assessments of technology advances for industry and NASA vehicle concepts, systems studies enabled the SCR team to assess the impact of vehicle sizing, wing/body blending, and large-payload supersonic airplanes.

The SCR Program encouraged innovation and new ideas from the disciplinary specialists. Research stimulated at Langley included several examples of novel technology, such as active control of aeroelastic characteristics (especially flutter), resulting in the potential for significant structural weight savings and cost. Active landing gear technology was pursued by the structural dynamics group at Langley, providing concepts to reduce landing loads on the supersonic transport's structure during landing impact and ground taxi operations.

Unconventional supersonic cruise configurations were also examined, including the use of a novel twin-fuselage SST configuration for high-payload, high-productivity applications. John C. Houbolt and M. Leroy Spearman initially pursued the twin-fuselage configuration for enhanced performance and productivity for subsonic transports. Their approaches used dual fuselages as wingtip "end plates" and an unswept, long-chord wing section between the fuselages. Houbolt was interested in such designs for favorable trades between geometric internal volume and aerodynamic skin friction drag. Spearman's interest, on the other hand, was in larger capacity (900 passengers) transports with acceptable "footprint" and ground handling. The interest in supersonic transport applications was generated by results of computational and wind-tunnel studies conducted by Samuel M. Dollyhigh and other Langley researchers. The aerodynamic principle involved in the supersonic application is using a pressure field generated by one airplane component to favorably modify

Unconventional supersonic transport designs, such as this high-payload twin-fuselage concept, were explored in the Supersonic Cruise Research Program.

A 1981 conceptual Boeing design for a supersonic business jet.

the pressures on another component. In the twin-fuselage configuration arrangement, supersonic drag is reduced because of the positive fuselage forebody compression pressures emanating from each forebody and impinging on the rearward facing surface of the half portion of the adjacent body, thereby increasing the afterbody pressure and reducing the drag. Langley data indicated that a two-fuselage vehicle could have 25 to 30 percent lower pressure drag than a single fuselage configuration.

Finally, the SCR interactions with industry sparked the interest of a number of industry design teams for supersonic business jets. Several configurations were brought forward, and the potential benefits of a smaller airplane over a large supersonic transport in areas such as sonic boom and noise were recognized. Years later, in the early 2000s, the concept of supersonic business jets would again arise, as will be discussed.

Termination of the SCR Program

Nearly 1,000 reports and presentations resulted either directly or indirectly from research supported by the SCR Program. In addition to day-to-day contact with pertinent industry personnel, annual reviews of the disciplinary research were conducted, and two major NASA conferences were held at Langley in 1976 and 1979, both well attended by members of the aerospace, military, and academic communities. Technology transfer from the program was especially effective because of the intimate working relationships that existed with program participants and the focused nature of research efforts. The approach of active industry involvement in NASA research activities has been continually demonstrated as the most productive approach to technology maturation and application of advanced concepts, and the legacy of the SCR Program serves as an outstanding example of this success.

In the late 1970s, the anti-SST movement was still very active in the United States, worldwide fuel shortages and price increases were encountered, the Concorde had proven an economical disaster, and low-fare availability on wide-body subsonic transports attracted the interest of the flying public. NASA faced major funding issues in its Space Shuttle programs, and the Agency decided to terminate the SCR Program in 1982. Although the program did not result in a second-generation SST, it produced technology of immediate value to the subsonic transport industry, including SPF/DB, materials such as advanced metal matrix composite structures, and aerodynamic design tools such as advanced flow modeling.

The NASA High Speed Research Program

Program Genesis

After cancellation of the SCR Program, the sporadic NASA interest in supersonic transports once again greatly diminished. In November 1982, an interagency group under the direction of the White House Office of Science and Technology Policy (OSTP) conducted a study on the state of aeronautical research and the role of the government in supporting that research. The study concluded that revolutionary advances in aeronautics were possible and that industry and government must work together to realize the benefits. In 1985, a committee of government, industry, and academic experts reviewed the study and specified three goals for future research, one of which was Supersonic Goal: To Attain Long-Distance Efficiency. It was noted that from a strategic and economic perspective, the Pacific areas were constrained by distance, a factor adding significance to the supersonic goal. In February 1987, OSTP reinforced the supersonic research goals by adding emphasis on the need to resolve environmental issues and proposed a plan to achieve the goals.

Meanwhile, a new subject in high-speed flight abruptly burst on the international scene with widespread support and enthusiasm. Hypersonics, once an area of extremely low national interest (at one point in the 1970s, Langley was the only significant United States participant in hypersonic research), suddenly became a high priority target, as a result of emerging hypersonic propulsion concepts within NASA and DoD, with perceived military and civil benefits. The rising support for hypersonics even penetrated the White House, and in his 1986 State of the Union Address, President Ronald W. Reagan praised the virtues of an "Orient Express," a Mach 25 hypersonic transport that could fly from New York to Tokyo in 2 hours.

In response to the OSTP call for technology development to support a long-range high-speed transport as one of the national goals in aeronautics, NASA awarded 2-year contracts for market and technology feasibility studies of a high-speed civil transport (HSCT) to Boeing Commercial Airplanes and Douglas Aircraft Company in October 1986. The scope of the studies requested a broad consideration of civil supersonic, hypersonic, or transatmospheric vehicles for high-speed transportation. These studies were coupled with independent in-house NASA team assessments. The assessments included the market potential of an HSCT, candidate vehicle concepts and the critical technologies that each concept would require, and the environmental issues relevant to each concept. Environmental issues were the focus during final contract phases in order to verify feasible technological resolutions.

The in-house Langley systems studies were organized by Charlie M. Jackson (Chief of the High-Speed Research Division) and his deputy, Wallace C. Sawyer, and conducted by a team led by Samuel M. Dollyhigh and Langley retiree A. Warner Robins. The results of the industry studies, as well as the in-house study, dismissed the "Orient Express" hypersonic transport concept as infeasible due to its projected impact on airport infrastructure costs, technology shortcomings, and vehicle performance limitations. At the same time, other studies were underway including a major workshop at Langley in January 1988 to discuss the status of sonic boom methodology and understanding. Industry and NASA both concluded that an economically viable HSCT would be feasible if the environmental issues, such as noise and emission problems, could be resolved. The cruise speed recommended from the study results was in the range between Mach 2 and 3.2.

High-Speed Research Phase I

As a result of these systems studies, and growing U.S. industry and government concern over the threats of emerging European and Japanese interests in a second-generation SST, NASA formally initiated a High Speed Research (HSR) Program in 1990 to identify and develop technical and economically feasible solutions to the many environmental concerns surrounding a second-generation supersonic transport. In the phase I HSR studies, all efforts were directed toward resolution of the three great environmental demons that had devastated the first U.S. SST program: ozone depletion, airport and community noise, and sonic boom. Before the HSR activities could be assured of further support from Congress or industry, the HSR team had to prove that the HSCT would be ozone-neutral, that it could meet the current airport noise requirements (FAR 36 stage III), and that the sonic boom generated by the HSCT could be made acceptable for overland or overwater flight. The HSCT envisioned by the HSR participants would fly 300 passengers at 2.4 times the speed of sound—crossing the Pacific or Atlantic in less than half the time presently required on modern subsonic, wide-bodied jets—at an affordable ticket price (estimated at less than 20 percent above comparable subsonic flights), and be environmentally friendly. The Mach 2.4 cruise speed was selected by program participants because of a desire to use conventional jet fuel at worldwide locations instead of more exotic or expensive fuels required for higher cruise speeds. At Langley, Allen H. Whitehead, Jr., was assigned responsibility for the airframe-related research of phase I HSR studies, while Robert J. Shaw of Lewis Research Center managed the propulsion elements from Lewis. Langley took an aggressive lead in attacking the environmental issues of phase I. In the sonic boom area, for example, a major workshop had been held at Langley in January 1988 to discuss the status of sonic boom methodology and understanding.

The first HSR workshop was hosted by NASA Langley in May 1991 at Williamsburg, Virginia. Throughout the program's duration, NASA and its industry partners placed special emphasis on

Researcher David E. Hahne inspects a 1/10-scale model of a McDonnell
Douglas Mach 2.2 transport in the Full-Scale Tunnel in 1992.

the fact that technology included in the HSR Program was commercially sensitive, and all data and program results throughout the program's lifetime were protected by a special limited distribution system. Many research contributions of the HSR Program are still sensitive, and the discussion herein will only highlight the scope of activities.

Attendance at the first HSR workshop was by invitation only and included industry, academic, and government participants who were actively involved in HSR activities. The workshop sessions were organized around the major task elements, which addressed the environmental issues of atmospheric emissions, community noise, and sonic boom. Sessions included airframe systems studies, atmospheric effects, source noise, sonic boom, propulsion systems, emission

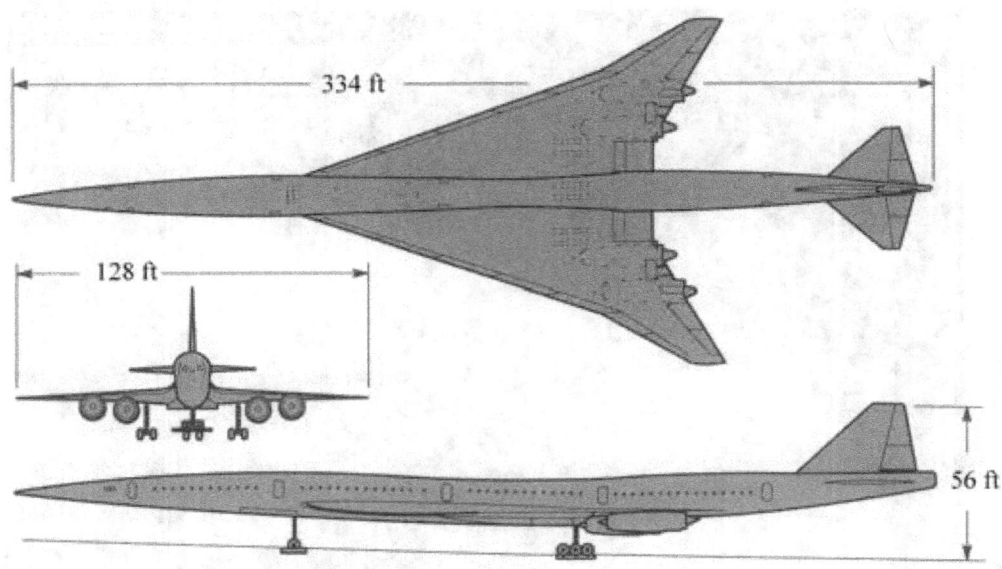

A McDonnell Douglas High-Speed Civil Transport concept during the early High Speed Research Program

reduction, aeroacoustics and community noise, airframe/propulsion integration, airframe and engine materials, high lift, and supersonic laminar flow control. Key Langley presenters included Samuel D. Dollyhigh, Ellis E. Remsberg, William L. Grose, Lamont R. Poole, John M. Seiner, Christine M. Darden, Langley retirees Domenic J. Maglieri and Percy J. (Bud) Bobbitt, Robert A. Golub, Gerry L. McAninch, Kevin P. Shepherd, Thomas T. Bales, Paul L. Coe, Jr., and Michael C. Fischer.

At the end of phase I studies, the industry and NASA team had provided convincing results that airport noise and emission problems could be conquered with advanced technology, but the lack of a solution to the sonic boom issue resulted in a restriction to overwater supersonic flight. Major questions remained as to the airplane's economic viability and the acceptability of the advanced technology costs.

High-Speed Research Phase II

Phase II of the HSR Program began in 1995 and was designed to assess and enhance the economic competitiveness of the HSCT. NASA and its industry team had begun to crystallize the advocacy and benefits of the HSCT. Industry experts predicted that the number of flights to the Pacific Rim would quadruple by the next century, spurring demand for over 500 next-generation supersonic passenger transports. Therefore, the first country to develop a supersonic transport that was competitive (i.e., less than a 30-percent ticket surcharge) with existing fares for subsonic transports would capture a significant portion of the long-haul intercontinental market.

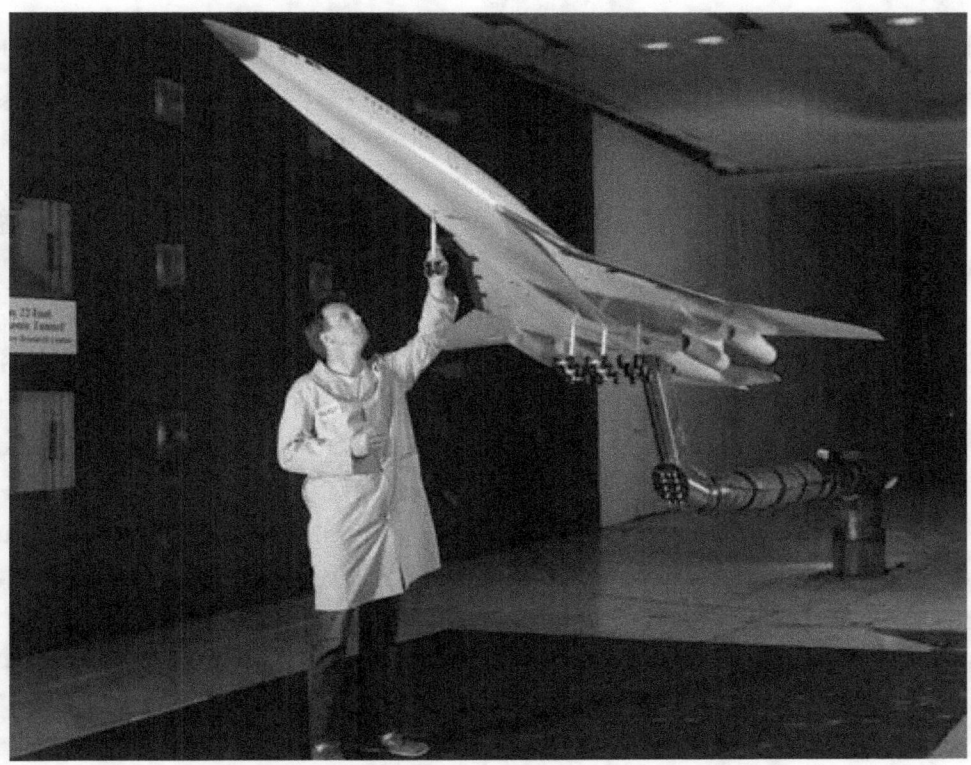

Technician Michael E. Ramsey inspects a 19-ft long model of the High-Speed Civil Transport Reference H configuration during tests of takeoff and landing characteristics in Langley's 14- by 22-Foot Subsonic Tunnel.

The NASA HSR phase II team was led by the Agency's HSR Program Office at Langley and was supported by Dryden Flight Research Center, Ames Research Center, and Lewis Research Center. Major industry partners in the HSR Program were Boeing Commercial Airplane Group, McDonnell Douglas Aerospace, Rockwell North American Aircraft Division, General Electric Aircraft Engines, and Pratt & Whitney.

The HSR Program Office was led by Wallace C. Sawyer, who began as the Agency's Director of the High-Speed Research Program in 1994. Sawyer was assisted by his deputy Alan W. Wilhite, and research tasks were accomplished by a unique teaming of NASA-industry team, with personnel divided into integrated disciplinary work units. As the program progressed, Boeing was declared the industry manager for HSR activities, and the program office at Langley instituted technology offices focused on topics that included technology integration (William P. Gilbert), aerodynamic performance (Robert L. Calloway), materials and structures (Rodney H. Ricketts), environmental impact (Allen H. Whitehead, Jr.), and flight deck technology (Daniel G. Baize).

Advanced technologies in the areas of aerodynamics, structures, propulsion, and flight deck systems were applied to a baseline Mach 2.4 vehicle concept called Reference H. In late 1995, the participants produced a new transport design known as the Technology-Concept Airplane (TCA). The TCA was also a Mach 2.4 aircraft with a 300-passenger capacity and a range of more than 5,000 nmi. During HSR phase II, NASA was spending a quarter of its annual $1 billion aeronautics budget for high-speed research.

Phase II technology advances needed for economic viability included weight reductions in every aspect of the baseline configuration because weight affects not only the aircraft's performance but also its acquisition cost, operating costs, and environmental compatibility. In materials and structures, the HSR team attempted to develop technology for trimming the baseline airframe by 30 to 40 percent; in aerodynamics, a major goal was to minimize drag to enable a substantial increase in range; propulsion research was directed at environment-related and general efficiency improvements in critical engine components, such as inlet systems. Other research involved ground and flight simulations aimed at the development of advanced control systems, flight deck instrumentation, and displays.

In addition to maintaining close coordination of industry and NASA research activities on a daily basis, the various disciplinary offices within the HSR Program Office organized and conducted extensive workshops in their technical areas, summarizing and disseminating the latest advances in technology and assessing the progress toward goals set for the HSR Program. For example, annual HSR sonic boom workshops from 1992 to 1995 brought together NASA's scientists and engineers with their counterparts in industry, other Government agencies, and academia to review the program's sonic boom element. Specific objectives of these workshops were to discuss theoretical aspects of sonic boom propagation, experimental results of boom propagation, results of boom acceptability studies for humans and animals (including sea life), and low-boom configuration design, analysis, and testing.

In the sonic boom area, the first low-boom designs developed by Christine Darden and Bob Mack as part of the HSR Program were updated to include more features of real airplanes than the simple flat wing-body designs that had been used in signature modification testing of the mid 1970s. In 1990, Mach 2 and Mach 3 twisted wing-body-nacelle model concepts were designed to produce tailored "flat-top" or "ramp" type signatures. During wind-tunnel tests of these models, large, unpredicted shocks emanating from the flow-through nacelles were encountered (the first wind-tunnel models had no flow-through nacelles). The next generation of low-sonic boom designs, begun in 1991, had two objectives: to correct the nacelle integration concerns and to improve the overall aerodynamic performance of the low-boom concept. Analysis methods were modified to ensure that inlet shocks were predicted. Also, for the first time analysis was accomplished with powerful, nonlinear computational fluid dynamics (CFD) methods. Because the traditionally used sonic boom theory is only valid at mid- to far-field distances, CFD methods are the only means of generating a near-field signature, one that can be compared directly with wind-tunnel data, and one wherein signature features can be directly correlated with configuration features.

For several models in this cycle, CFD methods were used to iteratively design the desired signature. The use of CFD had also become more imperative as wind-tunnel models became larger in order to incorporate the increasingly realistic features, such as twist and camber, and nacelles. Larger models necessitated measuring the signatures at closer and closer distances. All sonic boom models built during the HSR Program were 12 in. long and measurements were taken at 2 to 3 body-lengths away. Test results on this generation of models met with moderate success. Shocks from the nacelles were successfully embedded within the expansion wave of the vehicle and, while the predicted ground signature was not an N-wave, the slope of the pressure growth was much steeper than predicted. Because the initial signatures were now being measured quite close to the model, concerns for three-dimensional effects or uniform atmosphere effects began to arise. Because of the varying levels of systems analysis accompanying the low-boom designs begun in 1991, and because the impact of sonic-boom reduction techniques on the mission performance was a critical measure of success, an attempt was made to conduct a consistent analysis of the mission performance on all the designs. Extensive laboratory and field research on boom signature acceptability by J. D. Leatherwood, Kevin P. Shepherd, and Brenda M. Sullivan played a key role in the evolution of the target overpressure levels and boom shaping characteristics for low-boom aircraft.

Early in the HSR Program, Langley retiree Domenic Maglieri had pointed out that real-world demonstrations of boom signature modifications were needed to validate that a beneficial shaped boom signature (which had only been accomplished on wind-tunnel models out to about 10 to 30 body lengths) would persist for large distances in the atmosphere, for example, to 200 or more body lengths. Potential approaches to obtaining the necessary data were addressed, including the use of nonrecoverable supersonic target drones, missiles, full-scale airplane drones, very large wind tunnels, ballistic facilities, whirling-arm techniques, rocket sled tracks, and airplane nose probes.

Under Maglieri's leadership, it was found that the relatively large 28-ft supersonic Teledyne-Ryan BQM-34E Firebee II drone would be a suitable test vehicle in terms of its adaptability to geometric modifications, operational envelope capabilities, availability, and cost. A program was funded from 1989 through 1992 that included CFD analyses and wind-tunnel tests on models of the baseline Firebee II, including one in which the vehicle forebody was lengthened and reshaped so as to provide a "flat-top" sonic boom signature. Before funding was terminated when NASA's HSR

Program ended, a flight-test plan was developed that involved measurements read at ground level and also in the vehicle near-field using microphones mounted on a Pioneer uninhabited aerial vehicle (UAV).

Another flight demonstration that was proposed to prove the persistence of a "shaped" boom signature from an aircraft flying in a real atmosphere used a modification of the SR-71. Initial studies suggested that a significant amount of airframe modifications of the baseline airplane would be required in order to acquire the desired shaped boom signature. The SR-71 proposal also died due to the subsequent termination of the NASA HSR Program. However, interest in the SR-71 as a test article initiated a NASA flight test program to probe an SR-71 in flight to measure off-body pressures.

NASA's SR-71 and F-16XL during sonic boom research at Dryden in 1995.

During 1995, a NASA team at Dryden conducted a series of flights in the vicinity of Edwards Air Force Base with an SR-71 aircraft to study the characteristics of sonic booms. On each flight, two other NASA aircraft took in-flight sonic boom measurements to augment ground-based measurements and to aid in the analysis. A NASA YO-3A propeller-driven "quiet" airplane flew subsonically between the SR-71 and the ground, while a NASA F-16XL flew supersonically near the SR-71 to measure the sonic boom at distances as close as 85 ft below and to the rear of the generating aircraft. Speeds flown by the two aircraft ranged between Mach 1.25 and Mach 1.8 at altitudes of about 30,000 ft. This successful technique of in-flight probing was to be used later in a NASA-DARPA boom-shaping study, as will be discussed.

In 1997, Robert L. Calloway and Daniel G. Baize organized an HSR aerodynamic performance workshop that included sessions on configuration aerodynamics (transonic and supersonic cruise drag prediction and minimization), high-lift, flight controls, and supersonic laminar flow control. Single- and multipoint optimized HSCT configurations, HSCT high-lift system performance predictions, and HSCT motion simulator results were presented, along with executive summaries for all aerodynamic performance technology areas. Key Langley presenters were Steven X. S. Bauer, Richard A. Wahls, Steven E. Krist, Richard L. Campbell, Francis J. Capone, Guy Kemmerly, Wendy Lessard, Lewis Owens, and Bryan Campbell. Subsequent workshops were held in 1998 and 1999 simultaneously with annual reviews of activities in materials and structures, environmental impact, flight deck, and technology integration.

Another critical area of the HSR Program was materials durability. An objective of the activities was to demonstrate the ability of candidate materials to withstand mechanical and environmental factors that contribute to long-term degradation of properties under conditions simulating HSCT flight. Among the objectives was the development of associated predictive and accelerated tests,

Extensive testing of the High Speed Research Reference H configuration was conducted in many Langley tunnels, including the 16-Foot Transonic Tunnel (left) and the National Transonic Facility (right).

methods, and assessment of durability protocols for design. Langley's Thomas S. Gates led research in the area for the Center. The HSCT was intended to have a design life of 60,000 flight hours. At speeds of Mach 2.4, the maximum operating temperature could reach 350 °F under conditions of varying oxygen, moisture, solvents, and other environmental stress factors. Materials chosen for the vehicle would have to withstand at least one lifetime of this mechanical and environmental loading prior to production go-ahead. With the introduction of new materials and material forms, not all candidate materials could be subjected to adequate long-term tests; therefore, the necessity of accelerated and predictive methods requires development of validated techniques for screening materials for long-term durability. Gates directed efforts at predictive and accelerated test methods that use empirical and analytical relationships to simulate long-term durability performance and provide the engineer a standard and simple approach to screening new materials for long-term durability performance.

Langley also organized studies of appropriate materials for HSCT applications, including advanced aluminums and composite materials. Thomas T. Bales led the efforts on aluminums, and starting in 1991, Paul M. Hergenrother led a team in the development of new high temperature composites, adhesives, sealants and surface treatments. After sorting through over 200 candidate materials, in 1995 they arrived at a highly promising candidate known as PETI-5. PETI-5 is the commonly used term for phenylethynyl terminated imide oligomers. PETI-5 is a chemical material that can be used as both a composite resin and an adhesive. It also combines superb mechanical properties and extreme durability with easy processing and environmental stability. The material met the temperature/flight hour targets for the HSCT and was prepared from commercially available materials. Curing was done with the application of heat and mild pressure, which resulted in the formation of a strong, resistant polymer. Due to the nature of this polymer, it was fairly easy to create large, complex parts using PETI-5. Unacceptably high manufacturing cost was one of the major barriers to applications of the technology.

A series of highly successful fabrication demonstrations by Langley raised excitement over the material's potential use in the HSCT; however, developmental testing required for the application could not be accomplished within the lifetime of the HSR Program. Nonetheless, the material was recognized as a major materials breakthrough and quickly gained interest for other applications. PETI-5 won the NASA Commercial Invention of the Year award for 1998, and the material received Research and Development Magazine's R&D 100 award. Through licensing from Langley, companies have positioned themselves for markets in the areas of electronic components, jet engines, and high performance automotive applications.

Langley's research on improving cruise efficiency for HSCT configurations included wind-tunnel, computational, and flight studies of supersonic laminar flow control (SLFC). Efforts with industry in the development of SLFC in the HSR Program are discussed in detail in a separate section herein on laminar flow control. That joint effort by Langley, Dryden, Boeing, McDonnell Douglas and Rockwell had resulted in flight evaluations of the state of design theory and the robustness of SLFC using an active suction panel gloved to the wing of an F-16XL airplane in flight tests at Dryden. Michael C. Fischer was Langley's manager for the flight experiment. Several technical concerns had resulted during the flight tests, including contamination of flows by shocks emanating from the airplane's canopy and forebody. In addition, the F-16XL was deemed too small for convincing proof of the feasibility of SLFC applied to a large supersonic transport, and other potential airplane test beds were explored.

NASA's F-16XL was modified with an active suction laminar flow glove on its left wing for exploratory assessments of supersonic laminar flow control.

The Russian Tu-144LL takes off on a research flight sponsored by the NASA High Speed Research Program in 1998.

Several larger airplanes had been considered by NASA for follow-up experiments, including a Concorde. Late in 1992, Langley's Dennis Bushnell had discussed refurbishing one of the surviving Russian Tu-144 SSTs for use in supersonic flight research with the aircraft's manufacturer, Tupolev. Following several in-house staff discussions at Langley, Joseph R. Chambers, Chief of Langley's Flight Applications Division (sponsoring organization for the SLFC HSR effort), sent a formal proposal to NASA Headquarters (with the approval of Roy V. Harris, Langley's Director for Aeronautics) for using a Tu-144 for SLFC flight experiments. Meanwhile, Headquarters had independently funded a study contract to Rockwell to conduct a feasibility assessment of restoring a Tu-144 for use in the HSR program. Informal discussions between Headquarters and Tupolev had taken place at the Paris Air Show with positive results. The proposal met with approval at Headquarters, and a cooperative Tu-144 project was enabled by an agreement signed in June 1994 in Vancouver, Canada, by Vice President Albert A. Gore and Russian Prime Minister Viktor Chernomyrdin. This was the most significant post-Cold War joint aeronautics program to date between the two countries. Unfortunately, the original Langley interest in SLFC research on the Tu-144 was not accommodated within the ensuing activities.

A Tu-144 was modified by the Tupolev Aircraft Design Bureau in 1995 and 1996. The newly designated Tu-144LL Flying Laboratory performed flight experiments as part of NASA's HSR Program for studies of high temperature materials and structures, acoustics, supersonic aerodynamics, and supersonic propulsion. The Tu-144LL rolled out of its hangar on March 17, 1996, to begin a 6-month, 32-flight test program. Six flight and two ground experiments were conducted during the program's first flight phase, which began in June 1996 and concluded in February 1998 after 19 research flights. A shorter follow-up program involving seven flights began in September 1998 and concluded in April 1999. All flights were conducted in Russia from Tupolev's facility at the Zhukovsky Air Development Center near Moscow. Langley's Robert A. Rivers and Dryden's C. Gordon Fullerton became the first American pilots to fly the modified Tu-144LL during the 1998 experiments. during their evaluations, Rivers and Fullerton were primarily concerned with the Tu-144LL's handling qualities at a variety of subsonic and supersonic speeds and flight altitudes.

The HSR flight deck studies at Langley contributed enabling efforts in synthetic vision that might eliminate the need for the heavy drooped nose concept used on the first generation SSTs. The very significant accomplishments of this program are presented in another section of this document. Under the management of Daniel G. Baize, the synthetic vision program accelerated following the demise of the HSR Program as the technology was widely appreciated for its potential use for all types of aircraft. The work area became part of the new NASA Aviation Safety Program following the end of the HSR activities. The technology ultimately transitioned into a broader applications arena for the subsonic commercial airplane fleet, including aggressive applications by the business airplane community. This area is also covered in detail in a separate section herein.

Langley's HSR activities also included participation in efforts to resolve issues regarding the impact of emissions from commercial aircraft. A NASA Atmospheric Effects of Aviation Project (AEAP) was organized to develop scientific bases for assessing atmospheric impacts of the exhaust emissions discharged during cruise operations by fleets of subsonic and supersonic civil aircraft. The AEAP comprised two major entities, a Subsonic Assessment (SASS) project and an Atmospheric Effects of Supersonic Aircraft (AESA) project. The SASS project was conducted under the auspices of NASA's Advanced Subsonic Technology Program (ASTP), and the AESA project was conducted under the HSR Program. Because of the shared focus on environmental impact, program management of the two assessment efforts was consolidated into an overall program, the AEAP. The AESA project was designed to assess the impacts of a potential future fleet of HSCTs with cruise operations at midstratospheric altitudes.

HSR activities also focused on yet another environmental issue: the effects of radiation exposure to aircrew and passengers of an HSCT. The National Council on Radiation Protection and Measurement (NCRP) and the National Academy of Science (NAS) had concluded that the data and models associated with the high-altitude radiation environment needed refinement and validation. In response, NASA and the Department of Energy Environmental Measurements Laboratory (EML) created the Atmospheric Ionizing Radiation (AIR) project under the auspices of the HSR Program. In AIR, international investigators were solicited to contribute instruments to fly on an ER-2 aircraft at altitudes similar to those proposed for the HSCT. The flight series took place at solar minimum (radiation maximum) with northern, southern, and east/west flights. The investigators analyzed their data and presented preliminary results at an AIR workshop in March 1998.

Termination of the High-Speed Research Program

In the late 1990s, Boeing assessed its seriousness in future commitments to pursuing the HSCT. The development cost of this relatively high-risk airplane was estimated to be over $13 billion. Key technologies had not advanced to an acceptable level, especially in the areas of propulsion, noise, and fuel tank sealants. These technical shortcomings, together with production problems with its existing line of subsonic transports, major commitments to its new B777 transport, and other considerations led to a Boeing perspective that reflected critical technical obstacles that required solutions confirmed by full-scale component demonstrations and continuing study regarding the marketability and economics of supersonic aircraft.

The HSR Program was dependent on an active partnership between the government and industry. NASA terminated HSR in 1999 after Boeing dramatically reduced support for the project, shrinking its staff devoted to HSR from 300 to 50 and pushing the operational date for a high-speed supersonic transport from 2010 to at least 2020. Boeing's action was the result of market analysis and technology requirement assessments indicating that, from an economically and environmentally sound perspective, the introduction of a commercial HSCT could not reasonably occur prior to the year 2020. Industry and NASA also questioned whether the technologies being pursued would appropriately address environmental standards and other challenges in 2020. In response, NASA reduced activity in the HSR Program to a level commensurate with industry interest.

NASA terminated the program in 1999 in order to add $600 million to the budget for the International Space Station (ISS). The extra money was needed as part of a $2 billion, 5-year commitment to back up Russian delays on the ISS. Daniel S. Goldin, NASA Administrator, speaking to the House's Subcommittee on Space and Aeronautics on February 24, 1999, summarized the termination of the HSR activity:

"We are proud of our past accomplishments in two focused programs, High Speed Research and Advanced Subsonic Technology. Although dramatic advances were made against the original HSR Program goals, our industry partners indicated that product development would be significantly delayed, which led to the decision to terminate this program at the end of 1999."

Post–High-Speed Research Activities

Following the NASA HSR Program termination in 1999, the subsequent retirement of the Concorde fleet, rising fuel prices, and the demise of a large number of commercial airlines, worldwide interest in large commercial supersonic transports plummeted to new lows. Langley's research activities in supersonic civil aircraft were severely reduced in scope and funding, and the remaining funds and researchers were reoriented toward more fundamental research on a few critical technologies, with a limited focus on notional vehicles, such as supersonic business jets, and demonstrations of selected technologies, such as sonic boom shaping. Meanwhile, industry expertise and design teams from the previous supersonic transport programs were disbanded and reassigned to other programs, leaving a wide void nationally in experienced supersonic civil airplane capability.

In 2004, the NASA Headquarters Aeronautics Research Mission Directorate sponsored a Vehicle Systems Program (VSP) on notional advanced aeronautical vehicles, including studies of associated technology goals and development roadmaps to help focus NASA research efforts in civil aircraft.

The research community is developing new interest in supersonic business jets.

As part of its VSP responsibilities, Langley participated in vehicle-specific studies to identify breakthrough capabilities and resolve barrier issues for subsonic and supersonic civil vehicles. Coordinated by Langley's Peter G. Cohen, the supersonic segment of the program included three major thrusts: fundamental research in propulsion, aerodynamics, emissions, and noise; participation in cooperative programs with the military and industry; and participation in advocacy planning for new initiatives in supersonics. At Langley, specific tasks included aerodynamic design optimization, sonic boom, low-boom configurations, systems studies, and aeroelastic phenomena. In addition to Langley activities, the research program supported research efforts at the Glenn and Dryden Centers, as well as cooperative investigations with industry partners.

Notional supersonic civil aircraft studies in the VSP effort included business jet and large commercial transport applications. Compared with Concorde state-of-the-art characteristics—Mach 2.0, 400,000 lbs, 100 passengers, overwater cruise—Langley investigated a near-term (5 to 10 years) supersonic business jet—designed for Mach 1.6 to 2.0, 100,000 lbs, 6 to 10 passengers, and overland cruise—known as the Silent Small Supersonic Transport. The studies also included a far-term (15 years), second-generation transport (Mach 0.95-2.0, 400,000 lbs, 150 to 200 passengers, "corridor flight path"). The Langley VSP sponsored basic research in a number of areas, including advanced supersonic design methods and assessments of unconventional supersonic transport configurations, such as multiple-body designs.

Sonic Boom Shaping

One key technology target addressed by the VSP Program is the eternal sonic boom issue. Widely accepted as the single largest barrier to economically feasible supersonic civil flight operations, the boom continues to receive research focus. Perhaps no other aspect of supersonic flight has created such a polarization of attitudes regarding the possibility of solutions. Many within the aviation community regard the boom as a fact of life— generated by fundamental laws of physics unyielding to modification. However, Langley's past research on methodology to modify sonic boom signatures and make them more acceptable is providing optimism and guidance to today's efforts. The extensive results of computational analysis, wind-tunnel tests, flight tests, and human response evaluations have resulted in a relatively mature understanding of key sonic boom physical characteristics and the objectionable nature of certain critical boom phenomena.

While it is universally agreed that lifting the ban on commercial supersonic overland flight would represent a significant breakthrough in aviation history, the technology to do so has not yet been demonstrated. Mitigation of the sonic boom via specialized shaping techniques was theorized nearly four decades ago; but, until recently, this theory had never been tested with a flight vehicle subjected to actual flight conditions in a real atmosphere.

A major finding of sonic boom research from the 1960s is that noise reductions in the boom signature could potentially be achieved by the use of boom shaping. In particular, if the classical sharp-edged "N" waveform could be modified so that the onset of the initial shock was systematically reduced and shaped (lower initial pressure increase, resulting in a longer rise time and a less abrupt change in pressure), the booms might be less objectionable, as confirmed by Langley research using human test subjects experiencing sonic booms in a high-fidelity simulation facility.

The overpressure level of sonic boom that would be acceptable to the general population has been investigated in ground-based simulation as well as actual supersonic overflights of populated areas. Based on these results and studies of community response to other high-level impulsive sounds (artillery, explosions, etc.), criteria were developed for acceptability of sonic booms. Analysis has indicated that overpressure levels for a representative large HSCT-type supersonic transport would be about 3 lb/ft^2, and that significant reductions would be necessary to achieve acceptance by the bulk of the population (the smaller Concorde generated an unacceptable level of 2 lb/ft^2).

The aerospace community has, of course, been aware that sonic boom overpressures are directly relatable to size, weight, and length of the vehicle under consideration, and that smaller aircraft, such as business jets, could have a lower level of sonic boom than large transports. In addition, the Langley results discussed earlier in other supersonic transport programs had indicated that shaping of the boom signature might be accomplished through aircraft geometric shaping. In particular, wind-tunnel and computational results showed that features—such as placement of engine nacelles, wing planform shape and lift distribution, and fuselage forebody geometry—could modify boom characteristics in a favorable matter.

Stimulated by the possibility that sonic boom minimization might be achievable for supersonic business jets, thereby permitting overland supersonic cruise with its favorable economics, industry has continued its interest in the design methodology for this application, but the lack of full-scale flight data to validate approaches to boom minimization has continued to block applications.

After NASA's termination of the HSR effort, essentially all activity on sonic boom minimization ceased with the exception of small Langley contracts with industry. In 1999, a small cooperative research program was conducted with Lockheed to study the feasibility of a supersonic business jet. Both Langley and the Ames Research Center conducted design and testing in this effort. In 2003 a contract was awarded to Raytheon to study the feasibility and technology requirements for supersonic business jets.

In 2000, the Defense Advanced Research Projects Agency (DARPA) initiated a Quiet Supersonic Platform (QSP) Program directed toward the development and validation of critical technology for long-range advanced supersonic aircraft with substantially reduced sonic boom, reduced takeoff and landing noise, and increased efficiency relative to current technology supersonic aircraft. Improved capabilities would include supersonic flight over land without adverse sonic boom consequences, with an initial boom overpressure rise less than 0.3 lb/ft^2, unrefueled range approaching 6,000 nmi, and lower overall operational cost. Advanced airframe technologies would be explored to minimize sonic boom and vehicle drag, including natural laminar flow, aerodynamic minimization (aircraft shaping), exotic concepts (plasma, heat and particle injection), and low weight structures.

DARPA's Program Manager for QSP was Richard W. Wlezien, a Langley researcher on assignment to DARPA. In formulating the advocacy and content of his program, Wlezien had received major briefings on the state of low-boom research by Peter Coen, William P. Gilbert, and Langley retirees Domenic J. Maglieri and Percy J. (Bud) Bobbitt. These briefings played a major influence on DARPA's management in the final program approval.

Northrop Grumman, Boeing, and Lockheed Martin all performed phase I design studies for DARPA. In phase II, DARPA focused on military variants and attached particular significance to a flight demonstration of boom shaping and validation of propagation theories in real atmospheric conditions as had been advocated by Maglieri. DARPA subsequently initiated a Shaped Sonic Boom Demonstration (SSBD) Program with Northrop Grumman Corporation, including major participation from NASA Langley and Dryden, Naval Air Systems Command, Lockheed Martin, General Electric, Boeing, Gulfstream, Wyle Laboratories, and Eagle Aeronautics. Eagle Aeronautics' participation was particularly valuable to the team. Maglieri and Bobbitt represented Eagle, bringing with them decades of valuable experience and expertise in the areas of sonic boom and applied aerodynamics. Maglieri's leadership and creativity in sonic boom technology played a key role in the formulation and approach used in the program; while Bobbitt's extensive knowledge of applied aerodynamics provided guidance in aerodynamic integration and wind-tunnel testing.

The SSBD project was yet another example of close working relationships between NASA research centers. In addition to wind-tunnel tests conducted by the Glenn Research Center, the effort included valuable contributions by Dryden personnel in the planning and conduct of the flight experiment. Dryden's Edward A. Haering, Jr. was a critical participant in the project working with Peter Coen and others to create valid and successful sonic boom experiments. Haering also conceived, designed, and led an inlet shock spillage measurement test that was used to calibrate Northrop Grumman CFD methods for the design of the SSBD.

The objective of the SSBD Program was to demonstrate the validity of sonic-boom shaping theory in real flight conditions. For the demonstration, Northrop Grumman modified an F-5E fighter aircraft that was provided by the U.S. Navy's Naval Air Systems Command. The company designed and installed a specially shaped "nose glove" substructure and a composite skin to the underside of the fuselage. The final nose shape designed by Northrop Grumman met boom, pilot vision, wave drag, trim drag, and stability and control requirements.

The F-5 shaped-boom demonstrator aircraft.

Aircraft used in the shaped-boom tests. From top: the NASA F-15B survey airplane, the F-5 shaped-boom demonstrator, and the baseline F-5E.

On August 27, 2003, the first ever in-flight demonstration of boom shaping occurred at Dryden. During the experiment, the modified F-5E aircraft flew through a test range at supersonic speeds. Dryden sensors and industry sensors on the ground and in a Dryden F-15B research plane measured the shape and magnitude of the sonic boom's atmospheric characteristics. During some of the demonstrations, a NASA F-15B flew behind the modified F-5E in order to measure that

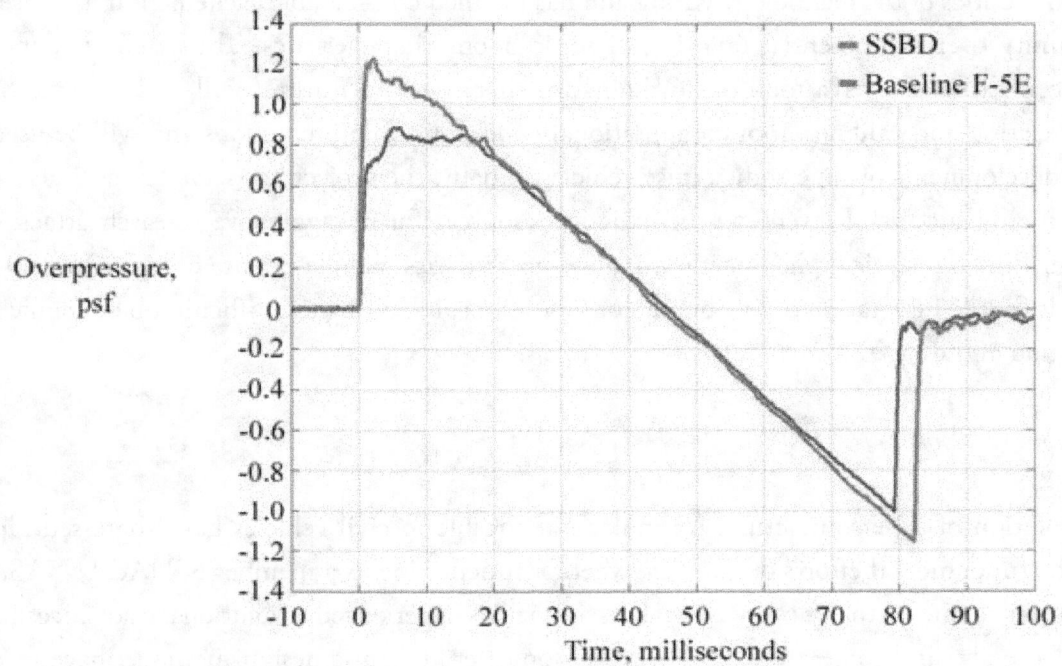

First proof that the sonic boom signature can be altered in actual flight conditions. The signature of the shaped sonic boom demonstrator clearly shows a beneficial softening of the bow shock signature compared with the baseline F-5E.

aircraft's near-field shockwave patterns. The F-15B's specially instrumented nose boom recorded static pressure measurements while flying behind and below the modified F-5E at a speed of about Mach 1.4. As previously discussed, this innovative in-flight data gathering technique had been developed by Dryden in 1995, when a Dryden F-16XL aircraft probed the shockwave patterns of a SR-71 research aircraft.

On three other flights, the modified F-5E SSBD was followed about 45 seconds later by an unmodified F-5E to determine the effect of aircraft shaping on sonic booms. The ground-based and airborne instruments measured the world's first shaped (flat top) sonic boom, followed by the normal, unshaped (N-wave) sonic boom from the unmodified F-5E. The data comparison of the two aircraft signatures clearly showed the persistence of the shaped boom, which was the highest priority of the study. A reduction in the sonic boom intensity was an attendant benefit. Another significant result was a demonstration of the robust nature of the intensity reduction. An identical test performed later that day confirmed the original results.

The results of this remarkable flight project have made history as the first demonstration of the persistence of a shaped sonic boom waveform. The impact of the demonstration has proven numerous critics of the technology wrong and has instilled considerable excitement in the research community over the potential tailoring of sonic boom characteristics. The extensive data that have been gathered over a total of 26 flights for shaped-boom features will be used extensively for correlation and calibration of computational codes for boom predictions and will be used for future developments of supersonic cruise vehicles. Finally, the project has served as an illustration of the highly successful leveraging of national resources for an aggressive research attack on a very difficult aeronautical phenomenon. Follow-up activities include the use of higher order CFD methods for the integration of low-boom configurations, including the difficult job of engine inlet integration and drag minimization.

Status and Outlook

The evolution of supersonic cruise technology applicable to civil vehicles has progressed rapidly from the fundamental efforts of Langley, whether functioning as a member of NACA, NASA, or an industry team. Virtually every critical discipline has experienced revolutionary advances in the state of the art, and some areas, such as supersonic aerodynamic design methods, have reached a remarkable level of maturity. Focused efforts of the U.S. SST Program, and the Langley-led SCR and HSR Programs, have accelerated technology development in highly relevant national efforts. Although tremendous progress has been made toward providing the technology required for economically feasible and environmentally compatible civil supersonic aircraft, several major barriers—both technical and nontechnical—still limit the introduction of this class of vehicle.

The sonic boom issue remains a most formidable challenge to supersonic overland flight, along with airport noise and reaching closure for a supersonic transport airplane with acceptable economic return. Skyrocketing fuel costs, increased bankruptcy by major airlines, low-fare competition from very efficient advanced subsonic transports, and emerging alternatives to business travel also stand in the way of progress. Finally, the high risk and astounding development costs associated with developing and certifying such a revolutionary airplane in today's climate of economic uncertainty and environmental sensitivity closes the door for further developments. In view of these considerations, large supersonic transport aircraft are far removed from the aerospace community's venue for near-term commercial applications.

In early 2000, NASA requested the Aeronautics and Space Engineering Board (ASEB) of the National Research Council to conduct an 18-month study to identify breakthrough technologies for overcoming key barriers to the development of an environmentally acceptable and economically viable commercial supersonic aircraft. The ASEB responded with a Committee on Breakthrough Technology for Commercial Supersonic Aircraft, composed of ex-NASA, industry, DoD, and academic experts in virtually every critical discipline involved in supersonic cruise vehicles, many of which had participated in past supersonic transport efforts. The committee's job was made more difficult by the fact that breakthrough technologies were guarded as proprietary and competition-sensitive for most of the HSR Program's results.

The committee members concluded that an economically viable supersonic aircraft will require new focused efforts in several areas, as well as continued development of technology on a broad front. Furthermore, NASA must advance key technologies to a readiness level high enough to facilitate the handoff of research results to the aerospace industry for commercial development. The committee concluded that maturation of key technologies could enable operational deployment in 25 years or less of an environmentally acceptable, economically viable commercial supersonic aircraft with a cruise speed of less than approximately Mach 2. They concluded that the time required for deployment could be considerably less if an aggressive technology development program were focused on smaller supersonic aircraft. The committee's rationale was that goals in many critical areas would be easier to achieve with smaller aircraft; however, it will certainly take much longer to overcome the more difficult technological and environmental challenges associated with building a large commercial supersonic aircraft with a cruise speed in excess of Mach 2.

Despite the challenges, a glimmer of hope exists in the business aircraft sector where a continuation of interest and sporadic projects continue to flourish for potential supersonic business jets (SBJ). SBJ aircraft have been discussed since the days of the NASA SCR Program, and several aborted industry programs arose in the 1980s, including a collaboration between Gulfstream and Sukhoi. Now, new SBJ efforts are appearing in the United States and abroad. These upstart activities reason that the smaller SBJ will present a lower technology challenge in the areas of sonic boom and airport noise requirements, and several proposed new SBJ projects have surfaced for a foothold in what might become a revolutionary sector of aeronautics. Also, the economic factors for the SBJ market may not be as critical as those for large transports, since a smaller number of high income individuals, companies, and even governments would be willing to pay the price for a business time advantage.

In October 2004, at the National Business Aircraft Association (NBAA) convention, Aerion Corporation and Supersonic Aerospace International (SAI) announced plans to launch full-scale development of supersonic business jets. Aerion's design makes use of an unswept, sharp leading-edge wing, a T-tail, and a novel natural laminar flow wing. SAI teamed with Lockheed Martin's Skunk Works, with its design incorporating low sonic boom (overpressures of less than 0.3 lb/ft^2) design features, such as an inverted tail and careful integration of the fuselage, wing, and engine components, with an arrow wing. Both teams are investigating aircraft with a stand-up cabin, a range of 4,000 nmi (one-stop transpacific and nonstop United States to Europe) at a cruise speed of Mach 1.6 to 1.8.

In 2005 NASA refocused its Vehicle Systems Program for aeronautics as the Agency directed its priorities on a new Vision for Space Exploration approved by President George W. Bush. The revised VSP places priority on four flight demonstrations of breakthrough aeronautical technologies, with one of the flight projects directed to the mitigation of sonic boom phenomena. The NASA Sonic Boom Mitigation Demonstration Project is managed by the Dryden Flight Research Center. In addition to NASA Intercenter participation and efforts, industry teams have been tasked to define technology and design requirements in preparation for a request for proposals for a low-sonic-boom demonstration aircraft. During a five-month study, the team's research will determine the feasibility of modifying an existing aircraft or whether a new design will be required as the quiet boom demonstrator.

NASA awarded a grant to American Technology Alliances (AmTech) to fund the industry studies which are being conducted by four industry teams. The teams include individual efforts by Boeing Phantom Works and Raytheon Aircraft, while Northrop Grumman has teamed with Gulfstream Aerospace and Lockheed Martin has teamed with Cessna Aircraft Company. The same grant also funded Allison Advanced Development Company, GE Transportation, and Pratt and Whitney to support the teams with engine-related data.

In summary, past contributions and leadership of NASA Langley Research Center in supersonic cruise technology have led the way toward potential application to civil aircraft. Much of the technology intended for supersonic aircraft has already been applied to existing and emerging subsonic aircraft. Together with its industry partners, the Center has provided disciplinary design methodology, concepts, and integrated assessments that have raised technology to relatively high levels of maturity. It remains to be seen if the existing technical hurdles, economic factors, and environmental obstacles can be removed for future applications to the civil aircraft transportation system.

The Blended Wing Body: Changing the Paradigm

Concept and Benefits

One highly motivating factor in innovative research for aeronautics is a desire to achieve the most efficient aircraft configuration for specific missions. For a high-performance military fighter, this challenge entails providing extreme maneuverability, low-observable radar signature, and sustained cruise performance. To a large extent in the highly competitive world of commercial air transports, efficiencies and economic considerations govern the introduction of innovative designs that change the standard paradigm.

The history of aircraft design evolution since the first powered controlled flight by the Wright brothers depicts an interesting perspective on an emerging focus toward the most efficient layout for airplane configurations. In the years following the revolutionary events of 1903, an amazing injection of innovative configurations occurred involving the evolution from biplanes to monoplanes, propellers to turbojets, and open cockpits to pressurized passenger cabins. Technical contributions rapidly accelerated within the key technical disciplines of aerodynamics, structures and materials, flight systems, and propulsion, which led to the development and first flight of the swept-wing Boeing B-47 exactly 44 years later on December 17, 1947. In contrast to earlier airplanes, the B-47 configuration included unconventional external features, such as a swept wing and empennage as well as podded turbojet engines mounted to pylons beneath and forward of the wing.

Progress in the mission capabilities of civil and military aircraft has, of course, improved significantly since the days of the B-47. The continuous application of new technologies—such as

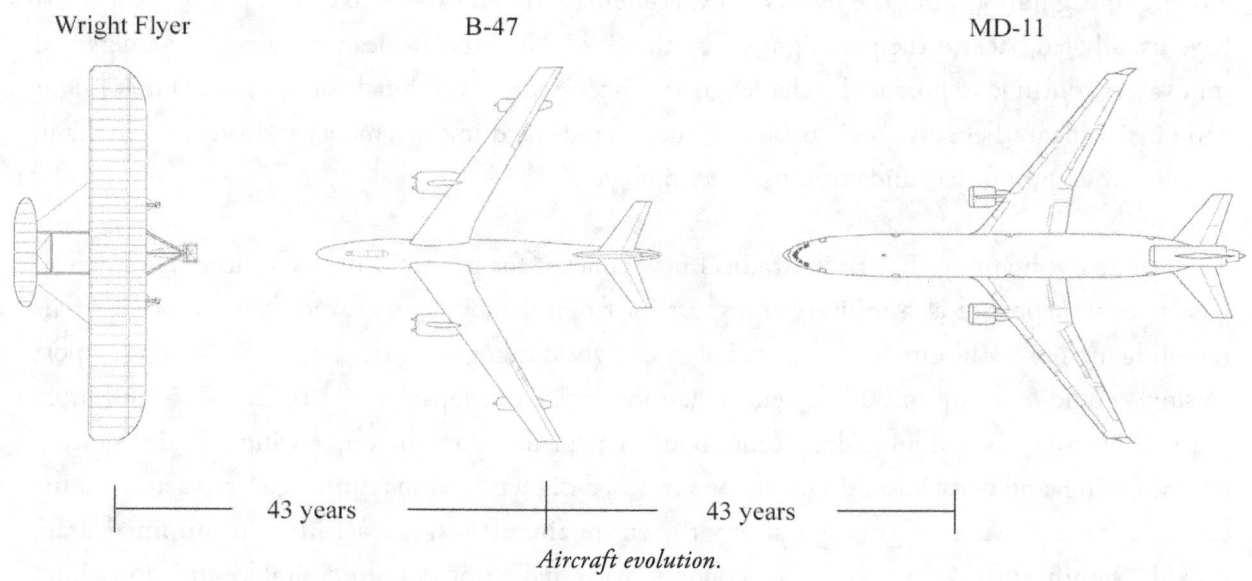

Aircraft evolution.

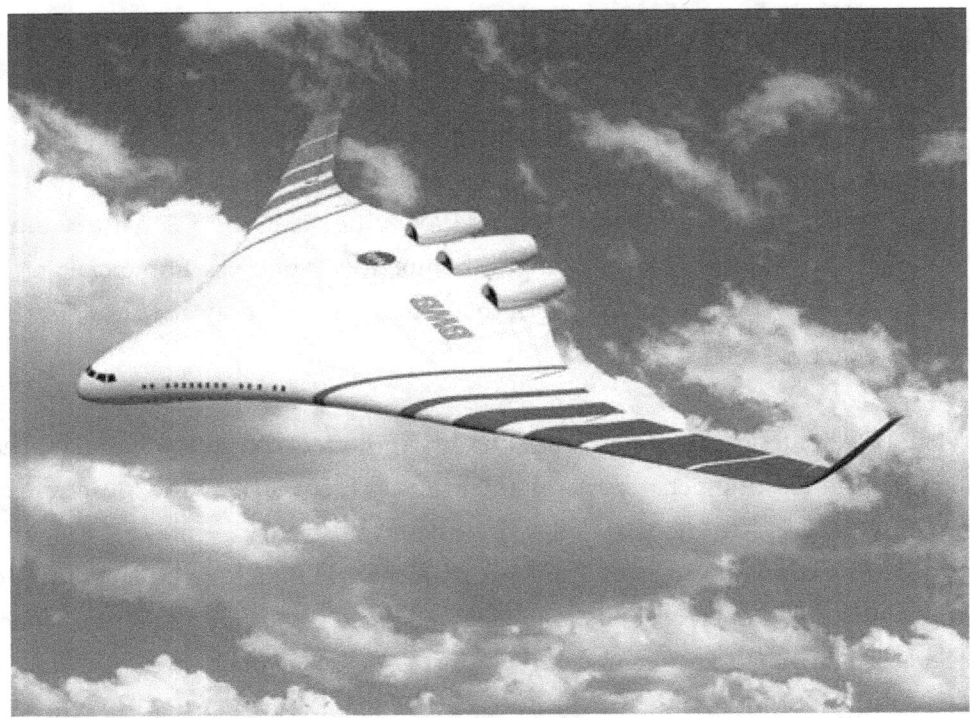

Artist's concept of a blended wing body commercial transport.

supercritical wings, advanced structures and materials, and high-bypass turbofans—has permitted the introduction of wide-body transports and extended-range missions. However, a cursory examination of the external features of current-day subsonic commercial transports indicates configuration features not unlike those used by the B-47 over 57 years ago. The historical attraction to the "tube and wing" arrangement has involved numerous flirtations with novel, unorthodox aircraft configurations, but the results of experimentation and experiences with prototype aircraft have usually led back to the paradigm set by the B-47. Nonetheless, leading-edge researchers and innovators continue to pursue the challenge to conceive unconventional configurations that might exhibit significant benefits and capabilities due to continued improvements and breakthroughs in aerodynamic, propulsive, and structural technologies.

The blended wing body (BWB) is a radical new concept for potential use as a future commercial passenger transport or as a military transport for troop deployments, cargo delivery, or air-to-air refueling missions. The airplane concept is a thick hybrid flying wing that, in commercial transport versions, could carry up to 800 passengers (almost twice the capacity of a Boeing 747-400) more than 7,000 miles in a double-deck centerbody that blends into the outer wings. By integrating engines, wing, and body into a single lifting surface, designers have maximized overall aerodynamic efficiency. The BWB configuration allows the entire aircraft to generate lift with minimal drag, provide significantly improved fuel economy with reductions in undesirable emissions, and potentially lower passenger-seat-mile costs, especially for large transports. In addition, the unique engine placement provides further operational benefits, such as reduced community noise during takeoff and landing operations.

Flying-wing aircraft configurations are not new, having been introduced and evaluated as early as the 1930s and 1940s within the United States and abroad. Today, the highly successful military B-2 bomber provides an example of a change in the paradigm, necessitated in its case by the need for low observables and international range. Like the B-2, the BWB incorporates advanced structural materials and flight control technologies to ensure safe and efficient operations across the flight envelope. Engineering analyses conducted to date by industry and NASA indicate that an advanced aircraft of this type could weigh less, generate less noise and fewer omissions, and operate at less cost than a conventional transport configuration using similar levels of technologies. The unprecedented levels of internal volume provided by the configuration offers revolutionary capabilities, such as passenger sleeper berths for transpacific flights and large weapon or troop loads for the military.

Challenges and Barriers

The unconventional configuration of the BWB design results in significant challenges and potential barriers in the technical disciplines, operational and regulatory requirements, and economic strategies for industry and commercial airlines.

Disciplinary Challenges

The concept of carrying as many as 800 passengers within a BWB configuration requires extensive research and development in the areas of aerodynamics, stability and control, structures and materials, and propulsion integration.

Aerodynamic design challenges for the BWB are present in all phases of flight operations: from takeoff, through cruise, and to landing. Many of these aerodynamic issues have proven difficult or impossible to solve for flying-wing type aircraft of the past. For example, the absence of a conventional aft-fuselage-mounted horizontal tail results in the need to deflect the wing trailing edge upward to provide trim for the nose-up attitudes required to produce lift for takeoff and landing. This trailing-edge deflection degrades the wing's lift potential and has resulted in unacceptable lift penalties for many past flying-wing aircraft. At high-subsonic cruise speeds, a particularly challenging requirement for the BWB is to ensure an acceptable level of aerodynamic drag for the large, thick wing centerbody.

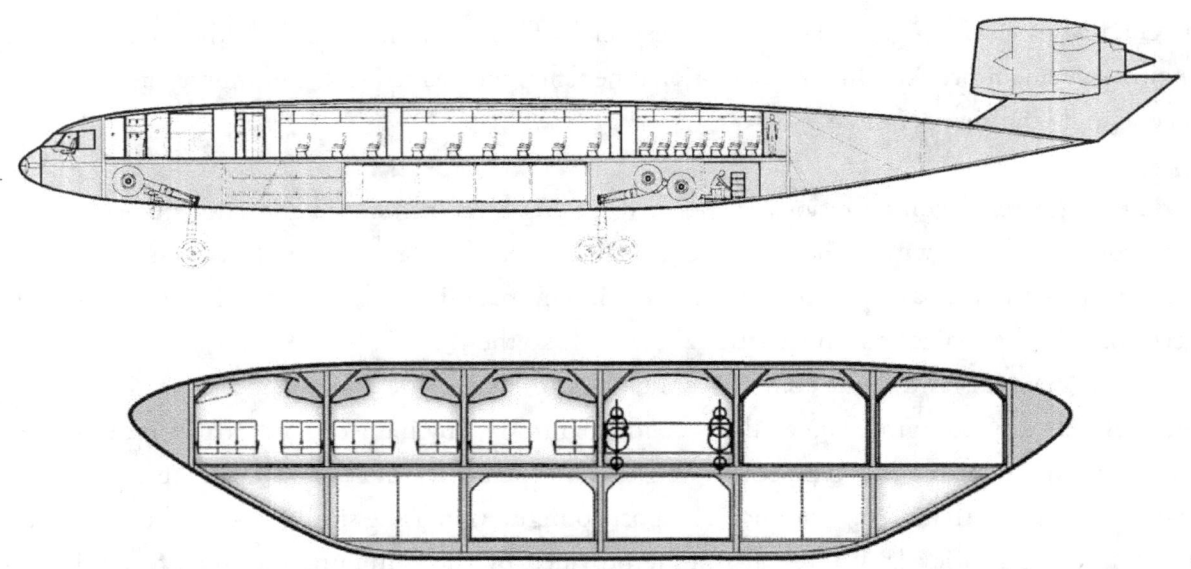

Schematic of Blended Wing Body passenger cabin showing side view
(upper sketch) and front view (lower sketch) of layout.

Stability and control problems have proven to be some of the most formidable barriers for past flying-wing aircraft configurations. The lack of a conventional aft tail results in potential longitudinal and directional stability and control deficiencies. Without a tail, aerodynamic stability variations with increasing angle of attack must be carefully assessed and designed to ensure that uncontrollable phenomena, such as "pitch up" or "tumbling," will not occur. Also, the absence of a conventional vertical tail results in directional stability and control requirements that require the use of wingtip-mounted fin or rudder configurations, as well as symmetric deflections of split upper and lower outer-wing trailing-edge segments for yaw control. The highly successful application of advanced digital flight controls and envelope limiting concepts to aircraft such as the B-2 has demonstrated that proper anticipation and design can meet these challenges.

Another daunting challenge to the successful design of BWB configurations involves the integration of engine and engine pylon concepts on the upper-aft surface of the blended fuselage centerbody. Shock-induced inflow conditions can severely degrade the operational efficiencies of large bypass ratio turbofans in proximity to uppersurface flows, and the engine installations can, in turn, lead to highly undesirable interference effects for the wing-body components and trailing-edge control surfaces.

Finally, of all the disciplinary design challenges facing the BWB concept, perhaps none is as important as the design of a highly noncircular pressurized cabin. The structural weight advantage of circular fuselage shapes for airplanes has been exploited since the earliest days of pressurized structures, and the experience base, design and evaluation methodology, and regulatory requirements are well known within the aviation community. In contrast, very little experience has been accumulated with noncircular pressurized structures and the integration of passenger and payload structural accommodations will require careful consideration if the advantages of span loading are not to be lost in an inefficient and unacceptably heavy structure. Possible solutions to these formidable requirements might be provided by current and near-term advanced composite materials and innovative approaches to structural design.

Operational Challenges

To date, studies of the BWB configuration concept's advantages, when compared with a conventional airplane, indicate that the concept becomes especially superior for very large (450 to 800 passenger) aircraft designs. Thus, in addition to operational challenges related to its unique configuration, the BWB is challenged by considerations of the safe and efficient accommodation of large numbers of passengers and payload.

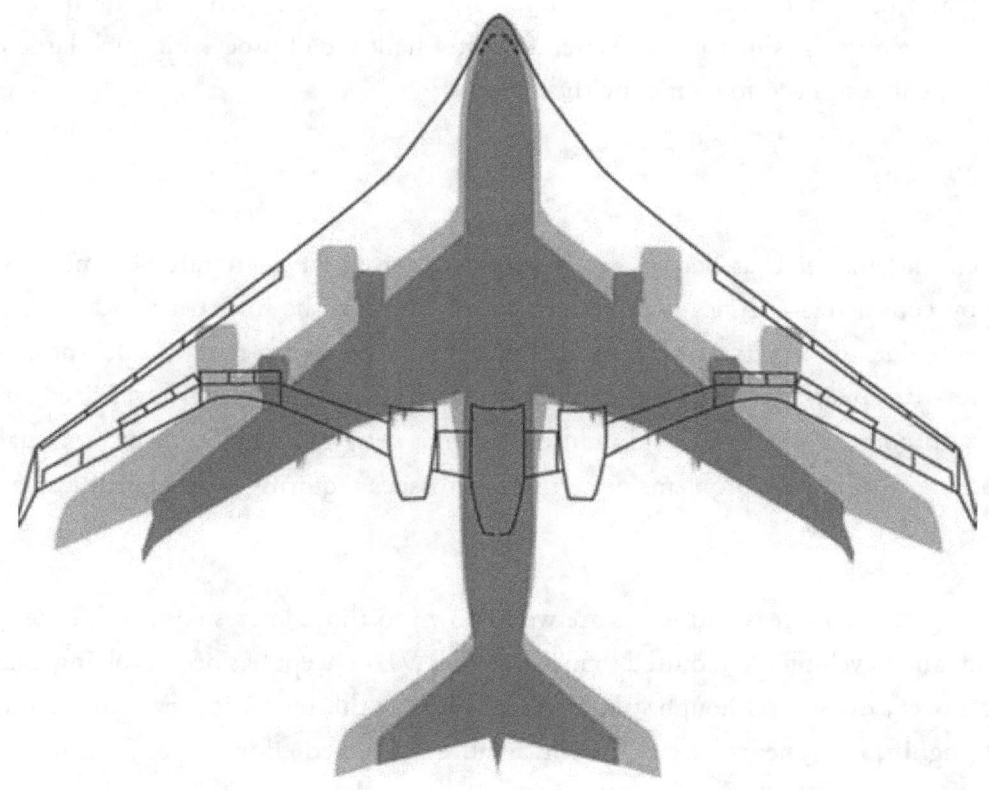

Planform view of an 800-passenger Blended Wing Body compared with an Airbus A380 (orange) and a Boeing B747-400 (green).

From a passenger's perspective, a unique characteristic of the BWB configuration is the provision of passenger volume within the centerbody structure. Versions of the BWB envision upper and lower seat accommodations in multiple side-by-side seating bays (150 in. wide) similar to Boeing 737 and 757 arrangements. The lack of windows for orientation and interest must be countered with the provision of appropriate systems for passenger comfort and relief of anxieties. Because some passengers will be distributed in a spanwise direction rather than the conventional arrangement near the center line of the airplane, the influence of rolling motions on passenger comfort during turbulence or intentional maneuvers must be resolved and satisfactory ride qualities must be ensured.

Provision for timely enplanement and deplaning procedures as well as emergency egress of such large numbers of passengers must be anticipated and provided within the airplane's structural layout and passageways.

Operational considerations, such as the dimensions of its wheel track and its operational weight for runway and taxiway operations will impose constraints on the span and length of the BWB (and other conventional "super jumbo" aircraft). The challenge of processing such large numbers of passengers will also need to be met by the airports.

Economic Challenges

The bottom-line answer that determines whether unconventional aircraft become feasible and procured by commercial airlines is economic payoff. Despite the inherent appeal of an aircraft configuration due to novelty, airlines will not acquire transports that are not profitable and competitive, passengers will not procure tickets that are not competitively priced or ride in airplanes with poor ride qualities, and industry will not design and build unconventional designs that expose significant new problems or certification issues requiring large expenditures of funds and market time to solve.

The foregoing problem areas and issues are well known to those interested in BWB designs, and the research and development required to mature the BWB concept has been evolving and making progress for over a decade. Although studies to date indicate this revolutionary configuration might contribute significant benefits, the BWB will require continued efforts to bring it to a sufficient level of technical maturity before it can begin to challenge the entrenched paradigm.

Langley Activities

Initial Activities

Dennis M. Bushnell, Senior Scientist of NASA Langley Research Center, is internationally recognized as one of the most credible and brilliant free-thinking innovators in the aerospace community. Throughout his career, Bushnell has personally challenged and encouraged researchers to embark on studies of fresh ideas and revolutionary approaches to age-old problems and barriers. The genesis of Langley's activities on the BWB concept began with a series of invitational workshops planned and conducted by Bushnell during the 1980s to stimulate aeronautical innovators across the Nation toward a vision of the unthinkable, the undoable, and the revolutionary. In 1988, he addressed a gathering of innovative leaders within the aeronautical community at Langley with this question: "Is there a renaissance for the long-haul transport?" Bushnell's question was stimulated by his perspective that advances in the aerodynamic performance of commercial transports had declined from revolutionary leaps to evolutionary gains wherein incremental benefits were becoming smaller and smaller. In particular, metrics such as airplane L/D ratio for subsonic transports had not increased significantly in modern times.

Bushnell encouraged the community to explore novel approaches for aircraft designs that might provide a breakthrough in what was rapidly becoming a stagnant area with diminishing advances. He especially suggested that unorthodox aircraft configurations deviating from the normal tube with wings might provide breakthrough performance if solutions to potential problems could be achieved within the technical disciplines.

One of the individuals accepting Bushnell's challenge in 1988 was Robert H. Liebeck of the Long Beach, California, division of the McDonnell Douglas Corporation (now The Boeing Company). Stimulated by his discussions with Bushnell, Liebeck and his associates conducted a "clean piece of paper" brief study and arrived at a revolutionary configuration that used adjacent pressurized passenger tubes aligned in a lateral plane and joined with a wing in an arrangement that vaguely resembled a tadpole. Comparisons made by the McDonnell Douglas team with a conventional configuration airplane, sized for the same design mission, indicated that the blended configuration was significantly lighter, had a higher L/D ratio, and had a substantially lower fuel burn. This first rudimentary design was the embryonic beginning of the BWB configuration.

Liebeck and his team held briefings that inspired immediate excitement and interest from the Langley researchers based on the performance potential and the clear technical disciplinary challenges that would have to be addressed by researchers in Langley's areas of expertise. Special

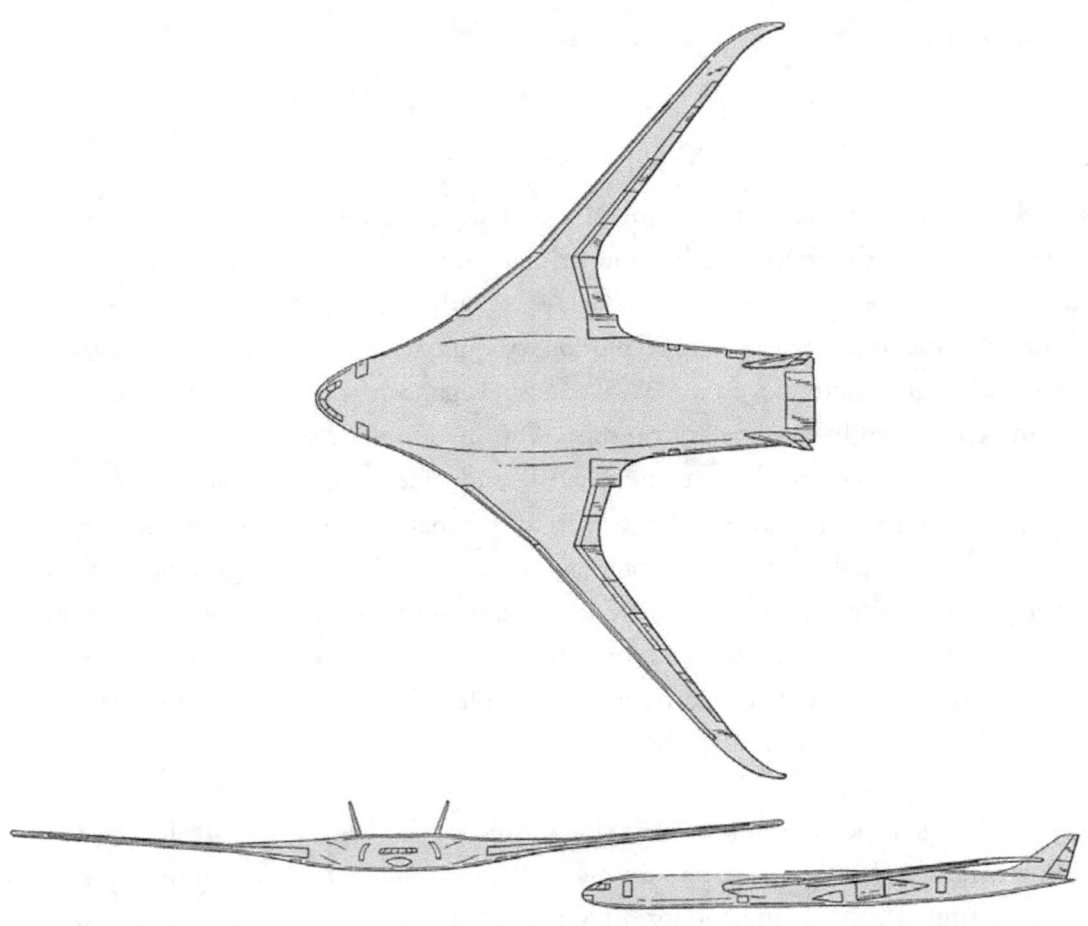

First Blended Wing Body configuration conceived by McDonnell Douglas team.

interest was especially exhibited by a Langley group responsible for the systems analysis of advanced aircraft configurations. This group specialized in assessing the ability of advanced configurations to meet specified mission requirements within current and future technology capabilities. William J. Small, Head of Langley's Mission Analysis Branch, sponsored a small study contract in April 1993 to allow Liebeck and his associates to further refine the configuration and compare its capabilities with those of a conventional subsonic transport configuration having similar advanced technologies (such as composite structures and advanced technology turbofans).

In this study, Liebeck and his associates made a critical decision when they decided that the unconventional transport would not use conventional tubular passenger compartments and that an advanced structural approach would permit more flexibility in designing the fuselage shape. With the assumption that the structures issue could be solved, the team moved forward into an assessment of various blended fuselage shapes that might result in an integrated wing-fuselage with minimal exposed (wetted) area design for minimum aerodynamic drag. The refined design incorporated an advanced supercritical wing, winglets with rudders for directional control, and "mailbox slot" engine intakes designed to swallow the boundary layer flow from the wing upper surface for enhanced performance. A double-deck pressurized passenger structure was designed, and all aspects of this unique configuration began to be addressed. Future development of BWB designs would begin to undertake issues not addressed by this initial concept, but the basic character of this first generation configuration persists to this day.

System Studies

Interactions between McDonnell Douglas and Langley on BWB-type configurations intensified in the middle 1990s. Langley was particularly interested in new configurations for very large subsonic transports (VLSTs) that might offer significant reductions in cost per passenger mile, and the BWB seemed to be a very promising approach. Langley's interest was stimulated by the state of air travel and projections of its growth in the future. Worldwide air travel passenger demand, measured in revenue passenger miles, was expected to triple within the next 15 to 20 years. Historically, the number of aircraft, aircraft size (i.e., passenger capacity), and the number of aircraft operations have all increased to accommodate the growing number of passengers; however, fewer new airports are being constructed, and the current airspace operations system is becoming saturated, thus making larger aircraft more attractive. Besides the beneficial effect on the air traffic control system, larger aircraft have also been one of the airlines' main means of reducing operating costs. Carrying more passengers on fewer planes is a proven way of reducing costs, assuming load factors (i.e., percentage of seats filled per aircraft) remain constant, as was experienced by the introduction of wide-body aircraft such as the Boeing B747. VLST concepts like the BWB are defined as intercontinental-range aircraft that carry more than 600 passengers.

The focal point of BWB interest at Langley was the Aeronautics Systems Analysis Division under the direction of Joseph R. Chambers. In 1995, several researchers led by Samuel M. Dollyhigh, Head of the Systems Analysis Branch, conducted independent research analyses of the BWB concept. Lead researcher for the group was Henri D. Fuhrmann, who conducted analyses of the mission performance capabilities of the McDonnell Douglas airplane and arrived at results that generally confirmed the benefit projections of Liebeck's team. At that time, the vehicle design focused on an 800-passenger BWB transport designed for a 7,000-nmi mission at a cruise Mach number of 0.85. The performance estimates by both the McDonnell Douglas group and the independent NASA team indicated extraordinary capabilities. McDonnell Douglas estimated that, compared with an 800-passenger conventional transport with the same advanced technologies, the BWB exhibited 20 percent lower fuel burned, 10 to15 percent less weight, and 20 percent lower direct operating costs. Both groups, however, recognized that many technical challenges would have to be met to reach the maturity levels required for a feasible configuration. Aerodynamics, structures and materials, flight controls, and propulsion technologies all presented formidable problems. Obviously, many research projects and data would be required to address these issues.

The Advanced Concepts Program

In 1994, NASA initiated a new program named the Advanced Concepts for Aeronautics Program (ACP) to stimulate revolutionary research in aeronautics, encouraging the participation of NASA researchers, industry, and academia in teamed efforts to investigate precompetitive and potentially high payoff aeronautical concepts. The systems analysis studies and interactions that had already occurred between Langley and McDonnell Douglas offered an excellent technical foundation for a potential ACP project for the BWB. After Joe Chambers presented the advocacy for the project proposal to managers at NASA Headquarters, a 3-year project on BWB configurations was awarded to a team composed of McDonnell Douglas (team leader), NASA Langley, Stanford University, NASA Lewis (now NASA Glenn), the University of Southern California, the University of Florida, and Clark-Atlanta University. The latest version of the McDonnell Douglas 800-passenger, 7,000-nmi configuration was used for the study.

Langley members of the ACP effort in 1995 included Cheryl A. Rose, Daniel G. Murri, Vivek Mukhopadhyay, Thomas M. Moul, Robert E. McKinley, Marcus O. McElroy, Ty V. Marien, Henri D. Fuhrmann, James R. Elliott, Dana J. Dunham, and Julio Chu. Elliott, who was initially ACP team lead was followed by McKinley in 1997.

The BWB concept for the ACP study had two full passenger decks in a typical long-range, three-class arrangement within a thick (about 17-percent thickness) centerbody. The seating was laid out in five parallel single aisle compartments on each deck. Each compartment was approximately equivalent to a very short narrow body aircraft, and even though the passenger complement was relatively high, the overall egress paths for passengers were shorter than most large conventional configurations. The estimated takeoff gross weight of the aircraft was 823,000 lbs (about three-quarters composites and one-quarter metal), and it used three 60,000-lb class turbofan engines. The engines were located on top of the wing, aft of the passenger compartment. This arrangement worked well for balance and had several beneficial side effects. The turbines and compressors were completely clear of the main structural elements, pressurized compartments, and fuel, which could improve safety. Also, the large fans on the high-bypass ratio engines were shielded from the ground by the centerbody, which was expected to improve the community noise characteristics.

The traditional low-speed, high-lift challenge associated with deflecting trailing-edge flaps (trim and net lift) was recognized in this design. Trailing-edge surfaces were not used as flaps, resulting in a maximum lifting capability less than that of a conventional flapped-wing design. To provide sufficient lift, the BWB wing loading was made substantially lower than the conventional norm by increasing the wing area. Leading-edge slats were used for additional lift at high angles of attack.

One of the numerous highlights of the ACP study was the highly successful development and flight testing of a large-scale, remotely controlled model of the BWB configuration by a team at Stanford under the direction of Professor Ilan Kroo. The Stanford group designed, fabricated, instrumented, and flew a large, 6-percent scale (17-ft wing span) model to explore the low-speed flight mechanics of the BWB as a relatively low-cost first step to define the stability and control of the configuration, especially at high angles of attack. Powered by a pair of pusher propellers, the "BWB-17" was dynamically scaled to predict the flight characteristics of a full size BWB. Stability augmentation and control laws were provided by an onboard computer that also recorded flight test parameters.

Initially, Kroo's students constructed and conducted exploratory flight tests of a 6-ft span model, which was flown in 1995 as a glider and later under powered conditions (using pusher propellers). They also conducted innovative semiconstrained car-top testing to check out the model prior to free flight. In this technique, they mounted the model to a test rig above an automobile, and the

 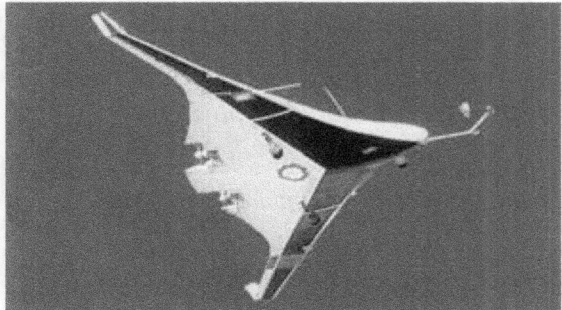

The 6-percent remotely controlled blended wing body model flown by the Stanford University Advanced Concepts for Aeronautics Program team.

model was free to rotate about all three axes. By releasing the model from a fixed angle of attack and observing the ensuing motions, the students were able to assess the effectiveness of an angle-of-attack limiter that guards against possible stall departure. The students tested the limiter using increasingly aggressive inputs and found it to be highly effective at preventing unwanted high angle-of-attack excursions. The students also developed a piloted simulator and studied several aspects of stability augmentation, including gain scheduling.

In 1997, the BWB-17 model underwent numerous research flights demonstrating the satisfactory flying characteristics of the BWB configuration. It was also flown before a highly impressed VIP audience of industry, NASA, and DoD representatives at El Mirage, California.

Meanwhile, other ACP team members were contributing additional data and results. The Langley team continued to conduct systems analyses of the mission capability of the BWB, estimates of aerodynamic performance for the airframe, and analyses of structural concepts that might meet the demanding requirements of the pressurized noncircular centerbody.

The NASA ACP-sponsored BWB study ended in 1998. At the end of the study, refined analysis had concluded that the performance of the BWB relative to a conventional configuration for the mission selected was indeed revolutionary. In comparison with the conventional design, the McDonnell Douglas estimates for BWB benefits included a reduction in takeoff gross weight of 15.2 percent, an increase in cruise L/D ratio of about 20 percent, a reduction in fuel burned of about 28 percent, and a reduction in direct operating costs of about 13 percent. Langley's independent analysis projected benefits about half as large as the McDonnell Douglas values, with differences attributed to higher NASA estimates of centerbody structural weight.

In comparison with the operational Boeing B747-400 airplane, the BWB had a 60-ft wider wing span, was 70-ft shorter in length, carried twice as many passengers, weighed about 7 percent less, and used fewer engines. Despite the excitement of the study projections, the team cited that many challenges and technology verification demonstrations would be required in the areas of structures and materials, aerostructural integration, aerodynamics, controls, propulsion-airframe integration, systems integration, and airport infrastructure.

Advanced Subsonic Technology Program

In 1997, Richard J. Re conducted wind-tunnel tests of a rudimentary BWB model (no engines or winglets) in the Langley National Transonic Facility (NTF) as part of the NASA-industry Advanced Subsonic Technology (AST) Program led by Samuel A. Morello. Part of the AST Integrated Wing Design element, the project's objective was to evaluate the capability of advanced CFD methods to predict the aerodynamic characteristics of an advanced, unconventional configuration.

Boeing personnel generated CFD solutions using several codes, including the Langley-developed CDISC code, developed by Richard L. Campbell. Separate but coordinated CFD efforts were also conducted at Langley and results used to assess engine inlet conditions and wind-tunnel sting interference effects. The data and detailed pressure distributions guided many design refinements to the BWB configuration. The CFD tools accurately predicted the aerodynamic data from cruise to buffet onset. This test was the first high-speed wind-tunnel study of the BWB configuration, and it was deemed particularly important because test data indicated that a configuration with a relatively thick centerbody could be designed for efficient cruise at high subsonic speeds (Mach 0.85).

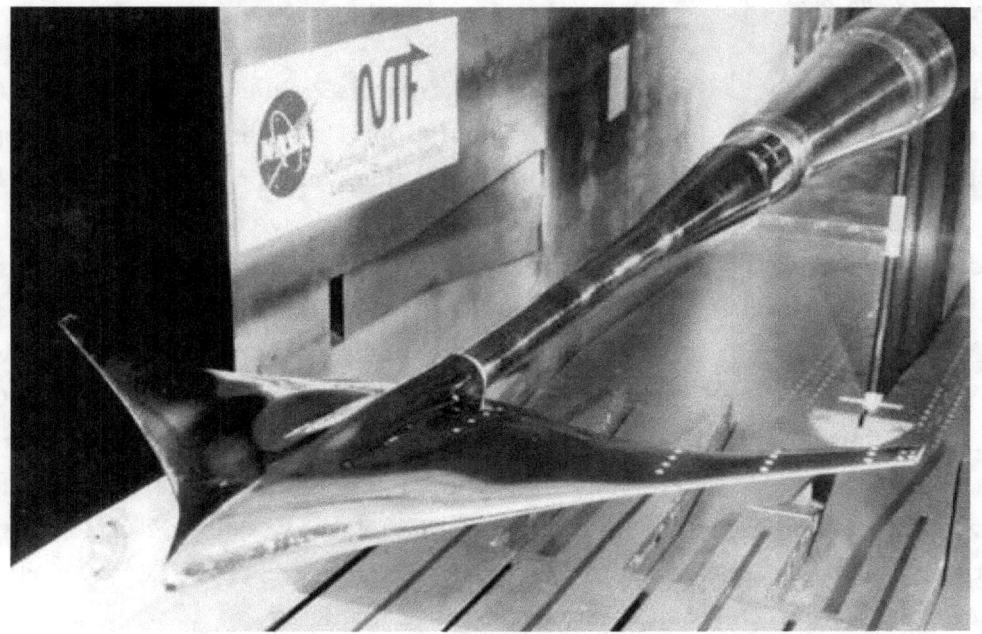

Blended Wing Body model undergoing tests in the Langley National Transonic Facility.

Interest in X-Planes

In the late 1990s, NASA expressed a renewed level of interest in X-plane projects, such as those that had generated so much enthusiasm and interest in past NACA and NASA aeronautics programs. Serious discussions occurred within industry and NASA for a manned, subscale jet-powered BWB airplane that would be capable of high-subsonic cruise evaluations during flight tests at NASA Dryden. In October 1996, a proposal white paper from industry and Langley BWB enthusiasts reached senior industry management and NASA Administrator Daniel S. Goldin. The proposed X-plane BWB configuration was for a 26-percent scale vehicle having a wingspan of 74 ft and a projected program cost of about $130 million. The NASA Administrator initially appeared to be supportive, but did not approve the proposal for go ahead.

McDonnell Douglas merged with Boeing in August 1997, and an immediate activity was undertaken by Boeing to reevaluate the BWB concept for its maturity, mission suitability, and its ability to conform to company strategies and outlook for future air travel requirements. Results of the Boeing study indicated that a very large 800-passenger airplane was not an appropriate vision, and that the BWB studies should refocus on a smaller, 450-passenger (240-ft wingspan) airplane. Also, certain configuration features, such as the propulsion installation, changed and the engines were mounted on upper surface pylons. Issues such as X-planes were put on hold.

Flight Research Revisited

In 1997, Darrel R. Tenney, Director of the Airframe Systems Program Office, initiated a series of in-house team studies to determine if Langley could support the fabrication and development of a series of unmanned, remotely controlled air vehicles that could be flown to support Langley's interest in revolutionary configurations. Langley's Director, Jeremiah F. Creedon, strongly supported the studies. At the same time, Joseph R. Chambers' staff in the Aeronautics Systems Analysis Division proposed a related new program based on the selection of precompetitive advanced configurations that would be designed, evaluated, fabricated, and test flown using remotely piloted vehicle technology at Dryden. The program, known as Revolutionary Concepts for Aeronautics (RevCon), would be based on a 4-year life cycle of support for concepts selected. Initial reactions to the proposed program from NASA Headquarters and Dryden were favorable, and following intercenter discussions, a formal RevCon Program was initiated in 2000, which was led by Dryden. Following a review of other advanced vehicle concepts by an intercenter team, the team selected the BWB as one of the concepts for further studies.

Under the Revolutionary Airframe Concept Research and System Studies (RACRSS) element of Airframe Systems, Robert E. McKinley led the interactions among Langley, Dryden, and Boeing. Initially, Boeing had proposed to fabricate a low-speed flight model of the BWB and Langley was planning to commit to fabrication and testing of a high-speed unpowered drop model. However, when Boeing could not support the low-speed model, Langley assumed responsibility for the design and fabrication of a model with Dryden supporting development of the flight control system. The 14.2-percent scale BWB low-speed vehicle (LSV) was to have a wingspan of 35 ft, a maximum weight of about 2,500 lbs, and be powered by three jet engines of the 200-lb class. The BWB configuration would be based on the BWB-450 design from the Boeing studies and was given the formal designation of X-48A. Project Manager of the RACRSS program was Bob McKinley, with the assistance of LSV Project Managers Wendy F. Pennington and Kurt N. Detweiler, and Chief Designer William M. Langford.

The LSV Program encountered major problems as flight control system development had to be put on hold when commitments to other programs changed the Agency's priorities and resource allotments. However, the program had successfully completed a preliminary design and review of

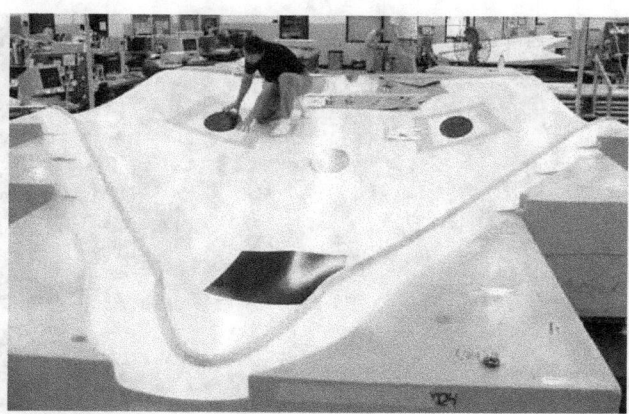

*Construction of the Blended Wing Body Low-Speed Vehicle
in a Langley shop.*

the vehicle's airframe, an initial round of structural material coupon and element testing, structural design of proof-of-concept wing box testing, and fabrication/assembly of the centerbody and wing molds for the composite LSV. The LSV Program was subsequently terminated by NASA because of higher priority program commitments.

After cancellation of the LSV, Langley moved toward the possibility of conducting lower cost low-speed free-flight tests of a smaller 5-percent scale model in the Langley Full-Scale Tunnel. In 2002, Boeing contracted Cranfield Aerospace Limited in England for the design and production of a smaller, 21-ft span LSV-type vehicle.

Flight Dynamics

Research of flight dynamics (stability, control, and flying qualities) is a particularly challenging area for unconventional configurations, such as the BWB. Thus, exhaustive studies have been conducted by the Langley staff to provide critical information for design refinements. The early efforts started with testing of a 4-percent scale model of the 800-passenger design in the Langley 14- by 22-Foot Tunnel during 1997. The research team, headed by Daniel G. Murri, obtained quantitative stability and control characteristics for low-speed flight conditions, including the effects of power and flow visualization.

In the later LSV commitment, Langley researchers conducted timely wind-tunnel tests and other analyses to support the project. Three different models were tested in three different wind tunnels. A team led by Dan Murri, Dan D. Vicroy, and Sue B. Grafton tested a 3-percent scale model of the BWB-450 in the Langley 14- by 22-Foot Tunnel in late 1999. The scope of testing included: conventional unpowered static tests to define performance and stability and control characteristics

BWB-450 model used for initial tests in the Langley 14- by 22-Foot Tunnel.

(including ground effects), unique forced-oscillation tests to determine dynamic aerodynamic characteristics, assessments of wing leading-edge slat configurations, and extremely large angle tests to determine aerodynamic characteristics over a complete 360° angle-of-attack range, giving a range of sideslip angle of −90° to +90°. The latter tests were conducted to obtain input data for analysis of tumbling characteristics, as will be discussed. During these tests, unexpected results were obtained on aerodynamic interference effects between the high-lift leading-edge slats and the rear-mounted engines, providing design information on how to alleviate a potential problem. The data also provided information on trailing-edge control allocation strategies for efficient lift, trim, and stability. The team conducted additional trailing-edge aerodynamic control interaction studies in a low-speed 12-ft tunnel at Langley.

Other models were used to examine the configuration's characteristics at extreme angles of attack. A team led by Charles M. (Mike) Fremaux and Dan Vicroy tested a 1-percent scale spin model of the BWB-450 in 1999 in the Langley 20-Foot Vertical Spin Tunnel, and they used a 2-percent scale model in rotary-balance testing in 2001 to determine the aerodynamic characteristics associated with spinning attitudes and angular rates. In free-spinning tests, the team assessed spin and recovery characteristics of the 1-percent scale model, determined the size of emergency spin recovery parachute required for the proposed LSV vehicle, and explored the configuration's tumble characteristics.

Experiences with flying-wing configurations during the 1940s and 1950s raised concern about an uncontrollable tumbling phenomenon, during which the vehicle would autorotate in pitch with

BWB-450 model used for unconventional large angle tests.
Model is positioned for airflow coming into the upper rear.

continuous 360° pitching motions. The tumbling motions were precipitated either by stalls or by flight at high angles of attack that resulted in massive separation of flow on the wing. Researchers in the Langley Spin Tunnel have conducted extensive studies up to the current day for predicting the phenomenon and identifying airplane characteristics that precipitate the problem. Even though it is generally agreed that a modern BWB transport configuration will use an angle-of-attack limiting concept in its flight control system to prevent tumbling, information on behavior during forced-tumble motions is of great value to the flight control designer. Mike Fremaux and Dan Vicroy conducted tumbling tests of the 1-percent scale model using a special 1-degree-of-freedom model mounting system in late 1999 and early 2000. These experimental results were augmented by theoretical calculations of the tumbling motions conducted by Dan Murri and Eugene Heim

Free-spinning tests of a Blended Wing Body model in the
Langley 20-Foot Vertical Spin Tunnel.

using aerodynamic data from static and dynamic wind-tunnel tests. The calculated motions were in close agreement with the observed free-tumbling results, providing good calibration for the mathematical model approach used in the simulated motions.

Langley researchers have also formulated and used piloted simulator studies of BWB configuration handling qualities to assess stability and control characteristics. Dan Murri led initial terminal-area (low-speed) studies of a BWB-800 configuration in the Langley Visual Motion Simulator (VMS), with Langley research pilot Robert A. Rivers and Boeing pilot Mike Norman providing assessments of the configuration's handling qualities. Other simulator studies have been implemented and checked out for use on the Generic Flight Deck (GFD) simulator. Objectives of these simulator efforts were to evaluate flying and handling qualities in the terminal area and to develop and refine control system elements and logic, including control allocation and envelope protection schemes.

Aerodynamic Performance

Aerodynamic performance refinements have continued to improve the BWB's predicted capabilities. Langley's expertise and tools have contributed valuable guidance and direction in these efforts. For example, Boeing's use of the Langley-developed CFD codes CDISC and CFL3D (developed by Richard L. Campbell and James L. Thomas, respectively) permitted extensive beneficial modifications of pressure distributions while maintaining constraints within cabin geometry, pitching moments, and span load distributions. These advances evolved the configuration's centerbody into a thinner section, resulting in a higher level of L/D ratio at cruise. Using these powerful design tools has also contributed to the sophistication of the wing airfoil characteristics. In the past, flying wings achieved longitudinal trim by using wing sweep and downloading the wingtips, resulting in a severe induced-drag penalty. With this approach, flying wings have not lived up to the performance potential. In contrast, the use of advanced wing design methods has permitted the BWB to be trimmed by careful distribution of wing trailing-edge camber and by the judicious use of wing twist (washout).

In 2004, Richard L. Campbell and Melissa B. Carter used CDISC, coupled with the USM3D unstructured grid Navier-Stokes flow solver, to improve the propulsion-airframe integration characteristics of a BWB configuration with boundary layer ingestion nacelles. Wind-tunnel models incorporating the baseline and redesigned geometry were built by Langley and tested in August 2004 at high Reynolds numbers in the NTF to verify the CFD design method. Campbell, Carter, and Odis C. Pendergraft, Jr., led the testing team. NTF force and moment data correlated well with CFD results in general, and matched predicted cruise drag reduction for redesign within the data's uncertainty. Experimental data also confirmed predicted changes in wing pressures in the design region near the nacelles. An intercenter team led by Langley's A. Neal Watkins, William K. Goad, Clifford J. Obara, and Ames' James H. Bell contributed a revolutionary flow visualization method for the unique environment of cryogenic flow in the NTF. The visualization results were used to further verify the CFD design method, indicating that flow features such as wing shock and boundary layer separation were accurately located in the computations.

Structures and Materials

Langley and Boeing interactions have addressed the formidable multidisciplinary challenges of designing a centerbody for the BWB with satisfactory structural weight, passenger accommodation, and pressurization within the configuration's aerodynamic lines. In a conventional circular fuselage section, a thin skin carries internal pressure efficiently via hoop tension. If the fuselage section is noncircular, internal pressure loads also induce large bending stresses. The structure must also withstand additional bending and compression loads from aerodynamic and gravitational forces.

Critical contributions from Langley staff in structures for the BWB have included development of a rapid structural code for preliminary design known as Equivalent Laminated Plate Solution (ELAPS). Developed by Gary L. Giles, this analysis method generates in a timely manner conceptual-level design data for aircraft wing structures that can be used effectively by multidisciplinary synthesis codes for performing systems studies. Samuel M. Dollyhigh's systems analysis groups used ELAPS for early conceptual studies during BWB assessments.

Researcher Vivek Mukhopadhyay conducted preliminary studies of structural concepts for noncircular fuselage configurations, focusing on multiple fuselage bays with noncircular sections. In his studies, flat and vaulted shell structural configurations were analyzed using deep honeycomb sandwich-shell and ribbed double-wall shell construction approaches. Combinations of these structural concepts were analyzed using both analytical and simple finite element models of isolated sections for a comparative conceptual study. Weight, stress, and deflection results were compared to identify a suitable configuration for detailed analyses. The flat sandwich-shell concept was found preferable to the vaulted shell concept due to its superior buckling stiffness. Vaulted double-skin ribbed shell configurations were found to be superior due to weight savings, load diffusion, and fail-safe features. The vaulted double-skin ribbed shell structure concept was also analyzed for an integrated wing-fuselage finite element model.

Additional problem areas such as wing-fuselage junction and pressure-bearing spar were identified. Mukhopadhyay also teamed with Jaroslaw Sobieski to analyze and develop a set of structural concepts for pressurized fuselages of BWB-type flight vehicles. A multibubble fuselage configuration concept was developed for balancing internal cabin pressure load efficiently using balanced membrane stress in inner cylindrical segment shells and intercabin walls. Additional cross-ribbed outer shell structures were also developed to provide buckling stability and carry spanwise bending loads. In this approach, it was advantageous to use the inner cylindrical shells for pressure containment and let the outer shells resist overall bending.

Support for the proposed LSV development challenged the Langley design and engineering group in the area of structures. William M. Langford designed and Regina L. Spellman analyzed an approach to LSV structural design. The dynamic scaling requirements for the LSV caused great difficulty in designing an airframe with the necessary strength and stiffness within the strict weight limits. Due to the vehicle's unusually large size, there were limitations to Langley's in-house composite fabrication capabilities and resources. These combined restrictions limited the design to a room temperature cure composite design. The design group also implemented an extensive materials characterization effort to identify materials and processes to meet the demanding requirements, and a proof-of-concept article was fabricated and tested.

*Acoustic tests of the BWB-800 configuration in the Langley Anechoic
Noise Research Facility.*

Noise

The unique layout of the BWB configuration places the noise generating engines above and at the rear of the wing-centerbody upper surfaces, suggesting that a significant reduction in projected noise might be obtained from structural shielding as compared with conventional configurations. To investigate and quantify any benefits of the BWB configuration, Langley researchers Lorenzo R. Clark and Carl H. Gerhold conducted acoustic tests of a 4-percent scale, 3-engine nacelle model in the Anechoic Noise Research Facility at NASA Langley Research Center. The test team placed a high-frequency wideband noise source inside the nacelles of the center engine and one of the side engines to simulate broadband engine noise. They also measured the model's sound field with a rotating microphone array that was moved to various stations along the model axis and with a fixed array of microphones that was erected behind the model. While no attempt was made to simulate the noise emission characteristics of an aircraft engine, the model source was intended to radiate sound in a frequency range typical of a full-scale engine. Clark and Gerhold found that the BWB configuration provided significant shielding of inlet noise. In particular, noise radiated downward into the forward sector was reduced by as much as 20 to 25 dB overall at certain full-scale frequencies.

Status and Outlook

NASA sponsorship of early BWB concept studies in 1993 has played a key role in the development of subsequent designs. Currently, disciplinary studies in several areas have continued at Langley, and the BWB concept as a future transport has become one of the focal points of NASA's vision of future air vehicles. Such focal points are being used to steer the direction of fundamental disciplinary research conducted within NASA toward high-payoff areas. Boeing has continued its close working relationship and cooperative studies with Langley while pursuing other markets for the BWB, including potential applications such as a large cargo transport or a military in-flight refueling transport to replace the aging KC-135 fleet. As a tanker, the BWB offers significant advantages, such as the ability to refuel more than one aircraft at a time. Many believe that the first applications of the BWB configuration will be for military uses with civil applications to follow. Dan D. Vicroy is leading an investigation of the free-flight characteristics of a 5-percent scale BWB model in the Langley Full-Scale Tunnel in 2005. The model flight tests are designed to provide further information on the flight dynamics of the configuration, including an assessment of stability and control characteristics for engine-out conditions at the edge of the low-speed envelope. One innovative example of this research will be the possibility of using thrust vectoring of the center line engine for auxiliary yaw control when one of the outer engines becomes inoperative. If successful, this application would be the first example of thrust vectoring used for a transport configuration. In addition to these low-speed tests, Boeing is also planning cooperative acoustic testing of a 3-percent model.

Under leadership of Robert M. Hall and Charles M. Fremaux, Langley is now assessing the ability of advanced CFD codes to predict static and dynamic stability characteristics of BWB configurations. The CFD team includes Paul S. Pao, Robert E. Bartels, Robert T. Biedron and Neal T. Frink.

Artist's concept of a military blended wing body application.

Langley is now considering variants of advanced versions of the BWB in continual system studies of the benefits of various configurations in future air transportation scenarios. In these studies, advanced technologies, such as hybrid laminar flow control and buried engines, result in BWB designs with even more potential than those researched thus far. Finally, piloted simulator studies of the BWB-450 configuration are planned for the Langley GFD simulator, and (with additional wind-tunnel data) far-term plans include eventual full-envelope simulation that will include the effects of compressibility and aeroelasticity.

The excitement and revolutionary capabilities of the BWB concept have not gone unnoticed by other international aircraft industries. France and Russia have conducted research on BWB configurations since the early 1990s. This interest has intensified and spread to other European nations, as evidenced by exhibits at the Berlin International Aerospace Exhibition held May 6–12, 2002. Airbus, ONERA/DLR, and TsAGI presented elaborate displays of BWB models and research activities at that international gathering of aerospace industry members. European collaboration on the BWB is obvious and accelerating, and the future global market for such configurations will be very competitive. In addition, the celebrated first flight of the super jumbo Airbus A380 in April 2005 has ushered in a new era of interest in very large transports.

In summary, the NASA Langley Research Center funded and stimulated the creation and initial development of the BWB concept, and the Center's interaction with industry and universities has resulted in valuable technical contributions that have brought continuing maturity to the concept. Langley's involvement in the BWB concept has been based on two traditional NASA roles in aeronautics: (1) the assessment of disciplinary design issues (especially off-design problems) for revolutionary vehicles, and (2) a credible independent assessment of revolutionary concepts that significantly advances the state of the art of aeronautics with the potential to change the paradigm.

Synthetic Vision: Enhancing Safety and Pilot Performance Through Virtual Vision

Concept and Benefits

Arguably, the most important physical human sense for piloting an aircraft is that of sight. Visibility is a key requirement for situational awareness, orientation, defensive warning and collision avoidance, and precision maneuvers. All aspects of aviation, including the airborne and ground operation of military, commercial, business, and general aviation airplanes, are severely impacted by limited visibility, resulting in degraded (perhaps catastrophic) safety and significant delays or cancellation of scheduled flights.

Restricted visibility is a major contributor to a class of accidents referred to as controlled flight into terrain (CFIT) wherein a fully functional aircraft collides with the ground, water, or other obstacles due to pilot disorientation, lack of awareness, or confusion. More than 1,750 people have died worldwide in airliner accidents due to CFIT since 1990, and a recent Boeing study indicates that a worldwide average of over 200 commercial jet CFIT fatalities occur per year. Such accidents may be precipitated by loss of orientation, unanticipated terrain features, and loss of key navigational and maneuver cues, especially under weather-related conditions or at night.

In addition to potentially catastrophic visibility-related conditions encountered during flight, a significant safety issue has been experienced during aircraft ground operations in low-visibility conditions. The world's worst aviation accident involved runway incursion, which involves potential collision hazards caused by the inadvertent intrusion of aircraft, vehicles, people, or other objects on runways. On March 27, 1977, a collision between a KLM 747-200 transport and a Pan Am 747-100 transport at Tenerife, Canary Islands, killed 578 passengers and crew members. Because of limited visibility and communications difficulties between air traffic control and the KLM aircraft, the KLM 747 started its takeoff while the Pan Am aircraft was on the same runway, resulting in a horrible fatal collision.

In reaction to an increasing number of nonfatal runway incursion incidents, the topic of runway incursion has ranked among the top five items on a list of high-priority issues identified by the National Transportation Safety Board (NTSB) for the last 6 years.

Today, unprecedented advances in modern computer capabilities and remote sensing technologies that can rapidly and accurately provide Earth-referenced models of geographic features—such as terrain, vertical obstructions, airport runways and taxiways, and advanced cockpit displays—have led to exciting new research on the development of cockpit display technologies. These technologies could eliminate low visibility accidents by providing the pilot with an accurate, realistic virtual image of the environment surrounding the aircraft, as well as guidance for safe maneuvers within physical constraints. Developments in this area have initially focused on enhanced vision systems (EVS), which use sensors such as forward looking infrared (FLIR) systems, systems, millimeter wave sensors, and other approaches based on information obtained from active onboard sensors. Results obtained from EVS studies and applications indicate improvements in awareness and safety; however, an innovative new technology involving "synthetic" vision promises to provide revolutionary all-weather operations without the loss of visibility.

Synthetic vision systems (SVS) differ from EVS technology in that they consist of a computer database-derived system (rather than sensors) that uses precise Global Positioning System (GPS) navigation, stored models of geospatial features, and integrity-monitoring sensors to provide an unrestricted synthetic view of the aircraft's external environment, including traffic depiction during airborne and ground operations. SVS provides an intuitive perspective view of the outside world in a manner sufficient for aircraft tactical guidance. The system is driven by airplane attitude from an inertial reference unit, GPS position, and a predetermined internal database that contains airport, terrain, obstacle, and path information for specific locales. The technology paints and displays a three-dimensional computer picture of the outside world so that pilots can see runways and obstacles, and they are given guidance information independent of weather conditions and time of day. In addition, SVS uses maneuver guidance features, such as "highway in the sky" concepts to reduce workload and potential disorientation in restricted visibility.

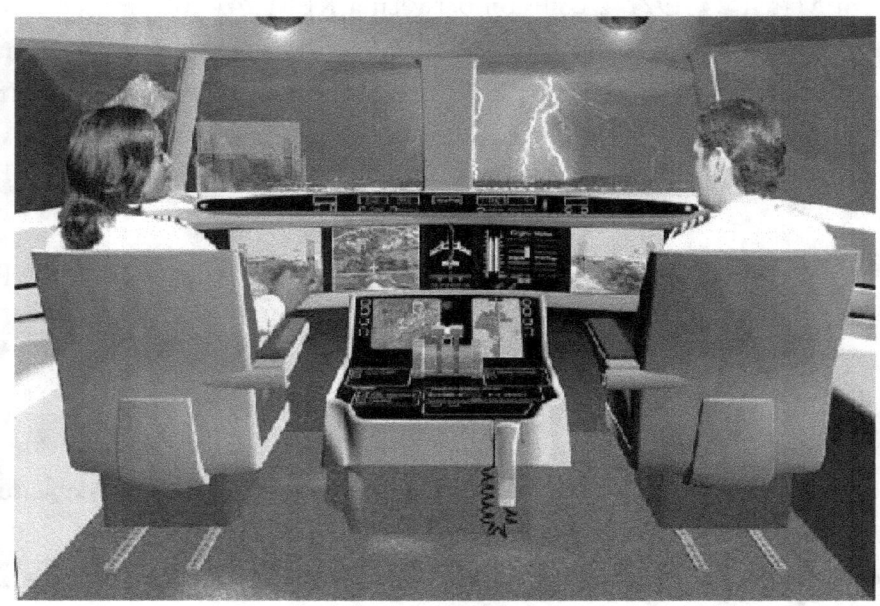

Artist's concept of advanced cockpit display with synthetic vision head-down and head-up displays.

For ground operations, SVS provides an electronic moving map of the airport surface— indications of obstacles, hold short locations, and other traffic—to help prevent inadvertent runway incursions. Other surface hazards can be avoided using SVS, such as wrong turns, taking off on the wrong runway, pilot confusion/disorientation at unfamiliar airports, and confusion resulting from language barriers.

The monumental challenge of building accurate airport and terrain databases for worldwide SVS operations has been met in large part thanks to data derived during dedicated Space Shuttle missions as well as other terrain measurement activities. During these focused missions, 80 percent of Earth's land surface (99.96 percent of land between 60° N and 56° S latitude) was mapped sufficiently for SVS enroute requirements. Shuttle Endeavor was launched in February 2000 (Shuttle Radar Topography Mission) on a 10-day effort to map the Earth, resulting in an 18-month data processing output. The primary objective of the DoD mission, conducted jointly with the National Imagery and Mapping Agency (NIMA), now called the National Geospatial-Intelligence Agency (NGA), was to acquire a high-resolution topographical map of the Earth's land mass. This activity represented a breakthrough in the science of remote sensing by producing topographical maps of Earth many times more precise than those previously available. Such databases (i.e., models) are currently being produced by both government and commercial entities on a global basis.

Challenges and Barriers

SVS promises unprecedented capability, but it also faces numerous operational issues and technical challenges, including the generation of reliable computer-based airport and terrain features, requirements for effective displays and onboard computational capability, certification issues, maintenance requirements, integration of flight-critical SVS into the airspace and transportation system, and the all-important issues of affordability and retrofitability. The following discussion identifies disciplinary and operational issues that have surfaced during NASA research on the technologies associated with SVS, most of which have been addressed and resolved by the efforts of Langley and its partners.

Disciplinary and Operational Challenges

Requirements for successful implementation of SVS technology are dominated by the fact that the concepts must meet flight-critical system requirements. The robustness of databases and computational systems is obviously critical, and failure modes and recovery strategies must be defined. Large deviations in database requirements will exist between various classes of aircraft (such as commercial transports versus general aviation), and the augmentation requirements for sensors of the various classes of vehicles must be addressed.

Development and demonstration of affordable, certifiable display configurations for aircraft involves major retrofit issues associated with this advanced display technology because to measurably impact safety and operations, a majority of the fleet has to be impacted positively. The actual and projected worldwide jet aircraft fleet shows that the majority of transports are now, and will remain, those equipped with glass cockpits. The display drivers, graphics drivers, and drawing capability necessary to host a synthetic vision display system are not in place for most aircraft. Additional equipment required for SVS includes GPS, an inertial reference unit (IRU), pilot controls/interfaces, several databases, and a computing host.

An opportunity may exist to use emerging head-up display (HUD) systems as a cost-effective retrofit path for SVS in HUD-equipped aircraft. This display concept is analogous in many respects to the EVS now certified and offered to customers on the Gulfstream V (GV) business jet, although the SVS raster image is synthetically derived rather than being a direct imaging sensor output. Also, evaluations of the potential advantages or disadvantages of head-down displays (HDD) must be conducted, particularly as a potential low-cost alternative for general aviation.

Display issues include the luminance, field of view, resolution, and display details required for satisfactory operations. Display content and computational onboard power requirements, cost, failure modes, and other details must also be studied. Ensuring accuracy and integrity of the terrain and global topology data used by SVS is a mandatory and critical task. Inadvertent use of inaccurate or misleading information will not be tolerated for potential applications.

Langley Activities

NASA Langley Research Center has conducted extensive research on pilot-vehicle display interfaces with a view toward safer, more efficient flight operations. In its early days as an NACA research laboratory, Langley led the world in studies of aircraft handling qualities and pilot-control system interface requirements, and explored IFR factors that result in accidents. As a NASA research

Conceptual head-up and head-down synthetic vision displays.

center, Langley has contributed some of the most significant technology ever produced on topics such as loss of control and orientation in limited visibility conditions; glass-cockpit technology; windshear detection and avoidance; advanced displays to simplify piloting tasks (particularly for general aviation pilots) such as "highway in the sky" and takeoff-performance monitoring displays; and fundamental human-display interface issues. Two past NASA programs, known as the Low Visibility Landing And Surface Operations (LVLASO) project and the HSR Program External Visibility System (XVS) element were crucial to start up and development of the SVS research conducted at the Center. Langley researchers have therefore exercised considerable expertise and experience in advancing the SVS concept.

In conducting its research mission in the SVS area, Langley is intensely integrating its efforts with those of the FAA, industry, DoD, and academia. In view of international interest and the significant potential payoff in reducing aviation accidents, this coordinated, teamed arrangement is accelerating the state of the art and maturing the technology base required for safe and effective applications in future civil and military aircraft.

The following discussion provides an overview of significant activities undertaken by Langley on SVS technology.

The NASA Aviation Safety and Security Program

In response to White House Commission on Aviation Safety and Security recommendations in 1997, NASA created an Aviation Safety Program (AvSP) to research and develop technologies focused toward a national goal of reducing the fatal aircraft accident rate by 80 percent by 2007. With aviation security taking on new importance, the program has been expanded to the Aviation

Safety and Security Program (AvSSP) to directly address the safety and security research and technology needs of the Nation's aviation system. Safety research in AvSSP develops prevention, intervention, and mitigation technologies and strategies aimed at one or more causal, contributory, or circumstantial factors associated with aviation accidents. Within the program, AvSSP contains an element called vehicle safety technologies, which includes synthetic vision systems. The SVS project started as a fully funded segment of AvSSP in 2000 with a planned project life of 5 years.

Langley's Michael S. Lewis initially served as the Director of the AvSP, with George B. Finelli serving as Deputy Director. Daniel G. Baize was assigned as Project Manager for SVS, and Russell V. Parrish was the project's Chief Scientist. Key Langley team members of the initial SVS project included James R. Comstock, Jr., Dave E. Eckhardt, Steven Harrah, Richard Hueschen, Denise R. Jones, Lynda J. Kramer, Raleigh (Brad) Perry, Jr., John J. White, III, and Steven D. Young. The team also included representatives from NASA Ames and Dryden Research Centers.

Synthetic vision research attempts to replicate the safety and operational benefits of clear-day flight operations for all weather conditions. The Aviation Safety Program envisions a system that would use new and existing technologies, such as GPS navigational information and terrain databases, to incorporate data into displays in aircraft cockpits. The displays would show terrain, ground obstacles, air traffic, landing and approach patterns, runway surfaces and other relevant information to the flight crew. The display includes critical flight path aids for reducing pilot workload and ensuring the appropriate and safe maneuver is followed.

Remarkable accomplishments of the SVS project have included critical technologies that enable key capabilities of the synthetic vision concept. With extensive in-house studies, as well as precisely planned cooperative activities with government, industry, and other partners, the Langley team has rapidly matured technologies and provided impressive demonstrations of the effectiveness of this revolutionary concept for both the commercial transport and general aviation communities.

The SVS project is composed of three elements or subprojects: commercial and business aircraft, enabling technologies, and general aviation. The element on commercial and business aircraft focuses on issues particular to large jet transport aircraft and considers more expensive sensor augmented systems. Research on enabling technologies focuses on required supporting technologies such as the geospatial feature databases; communication, navigation, and surveillance (CNS) technologies; hazard detection functions; sensors; and system integrity approaches. The general aviation element focuses on the particular needs and applications of general aviation aircraft, with sensitivity to relative expense of SVS.

Before reviewing the impressive accomplishments of the SVS project team, a brief review of various segments of past Langley research programs that have culminated in SVS technology is provided for background information.

"Highway in the Sky"

As early as the 1950s, researchers within civil and military communities began intense efforts to provide more intuitive and valuable artificial guidance concepts for use in cockpit displays. Efforts in the United States were led by the military, which conducted extensive ground and flight studies of concepts that depicted the intended and actual aircraft flight paths on electronic instruments. Much of this research was led by the Navy's George W. Hoover, who pioneered an early concept that became known as Highway in the Sky. Langley conducted efforts on conceptual pathway-in-the-sky technology in the 1970s, and although several research papers were written on the topic, little interest was shown from industry. In the early 1980s, Langley's research efforts on flight problems associated with poor visibility included studies of critical factors that precipitated inadvertent loss of control incidents for typical general aviation pilots under IFR. A key research activity known as the Single Pilot IFR (SPIFR) project, led by John D. Shaughnessy and Hugh P. Bergeron, used piloted simulator studies and actual flight tests of representative subjects to highlight the difficulty and potential dangers associated with such flight conditions. During the studies, a concept known as the "follow-me box" received considerable attention as an intuitive aid to inexperienced pilots during IFR flight. James J. Adams conducted in-depth simulator studies of the display, which used simplified rectangular lines to form a guidance box to be followed in flight by the human pilot on a projected display.

Later, the follow-me box concept was refined by Eric C. Stewart during simulator studies, where it was found that significant improvements in the piloting skills of relatively inexperienced general aviation pilots could be obtained using displays that provided the aircraft's current and future flight path in a visual depiction similar to driving an automobile. Stewart's research was conducted as part of the Langley General Aviation Stall/Spin Program, within a project known as "E-Z Fly." The project's objective was making piloting tasks more natural for new pilots. Stewart further developed the highway in the sky concept, using a graphic consisting of line segments that provided an automotive highway depiction, even to the point of providing objects such as telephone poles for improved orientation. Using this concept, Stewart evaluated the performance of numerous test subjects, including nonpilots who had no airplane flight training before flying the simulator. Results obtained with the display were remarkable, with significant improvements in pilot guidance and response. In fact, individuals with no exposure to pilot skills were able to rapidly adapt to the guidance system and successfully fly full operational missions. Despite the significant improvement observed in pilot performance, the fact that the concept was directed at the general aviation community's low end, where computational requirements and cost were immediate barriers, stifled further research.

In the mid 1990s, NASA formulated the Advanced General Aviation Transport Experiments (AGATE) Program to revitalize the ailing general aviation industry. With over $52 million and a consortium of NASA, FAA, and 80 industry members, the program included an element to develop more intuitive flight instruments. The AGATE team recognized the potential value of advanced graphical displays designed to enhance safety and pilot awareness. Such systems would depict all flight parameters, as well as aircraft position, attitude, and state vector on the flight display. Noting industry's movement toward a system that resembled a pathway-in-the-sky concept, AGATE planners decided to further develop such a system, reduce its cost, and enhance its commercial viability. The system was designed to be a low-cost replacement for instrument panels with graphical displays. In 1998, a competitive cost-sharing contract focused on development of the first phase of a highway-in-the-sky (HITS) display system was awarded to AvroTec Corporation. The HITS system consisted of a navigational display showing horizon and relative attitude and a weather display depicting weather information. Avidyne replaced AvroTec during further development of the HITS concept. In July 2000, the HITS system was installed and successfully tested on a production Lancair Columbia aircraft. Led by Walter S. Green, the Langley AGATE Flight Systems Work Package Leader, the project culminated in a highly successful demonstration of the HITS display at the EAA AirVenture 2001 activities in Oshkosh, Wisconsin.

The NASA High-Speed Research Program

While highway-in-the-sky concepts were being pursued in Langley's General Aviation Program, additional research was being conducted to guide efforts for future applications to more sophisticated commercial transports and business jets. From 1994 to 1996, Langley held a series of interactive workshops focused on highway-in-the-sky concepts for commercial transports. These workshops brought together government and industry display designers and pilots to discuss and fly various concepts in an iterative manner. The first workshops primarily focused on the utility and usability of pathways and the pros and cons of various features available. The final workshops focused on the specific high-speed research applications to the XVS. The NASA High-Speed Research Program was concerned with replacement of the forward windows in a second-generation high-speed civil transport with electronic displays and high-resolution video cameras to enable a "no-droop" nose configuration. By avoiding the Concorde-like drooped nose for low-speed operations, the future supersonic transport design could save considerable weight. Primary concerns in XVS application were preventing display clutter and obscuring of hazards as the camera image was the primary

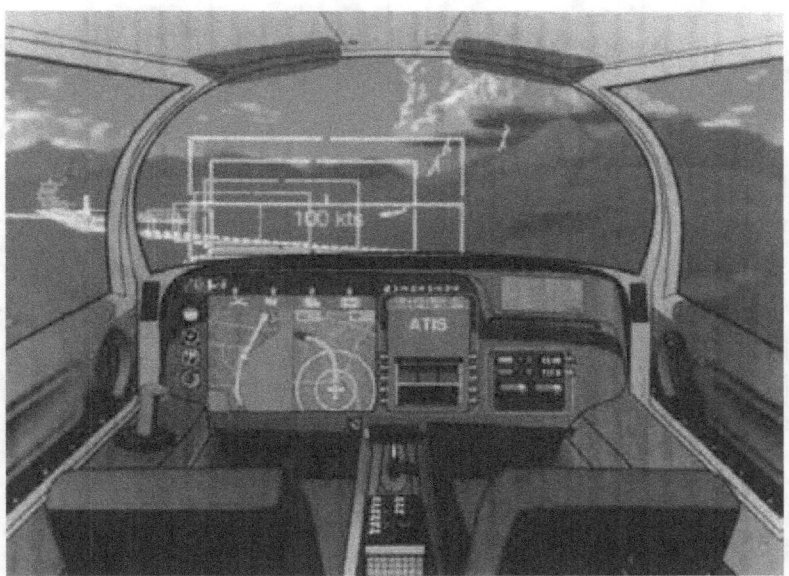

Highway-in-the-sky concept in NASA AGATE Program.

Highway-in-the-sky display in Cirrus SR-20 airplane.

means of traffic separation in clear visibility conditions. These concerns were not so prominent in the first workshops, which assumed an SVS application wherein hazard locations are known and obscuration is handled easily.

Langley researchers led by Russell V. Parrish, James R. Comstock, Jr., Dave E. Eckhardt, Lynda J. Kramer, Steve Williams, Jarvis J. (Trey) Arthur, III, and Louis J. Glaab contributed to XVS from 1993 to 1999. The XVS group conducted many simulation and flight test evaluations and established overall requirements for displays. Lynda Kramer and her Boeing team members developed a surveillance symbology system and proved the no-droop nose concept's feasibility. Early XVS flight tests used the U.S. Air Force's Total In Flight Simulator (TIFS) and Langley's B-737. The flight research range covered issues regarding the adequacy of computer-generated outside views with camera views, horizontal field of view requirements, conformal display and camera location issues, surveillance symbology, and guidance and flight control issues. The project culminated in validating the no-droop nose concept cockpit with a full 50-degree vertical by 40-degree horizontal field of view. With this highly successful accomplishment, the high-speed civil transport research transitioned to the SVS element of AvSP.

Conceptual synthetic vision system for a high-speed civil transport.

*The NASA Terminal Area Productivity Program and the Low Visibility Landing
and Surface Operations Project*

From 1993 to 2000, NASA conducted a Terminal Area Productivity (TAP) Program aimed at developing requirements and technologies for terminal area operations that would safely enable at least a 12-percent increase in capacity at major airports in instrument meteorological conditions (IMC). TAP research activities comprised four subelements: air traffic management, reduced separation operations, aircraft-air traffic control integration, and LVLASO. LVLASO research was aimed at investigating technology as a means to improve the safety and efficiency of aircraft movements on the surface during the operational phases of roll-out, turn-off, in-bound taxi, and out-bound taxi. This investigation was critical in the face of growing demands for air travel, the increasing number of reported surface incidents and fatal accidents, and the economic, environmental, and geographic infeasibility of constructing new airports or runways. During the LVLASO project, a prototype system was developed and demonstrated in order to validate an operational concept proposed to meet the project's objective. This system concept was used to help define international requirements for future advanced surface movement guidance and control systems defined by the International Civil Aviation Organization (ICAO).

The LVLASO system concept was built upon two display concepts called the Taxiway Navigation and Situational Awareness (T-NASA) system and the Roll-Out Turn-Off (ROTO) system. The T-NASA system provided head-up taxi guidance and head-down moving map functions, whereas the ROTO system provided head-up guidance to pilot-selected exits during landing roll-out. The underlying system used GPS; an accurate airport database, including the locations of runways, taxiways, and center lines; and data link technology that provided traffic information and controller-produced routing, or path instructions. The concept also included ground-based surface surveillance systems and controller displays.

Because the LVLASO concept was based on using an accurate airport database and precise positioning via GPS, it is considered a form of SVS constrained to the landing and surface operational phases. At the onset of the SVS project, it was apparent that the lessons learned during LVLASO were directly applicable to SVS operational goals and objectives. The SVS project challenge (2000 to 2005) became integrating the LVLASO surface display concepts such as taxi path, hold-short locations, and surface traffic with those of the in-flight display concepts such as tunnel, terrain, and obstacles. Detection of runway incursion events and subsequent alerting became an important add-on capability of the original LVLASO display suite.

Key researchers during the LVLASO project were Wayne H. Bryant (Project Manager), Steven D. Young, Denise R. Jones, Richard Hueschen, Dave Eckhardt, Dave Foyle, Rob McCann, and Tony Andre.

SVS Operational Concept Development

Langley research on crew-centered display concepts for synthetic vision systems includes activities to define concepts (tactical and strategic); retrofit analog systems; forward-fit (includes advanced display media); display information content, media, and format; operator interfaces; and simulation and flight evaluations. The display research is coupled with research on operational concepts, requirements, and integration. Critical elements in operational research include operations and requirements, procedures definition, operational enhancements, reversionary mode assessments, interface with the air traffic system, certification strategy, and operational benefits assessment. Finally, supporting research in crew response evaluation methodologies contributes information on situation awareness measurement techniques, crew performance assessments, and the application of measurement techniques in simulation and flight evaluations.

Key Langley research areas and lead individuals include: terrain database rendering (Trey Arthur and Steve Williams); pathway concepts (Russ Parrish, Lawrence J. Prinzel, III, Lynda Kramer, and Trey Arthur); runway incursion prevention systems (Denise R. Jones and Steven D. Young); CFIT avoidance using SVS (Trey Arthur); loss of control avoidance using SVS (Douglas T. Wong and Mohammad A. Takallu); database integrity (Steven D. Young); SVS sensors development (Steve Harrah); and SVS database development (Robert A. Kudlinski and Delwin R. Croom, Jr.).

Enabling Technologies for SVS

For the SVS concept to reach its fullest potential, several enabling technology developments are required. A large portion of the SVS project focused on concern over the availability and quality of terrain, obstacle, and airport databases. These issues were addressed through large-scale assessments, acquisitions, and development of representative prototypes. International standards were developed and published that established requirements for content, quality, and exchange. Sensor research, including FLIR and X-band radar, addressed integrity concerns by implementing data fusion, feature extraction, and monitoring techniques. Hazard detection algorithms were developed to address the runway incursion, runway alignment, and obstacle avoidance issues. Key researchers on the SVS team were Rob Kudlinski (lead), Steven D. Young, Denise R. Jones, Steve Harrah, and Delwin Croom, Jr.

Researchers Lynda Kramer (left) and Denise Jones conduct SVS study.

Partnerships

In addition to extensive in-house research, Langley is engaged in several partnerships and cooperative activities to develop synthetic vision systems. For commercial and business aircraft applications, an agreement partnership exists for a synthetic vision information systems implementation team led by Rockwell Collins, Inc., of Cedar Rapids, Iowa, with members from Jeppesen, The Boeing Company, American Airlines, Delft University of Technology, Embry-Riddle Aeronautical University, and Flight Dynamics, Inc. Another partnership on future flight deck information management and display systems is led by BAE Systems, Inc., CNI Division, Wayne, New Jersey, with members from BAE Systems Canada, Inc., and BAE Systems Astronics Company.

Enabling technology research was accomplished in large part through three cooperative research agreements. Database issues were addressed by a team led by Jeppesen and included members from American Airlines, the Technical University of Darmstadt, and Intermap Technologies Corporation. Runway incursion detection schemes were developed and tested by Rannoch Corporation. Lastly, Ohio University's Avionics Engineering Center contributed significant research results with respect to database integrity monitoring.

Specific research efforts are also directed at general aviation community needs where weather-induced loss of situational awareness has traditionally led to fatal accidents. In this area, Langley partners with an affordable, certifiable low end thrust synthetic vision system team led by AvroTec, Inc., Portland, Oregon, with members from B.F. Goodrich, Elite Software, Lancair/PAC USA, Massachusetts Institute of Technology, Raytheon Aircraft, Seagull Technologies, Inc., and FAA Civil Aerospace Medical Institute. In addition, a low cost synthetic vision display system capability for general aviation team is led by Research Triangle Institute, Research Triangle Park, North Carolina, with members from Archangel Systems, Inc., Flight International, Inc., Seagull Technologies, Inc., Dubbs & Severino, Inc., and FLIR Systems, Inc. Finally, a low-cost attitude and heading reference system (AHRS) to enable synthetic vision team is led by Seagull Technology, Inc., Los Gatos, California, with members from Dynamatt, BARTA, S-Tec Unmanned Technologies, Inc., Reichel Technology, Rockwell Collins, Inc., Stanford University, and Raytheon Aircraft.

Flight Demonstrations of SVS Technology

Throughout the SVS project's 5-year term, the rapidly advancing state of synthetic vision technology has been demonstrated by NASA in high-profile events with extensive participation by other government agencies, industry, airlines, and airport officials. Since 1999, the flight demonstrations have shown an increasingly sophisticated SVS capability with obvious benefits to piloting tasks and attendant safety in low-visibility conditions. The scope of the demonstrations has included airborne as well as ground operations designed to show the potential of SVS to minimize or eliminate CFIT and runway incursion accidents.

Asheville, North Carolina

As part of the AvSP, initial SVS flight efforts used the TIFS, a highly modified propeller-driven Convair 580 built in the 1950s and transformed into a flying simulator in the early 1970s. The TIFS aircraft is outfitted with two cockpits: a conventional safety cockpit that is always available to fly the plane and a forward research cockpit that is used to test advanced concepts. The simulation cockpit is equipped with special instruments and displays. Operated by the Veridian Corporation, the TIFS vehicle was configured to simulate flight characteristics of a representative high-speed civil transport for SVS flight research at Asheville, North Carolina, in 1999. In a demonstration of photorealistic synthetic vision, Langley researchers used an experimental terrain database of the Asheville Regional Airport that had been augmented by sophisticated computer rendering techniques. The use of photorealistic versus generic terrain texturing provided the researchers with an early look at display size and field of view issues.

The Total In-Flight Simulator research airplane during NASA flights at Asheville, North Carolina.

Display implementation for Asheville research flights

The Asheville demonstration was led by Langley's Russ Parrish, Lynda Kramer, Lou Glabb, and Veridian's Randy Bailey (who would later become a Langley researcher and join the SVS team), and it included the participation of Veridian, Boeing, Honeywell, the Research Triangle Institute, and various airlines. During October and November 1999, three evaluation pilots flew over 60 approaches, including three to final touchdown. Pilot comments on Asheville Regional Airport's nested database with photorealistic overlay were very favorable. The state-of-the-art terrain, obstacle, and airport databases worked extremely well in actual flight. Pilots said that tactical control of the aircraft using synthetic vision was intuitive and characterized by low workloads. NASA researchers gathered information data on transitioning from lower to higher resolution nested databases, as well as grid size requirements. They also assessed photorealistic terrain texture overlays compared with computer-generated terrain texture overlays.

Pilot comments during the demonstration were very impressive. The display's realism and the obvious impact on flights in low-visibility conditions during airline-type operations were cited as monumental contributions with tremendous potential for enhanced safety.

Armed with these very exciting early results, Langley researchers continued the pursuit of additional challenges and barriers, including SVS retrofit issues and technical questions, such as head-down display size and field-of-view requirements head-up opaque display concepts, terrain texturing issues, generic versus photorealistic views, and database integrity.

Dallas-Fort Worth International Airport, Texas

During September and October 2000, Langley researchers and their cooperative research partners—Rockwell-Collins, Rannoch, the FAA's Runway Incursion Reduction Program, Jeppesen, Ohio University, and the Dallas-Fort Worth Airport Authority—conducted an extensive flight demonstration/evaluation of synthetic vision at the Dallas-Fort Worth (DFW) International Airport. In this series of demonstrations, the Langley team used the Langley Boeing 757-200 Airborne Research Integrated Experiments System (ARIES) research aircraft with SVS displays.

The DFW demonstration was planned as part of the previously discussed NASA LVLASO project that was slated to end with this specific demonstration, culminating 7 years of Langley research on surface display concepts and systems for low visibility ground operations. The demonstration included an assessment of a Runway Incursion Prevention System (RIPS). A government study predicts runway collisions will be the single largest cause of aviation fatalities over the next 20 years unless something is done, and near misses on runways are up sharply in recent years. In one tragic example in 1991, 34 people died at Los Angeles International Airport during a runway incursion accident.

Langley's RIPS integrates several advanced technologies into a surface communication, navigation, and surveillance system for flight crews and air traffic controllers. RIPS combines a head-down cockpit display of an electronic moving map of airport runways and taxiways with a head-up display that gives the pilot real-time guidance.

In-flight SVS concepts were also included in the evaluation so that they could be assessed in a busy terminal environment. The in-flight SVS testing was decoupled from the RIPS testing; that is, both components of what would later be the integrated Langley SVS were tested and evaluated separately at DFW. Six pilots flew 76 landing approaches to evaluate the SVS concept, including about 18 research flight hours. RIPS was flown and assessed by four evaluation pilots. One of the other key investigations of this activity was an evaluation of a Langley opaque HUD concept during night operations.

Langley's B757 ARIES research aircraft at Dallas-Fort Worth International Airport.

Night scene during practice flights for Dallas-Fort Worth showing SVS display and out-the-window runway lights.

The evaluation pilots provided extremely favorable comments, including observations that synthetic vision appeared to be viable and effective, situational awareness was enhanced, and pilots liked the immersive feel of the HUD.

Two incursion events were emulated at DFW. For the first runway incursion event, the ARIES pulled onto an active runway upon which another aircraft (simulated by an FAA van) was landing. In another scenario, the ARIES performed a coupled instrument approach while traffic (the FAA van) proceeded across the hold-short line crossing the active runway. These emulated incursions demonstrated the effectiveness of RIPS.

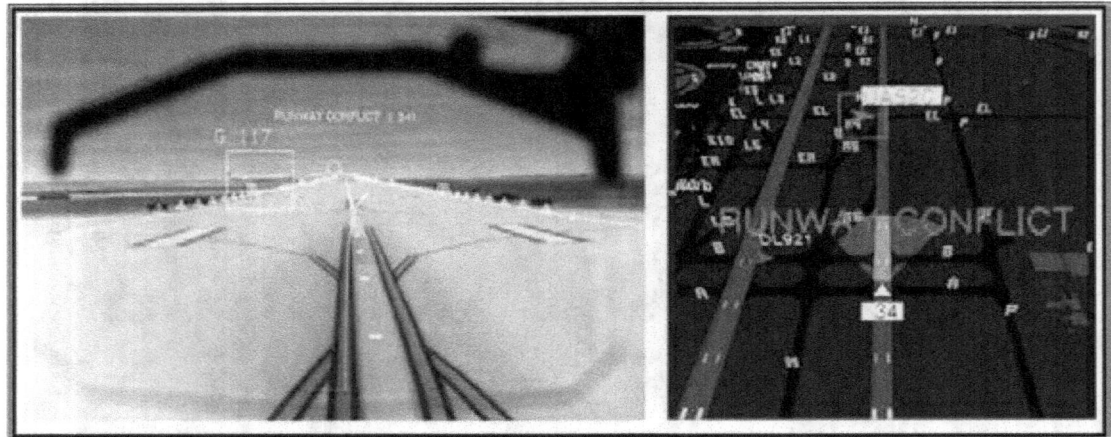

Runway incursion prevention displays.

Key researchers for the DFW demonstrations included Lou Glaab, Denise Jones, Richard Hueschen, Lynda Kramer, Trey Arthur, Steve Harrah and Russ Parrish, as well as Langley research pilots Harry A. Verstynen and Phillip Brown.

Eagle County Regional Airport, Colorado

In August and September of 2001, Langley and its partners provided demonstration flights of SVS for the third time in 3 years to NASA, airline, FAA, and Boeing pilots at the Eagle County Regional Airport near Vail, Colorado. Led by Randy Bailey, the project team's objective was to demonstrate SVS in an operationally realistic, terrain-challenged environment. Vail-Eagle is nestled in a valley with mountains close to the runway on three sides. Runway incursion prevention systems were not evaluated in this project. The vision of operational SVS concepts had matured to database-derived systems using precise GPS navigation and integrity monitoring sensors to provide an unrestricted synthetic view of the aircraft's current external environment, regardless of weather or time of day.

The Langley ARIES B757 research airplane was again used for the demonstration, which was flown by seven evaluation pilots for 11 research flights that included a total of 106 airport approaches. Specialized systems on the 757 aircraft included dedicated SVS pallets, special SVS display panels, a HUD, vision restriction capability for simulated instrument conditions, and an 18.1-in. (1280

Langley's ARIES B757 research airplane during flight at Vail-Eagle airport.

Head-up and head-down displays used during demonstration flights.

by 1024) flat panel display. Subject pilots included current B757 captains from American Airlines, United Airlines, Delta Airlines, FAA, Boeing, and NASA. The pilots were provided preflight training in Langley's simulators before the demonstration.

During the flights at Vail-Eagle, the evaluation pilots provided comments on two different HUD concepts and four different head-down concepts developed by NASA and Rockwell Collins. The pilots provided comments on the relative effectiveness of display sizes, fields of view, and computer graphic options.

In addition, EVS applications were investigated, both as a sensor for gathering images of the runway environment and as a database integrity monitor. Likewise, multiple radar altimeters and differential GPS receivers gathered height-above-terrain truth data for database integrity monitoring algorithm development.

Key Langley researchers included Principal Investigator Randy Bailey, Russ Parrish, Dan Williams, Lynda Kramer, Trey Arthur, Steve Harrah, Steve Young, Rob Kudlinski, Del Croom, and pilots Harry Verstynen and Leslie O. Kagey, III. Boeing, Jeppesen and Rockwell Collins, and Ohio University joined Langley as part of the synthetic vision project team. Approximately 70 representatives from the FAA, DoD, and the aviation industry participated in preflight briefings and in-flight demonstrations of SVS display concepts.

One impressive demonstration of the capabilities provided by SVS came during "circle-to-land" approaches on Vail-Eagle's runway 7. Commercial pilots get special training to land at mountain airports like Vail-Eagle, but until the NASA tests, pilots had never made the circle-to-land, which puts the jet very close to the surrounding terrain under instrument flight conditions. Published navigational charts require that the landing be flown with a visual approach. The technology provided by Langley and its partners worked flawlessly during the unprecedented approaches.

Reno-Tahoe International Airport, Nevada

In July 2004, Langley's consortium of government, industry, and university partners culminated an impressive 5-year demonstration of accelerated technologies for prototype SVS concepts with an in-depth series of flight tests at the Reno-Tahoe International Airport. The demonstration integrated a number of SVS elements to highlight the benefits to pilots in both airborne and ground operations in low visibility conditions. Dan Baize's team of government and industry participants used a GV business jet and a variety of Langley- and Rockwell Collins-derived SVS concepts that left indelible impressions on evaluation pilots for potential enhancement of flight safety. Using the

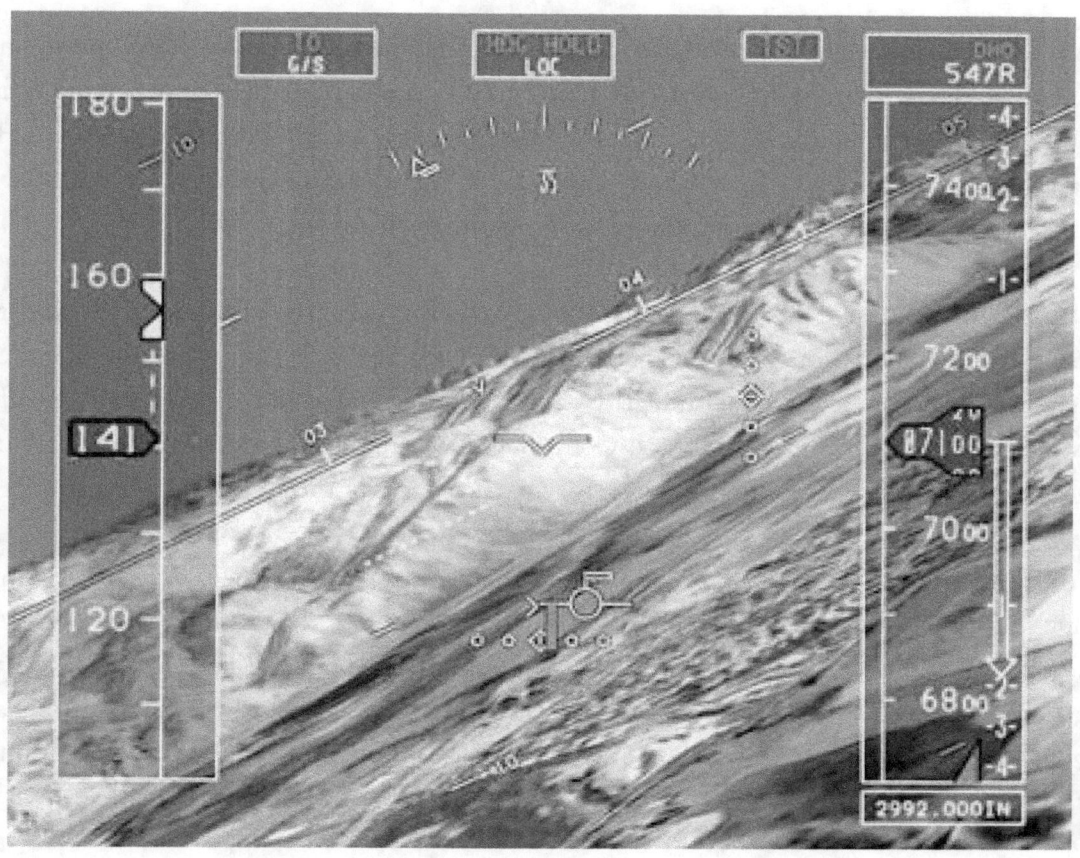

Details of the synthetic vision system display used in the demonstration flights.

Gulfstream V airplane used in demonstration flights at Reno International Airport.

GV was especially noteworthy because the aircraft was equipped with a production option known as the Gulfstream enhanced vision system, which uses a nose-mounted forward-looking infrared sensor to provide information depicted in terrain displays on a cockpit HUD panel and an HDD flat panel display, both of which provide symbology and flight path information. As of 2004, over 80 units of this concept had been produced for Gulfstream customers.

Langley's Randy Bailey led the project, while Denise Jones was Principal Investigator for the runway incursion technology, Lynda Kramer and Trey Arthur led the SVS display development, Steve Young and Del Croom led the data integrity efforts, and Steve Harrah led sensors-related efforts. Industry participants included Rockwell-Collins, Gulfstream, Northrop Grumman, Rannoch Corporation, and Jeppesen, and Ohio University participation in the area of data integrity.

Two different runway incursion prevention systems developed by NASA/Lockheed Martin and Rannoch Corporation, respectively, were evaluated. A dramatic demonstration of the NASA-Lockheed Martin concept's effectiveness involved a simulated potential collision with the team using a NASA aircraft acting as an intruder by accelerating on an intersecting runway during a GV takeoff. With unmistakable visual and aural cues, the RIPS quickly identified the situation and prompted the pilot for immediate action (in this case, deploying brakes and engine thrust reversers to stop the Gulfstream test vehicle).

Langley's SVS terrain and guidance displays used in the demonstration were intentionally conceived to be futuristic in terms of guidance displays, pathway-in-the-sky details, and terrain detail. As part of the SVS demonstration team, Rockwell Collins provided its version of a more near-term SVS/EVS display concept, highlighting the accelerated development that has occurred in synthetic vision capability and the potential for certification and widespread applications of the technology in the near future.

Terrain database integrity monitoring was also demonstrated during the Reno flights. In this element of the NASA program, Langley teamed with Ohio University and Rockwell Collins to use sensors—such as a radar altimeter, advanced weather radar, and forward-looking infrared information—to cross check the accuracy of the digital SVS database in real time during flight. The pilot was alerted if questionable data were detected during the correlation of sensor and digital information, and erroneous data were deleted from the displays. During the demonstration flights, the GV's standard infrared-based "all-weather window" produced by Kollsman, Inc., provided thermal imagery of features such as runway lights and terrain for cross checking of digital SVS data.

*View of the synthetic vision system head-up and head-down displays (left)
and the standard Gulfstream V display (right).*

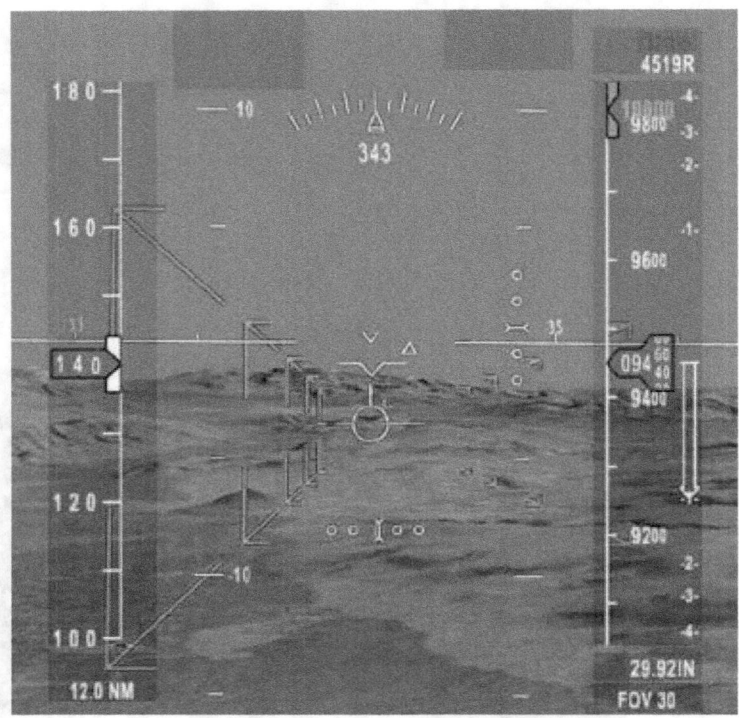

Details of synthetic vision system display evaluated in Reno tests.

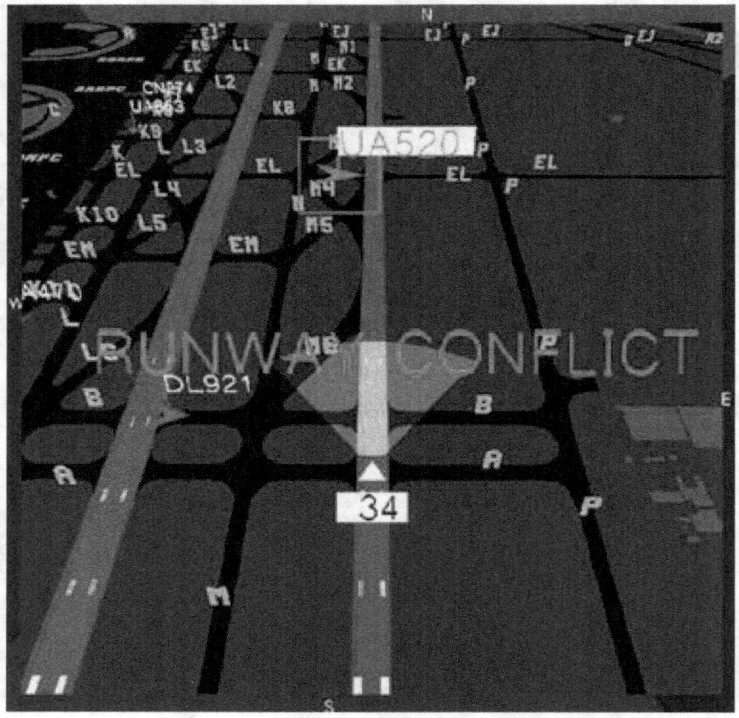

Runway incursion prevention display.

Simulated runway incursion by NASA King Air airplane (left) during Reno evaluations of incursion prevention systems on Gulfstream V airplane (right). Note the deployed engine thrust reverser doors on the Gulfstream V as the pilot takes corrective actions to avoid the potential collision.

The integrated SVS concept also included a voice recognition system for display control. Insensitive to individual voice or accent characteristics, the system proved to be extremely reliable during the flight demonstrations.

All evaluation pilots who assessed the SVS technology at Reno were extremely positive and enthusiastic over the capability of this revolutionary technology. The system's potential to ease pilot workload and provide a significant improvement in safety during low-visibility airborne and ground operations was readily apparent and appreciated by all participants.

General Aviation Activities and Flight Demonstrations

Langley researchers are also addressing technological and operational challenges facing the implementation of synthetic vision systems within the general aviation community. Low-visibility accidents are especially prevalent among inexperienced pilots. General aviation aircraft comprise about 85 percent of the total number of civil aircraft in the United States. In a recent NTSB accident database, general aviation accounted for 85 percent of all accidents and 65 percent of all fatalities. The combination of night and instrument flying increased the proportion of fatal to total accidents to 64.3 percent, making low-visibility conditions the most deadly general aviation flight environment.

Working with the FAA, industry, universities, and the general aviation pilot community, Langley is conducting extensive simulator and flight test evaluations to assess and demonstrate the benefits of SVS technology compared with current general aviation systems. As previously mentioned, the constraints on technical concepts directed toward the general aviation sector are unique, including low cost, no sensors, no HUD applications, and use of existing strategic terrain displays. Dominant in these considerations is the tremendous degree of variability in pilot background and capabilities.

Lead researchers for Langley's general aviation element of the SVS project include Louis J. Glaab and Monica F. Hughes. Initially, their team of researchers conducted piloted simulator studies using a generic general aviation workstation equipped with variants of SVS characteristics, such as terrain features and guidance information. Following the fundamental simulator studies, the more promising displays were incorporated in actual flight evaluations at two Virginia airports.

Flight Demonstration of Head-Down Displays

Glaab and Hughes conceived and conducted a series of flight evaluations to attack technical questions regarding terrain presentation realism and the resulting enhancements of pilot situational awareness and performance. Comprising coordinated simulation and flight test efforts, terrain portrayal for the head-down display (TP-HDD) test series examined the effects of digital resolution and terrain texturing. The TP-HDD test series was designed to provide comprehensive data to enable design trades to optimize all SVS applications, as well as develop requirements and recommendations to facilitate the implementation and certification of SVS displays. The TP-HDD flight experiment used the Langley Cessna 206 research aircraft and evaluated eight terrain portrayal concepts in an effort to confirm and extend results from a previously conducted TP-HDD simulation experiment. Fifteen evaluation pilots, of various qualifications, accumulated over 75 hours of dedicated research

Langley's Cessna 206 research airplane.

flight time at the Newport News-Williamsburg International and Roanoke Regional airports in Virginia from August through October, 2002.

Overall, a comprehensive evaluation of specific components of SVS terrain portrayal methods was conducted through an extensive simulation and flight-test effort. Project results indicated pilots were able to use SVS displays effectively with dramatically increased terrain awareness. In general, all SVS concepts tested provided results that correlated with other data produced by non-NASA researchers, suggesting that ultimate terrain portrayal fidelity (photorealism) might not be as important as effective terrain portrayal presentation (elevation-based generic).

The FAA's Alaskan Region Capstone Program

Langley has traditionally maintained a close partnership with the FAA and other government agencies in its aeronautical research and development activities, and this legacy has been maintained in the SVS project element for general aviation. In addition to frequent communication and partnership activities with the FAA, Langley has hosted two separate workshops, in 2002 and 2004, to ensure real-time transfer of its technology and to receive guidance and comments from appropriate FAA offices regarding future research.

An important FAA activity involving Langley participation has been the Alaskan Region Capstone Program. Capstone is an accelerated FAA effort to improve aviation safety in Alaska. No state relies as heavily upon aviation as Alaska does to provide many of life's bare essentials, yet Alaska ranks at or near the bottom in U.S. aviation safety because of its terrain, climate, and lack of such infrastructure as weather observation stations, communications, and radar coverage below 10,000 ft, where most general aviation and commercial carrier aircraft fly. The Capstone program was created to address Alaska's high accident rate for small aircraft (those weighing 12,500 lbs or less). These accidents occur at nearly five times the national average. Plans call for up to 200 aircraft to be voluntarily equipped with Capstone avionics. The program includes the installation of ground infrastructure, GPS-based avionics, and data link communications in commercial aircraft serving the Yukon-Kuskokwim Delta/Bethel area.

The Capstone Program equips over 150 aircraft used by commercial operators in Alaska with a combined data link and GPS-based avionics package designed to increase the situational awareness of pilots in averting mid-air collisions and CFIT accidents.

In 2001, the FAA released a Request for Proposal (RFP) for a Capstone phase II activity that would incorporate technologies matured in phase I, build on lessons learned, and explore other risk-mitigating technologies to reduce accidents and fatalities in the Southeast area of Alaska. The project would include technical elements representative of SVS concepts, such as the first certification of HITS technology for navigational guidance, use of forward-looking three-dimensional terrain, adjustable field of view on primary flight displays, and use of conformal perspective runway presentation and conformal obstructions on primary flight displays. In April 2002, Chelton Flight Systems of Boise, Idaho, received the phase II award for this revolutionary project. In March, 2003, Chelton Flight Systems received the first FAA approval for synthetic vision, highway-in-the-sky technology. Chelton's system incorporates synthetic vision, a flight path marker, and HITS technology, providing a three-dimensional series of boxes along the flight path from takeoff to touchdown. Selected for the FAA's Capstone Program, the certification of this technology was a groundbreaking partnership between industry and a progressive FAA.

Langley participated in the development of the RFP for Capstone II, including membership on the technical evaluation board. As part of the TP-HDD simulation experiment, the lead certification pilot for the FAA Capstone-II effort, as well as three potential Capstone-II equipment users from

Juneau, Alaska, participated in the TP-HDD data generation process as subject pilots. Due to this participation and extensive discussions during the previously mentioned workshops, Langley was able to effectively direct its research toward FAA-requested certification issues.

Status and Outlook

Research conducted to date on SVS addresses low-visibility-induced incidents and accidents with a visibility solution, making it possible for every flight to be nearly equivalent to clear-day operations. The key remaining challenges to actual commercial use appear to be: (1) certification and systems engineering, that is, how to map SVS functions to certified avionics and prove fail-safe operations; (2) operational approval for flight-critical uses, or how to prove that pilots can trust the system when flying close to the ground in low visibility; (3) effective crew procedures and interfaces; and (4) justifying investment in new technology in today's economic environment. Safety improvements are rarely implemented without operational benefits unless mandated. SVS may provide operational benefits by enabling lower landing minima and/or landing at unequipped runways and airports. Technical challenges such as database integrity and avoidance of hazardously misleading information also remain major concerns.

NASA strives to work cooperatively with the FAA, academia, and industry to ensure successful implementation of SVS. A large number of commercialization efforts are underway to bring the SVS concept to reality; certification is being pursued, and it is expected that the technology will be routinely used in commercial and general aviation in the near future.

Langley's SVS project is scheduled to end in 2005, completing an exceptionally productive program that has helped push this remarkable technology to the forefront of aviation. At this time, almost every general aviation manufacturer has a synthetic vision concept in its design pipeline for near-term application. The Gulfstream EVS system has set the mark for applications to business jets, and Chelton's FlightLogic™ system is leading the way for SVS.

Research conducted by Langley Research Center and its partners has had a profound effect on the state of the art in synthetic vision cockpit displays. The highly professional efforts contributed by Langley on fundamental pilot-display technology, enabling avionics technologies, computational requirements and methodology, and interactive demonstrations to the aviation community have greatly accelerated the implementation, database development, and confidence level required to pursue the certification and application of SVS. NASA has concentrated on futuristic systems and operations, becoming leaders in technology beyond that which is emerging today. The partnerships between Langley and industry have been especially productive, continuing a long Langley legacy of success in such arrangements. In addition, the accumulated expertise and experience of the Langley staff represents a national asset that is being applied to the solution of visibility-related safety issues for the U.S. aviation fleet for years to come.

Concept and Benefits

When an aircraft flies through the atmosphere, the effects of viscosity cause frictional forces as the flow in a thin layer of air next to the aircraft surface—known as the boundary layer—decelerates relative to the aircraft's speed. These frictional forces generate a significant amount of aerodynamic drag for all classes of aircraft. This drag, referred to as skin-friction drag, has a large effect on the amount of fuel consumed by most aircraft during cruise. For example, at subsonic cruise conditions, the skin-friction drag of a conventional subsonic transport accounts for about one-half the fuel required for flight, and, for a future supersonic transport, skin-friction drag could account for about one-third the fuel burned at cruise conditions. Very complex fluid dynamic interactions that occur within the boundary layer determine the specific level of skin-friction drag experienced by an aircraft. Three distinct types of boundary-layer flow states may occur: laminar, transitional, and turbulent. In the case of laminar boundary-layer flow, the fluid near the surface moves in smooth-flowing layers called laminae.

The sketch shows a representation of boundary-layer conditions (greatly magnified) resulting from

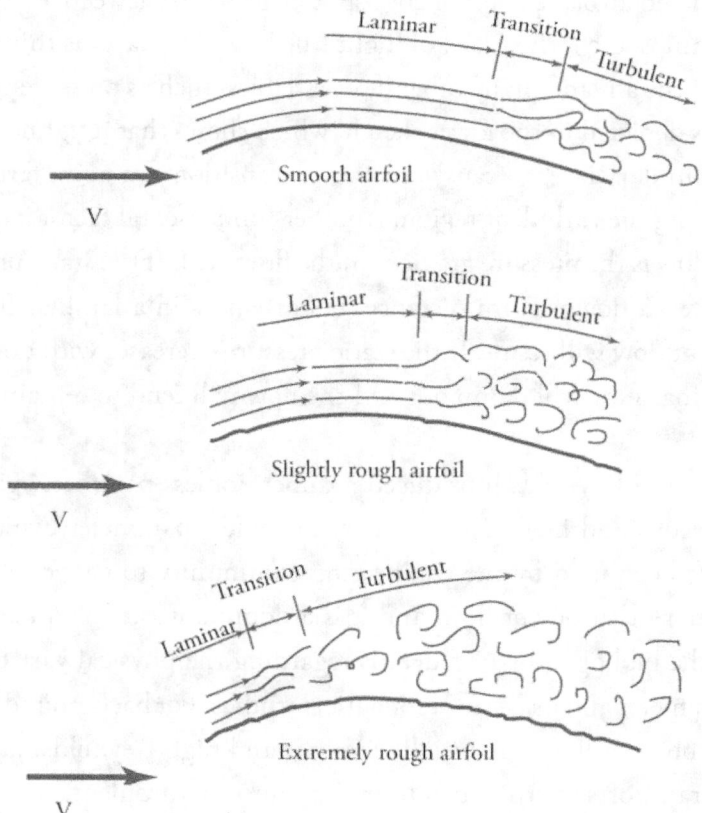

Effect of airfoil surface condition on boundary-layer flow.

air moving over an airfoil from left to right. For ideal conditions, the flow would follow the curved airfoil surface smoothly, in laminae. Unfortunately, for most full-scale aircraft flight conditions of interest, the thin boundary layer enters a state of transition wherein the adjacent layers of flow begin to intermix and destroy the desirable laminar condition. Following transition, the boundary layer is completely turbulent and laminar-flow features are completely lost. In turbulent flow, secondary random motions are superimposed on the principal flow within the thin boundary layer, such that slow-moving fluid particles speed up, and fast-moving particles give up momentum to the slower moving particles and slow down. The friction force between air and moving body dramatically increases when the boundary-layer flow changes from a laminar state to the turbulent condition. The reader is cautioned that the change from a laminar to a turbulent condition does not result in flow separation from the airfoil surface (as in a stalled condition). The eddies and turbulence shown in the sketch serve only to emphasize the chaotic nature of the turbulent flow within the thin boundary layer, which may extend from just a few fractions of an inch to inches above the surface.

Mechanisms causing boundary-layer flow change from laminar to turbulent are very complex depending on a considered airplane component's specific geometry (sweep, thickness, etc.), surface disturbances (gaps, bumps, etc.), the speed of flight (the boundary layer is thinner at high speeds), the relative viscosity of the air, and many other flow variables, such as pressure gradients. The effect of surface roughness is depicted in the lower sketch, which shows that roughness imparts sufficient disturbances to the boundary layer to cause premature transition at a point farther forward on the airfoil, thereby increasing the turbulent region. Another fundamental factor in the transition from laminar to turbulent flow is the pressure gradient in the flow field. If the static pressure encountered by the flow increases with downstream distance, disturbances in a laminar flow will tend to be amplified and turbulent flow will result. If the static pressure decreases with downstream distance, disturbances in a laminar flow will damp out and the flow will tend to remain laminar.

In 1883, scientist Osborne Reynolds introduced a dimensionless parameter giving a quantitative comparison of the viscous and inertial flow states. Reynolds' parameter, which is known as the Reynolds number, has been used by the engineering community to gauge relative probability of the existence of laminar or turbulent flow, and it is a dominant variable in aerodynamic design. Documents cited in the bibliography give details regarding the physical variables included in the Reynolds number parameter and its use in aeronautical studies. For background of the nontechnical reader, low Reynolds number flows are usually laminar and high Reynolds number flows (typical of large commercial transports at cruise conditions) are mostly turbulent.

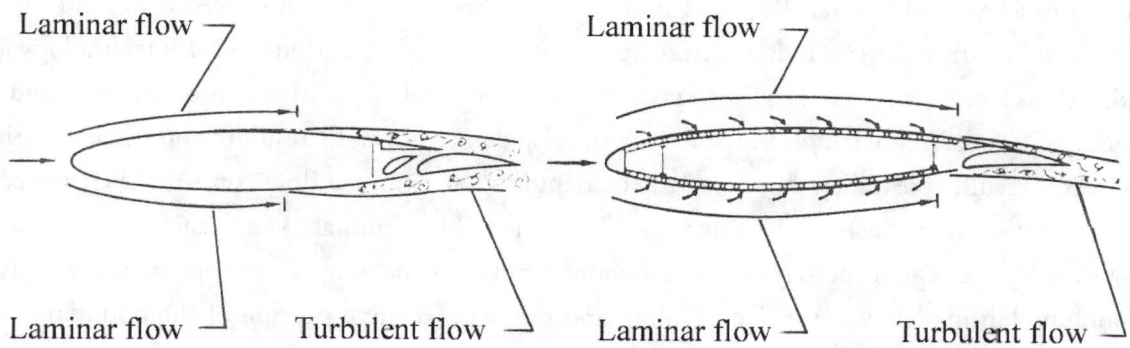

Laminar flow

Laminar flow

Laminar flow

Turbulent flow

Laminar flow

Turbulent flow

Natural laminar flow

Laminar flow control

Laminar flow obtained by airfoil shaping in the natural laminar flow concept (left) and active suction flow control in the laminar flow control concept (right).

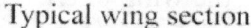

Typical wing section

Upper

c_p

x/c

Natural laminar flow

Suction laminar flow control

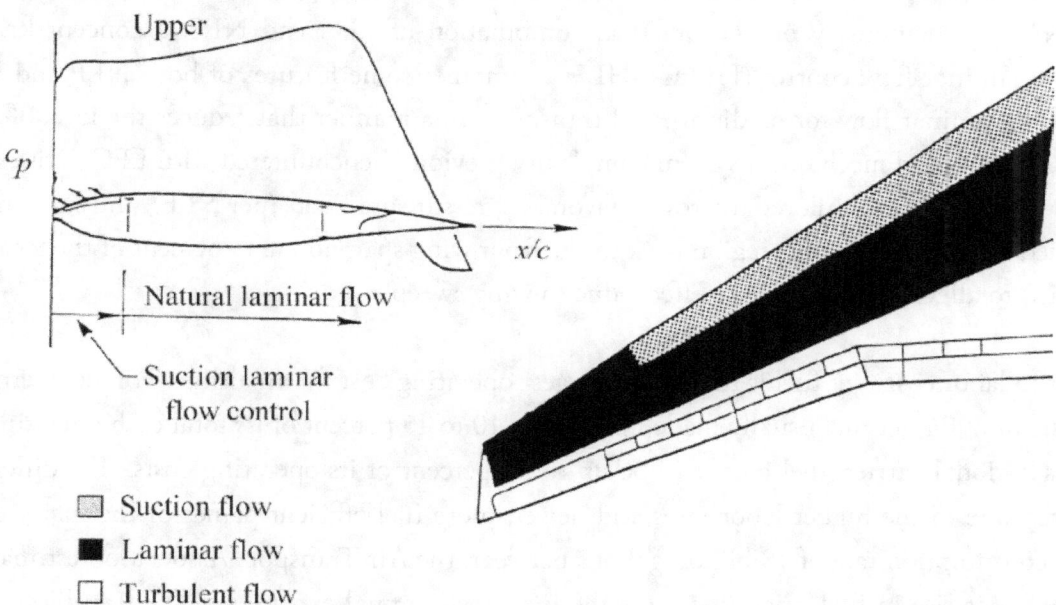

☐ Suction flow
■ Laminar flow
☐ Turbulent flow

The hybrid laminar flow control concept.

Successful analysis, prediction, and control of the boundary-layer transition process for improved aerodynamic efficiency has been the ultimate goal—the Holy Grail—of aerodynamicists since the earliest days of aviation. For some configurations with relatively small-chord unswept wings, designers have used shaping of the wing airfoil to promote favorable pressure gradients that, together with smooth composite wing structures, and, in some cases, high-altitude flight (low Reynolds number), result in extensive laminar flow. Examples include high-performance gliders and the military U-2 reconnaissance aircraft. In this passive manner, laminar flow is obtained in what is referred to as natural laminar flow (NLF). A previous book by the author, NASA SP-2003-4529, *Concept to Reality*, discusses Langley Research Center's past contributions to NLF technology and applications by industry. For larger aircraft, which cruise at relatively high subsonic speeds and are subject to surface irregularities, the principles of NLF do not normally result in appreciable laminar flow. As a result, researchers have turned their interests to laminar flow control (LFC) concepts that use artificial (active) mechanisms to delay or possibly eliminate the existence of turbulent flow over a large region (perhaps even full-chord length) of the wing. Two approaches to actively prolonging laminar flow are surface cooling and removing a small portion of the boundary-layer air by suction through slots or porous surfaces.

LFC concepts for aircraft applications from the late 1930s to the late 1980s focused on suction concepts, most of which used full-chord deployment of suction to encourage full-chord laminar flow. Unfortunately, LFC is a mechanically complex concept involving auxiliary power sources or engine bleed air for suction, extensive ductwork throughout the wing internal structure, and weight and maintenance penalties. However, in the late 1980s application studies were instituted by NASA and industry on the practical combination of LFC and NLF, a concept known as hybrid laminar flow control (HLFC). HLFC combines some features of both NLF and LFC to promote laminar flow for medium-sized transports in a manner that reduces the level of suction requirements and mechanical system complexity previously encountered with LFC. With HLFC, the wing geometry is tailored to provide favorable pressure gradients (per NLF), and suction is only applied in the leading-edge region back to the front wing spar (about 15 percent of the local chord length) to alleviate some adverse effects due to wing sweep.

Next to labor costs, jet fuel is the second largest operating cost for an airline. For an international airline in 2004, jet fuel usually made up roughly 10 to 15 percent of its total cash operating costs. For a regional carrier, fuel costs can be up to 25 percent of its operating costs. The difference is mainly due to the higher labor rates and newer, more fuel-efficient planes of the major carriers. At a consumption rate of 18 billion gallons per year, the Air Transport Association estimates that each 1-cent rise in fuel prices increases the industry's annual expenses by $180 million. Jet fuel price increases have consisted of large excursions that can abruptly occur over short time periods of months. For example, the average monthly cost of domestic jet fuel from January 1999 through July 2000 rose from a low of 43.8 cents per gallon to a high of 77.8 cents per gallon, a 78-percent increase. In early 2005 the rise of crude oil prices to over $55 per barrel caused financial chaos throughout the U.S. air and ground transportation systems.

Aerodynamic performance predictions by researchers indicate that LFC might decrease the fuel burned on long-range flights for transport-type aircraft by a phenomenal 30 percent. By reducing the amount of fuel burned, LFC would also reduce emissions and pollution caused by aircraft. Although the aerodynamic performance improvements of HLFC are not as great as LFC, the potential gains are substantial (up to about 15-percent reduction in fuel burned) and represent a significant benefit for cash-strapped airlines. Of course, the benefits of LFC and HLFC will be largest for fuel-intensive long-range missions rather than short routes. Studies also indicate that applications of LFC and HLFC at initial design phases of a new aircraft (versus retrofit applications to existing aircraft) may allow the airplane to be resized for weight reduction, smaller engines, and other benefits. For years the domestic and international aviation communities have widely recognized the benefits of LFC and the technology is currently being pursued for potential applications to civil and military aircraft.

Challenges and Barriers

Requirements for aircraft performance, economic viability, operational suitability, and safety must be addressed and demonstrated before active laminar flow concepts, such as LFC and HLFC, can be applied to operational aircraft. Obviously, technology maturation is required to a relatively high level, including demonstrations with actual aircraft. The following discussion highlights issues and concerns that surfaced early during research on LFC. NASA in-house, contracted, and cooperative research has addressed virtually all of these early concerns, but this section discusses these issues here to provide the reader with an appreciation of the breadth of research provided by NASA to solve potential problems and barriers to LFC and HLFC.

Disciplinary Challenges

Aerodynamicists face fundamental challenges in LFC technology. Developing reliable design methods that accurately predict characteristics of the inherently unstable boundary layer as affected by a myriad of aircraft and atmospheric parameters such as geometry, sweep, pressure gradients, surface conditions, suction levels and distribution, and Reynolds number is formidable. Inherent to the task is a full understanding of the physics and processes associated with the laminar to turbulent boundary-layer transition process.

Aerodynamicists must master transition mechanisms related to spanwise crossflows, eddies, and critical waves within the boundary-layer flow, and other aerodynamic phenomena that promote or inhibit transition. They must also develop and validate predictive tools for design. Within this effort, the transition-inducing effects of surface roughness, waviness, steps, and gaps must be determined and tolerances defined for manufacturing. Aerodynamic advantages of slotted, porous, and perforated suction distribution concepts must be evaluated and demonstrated.

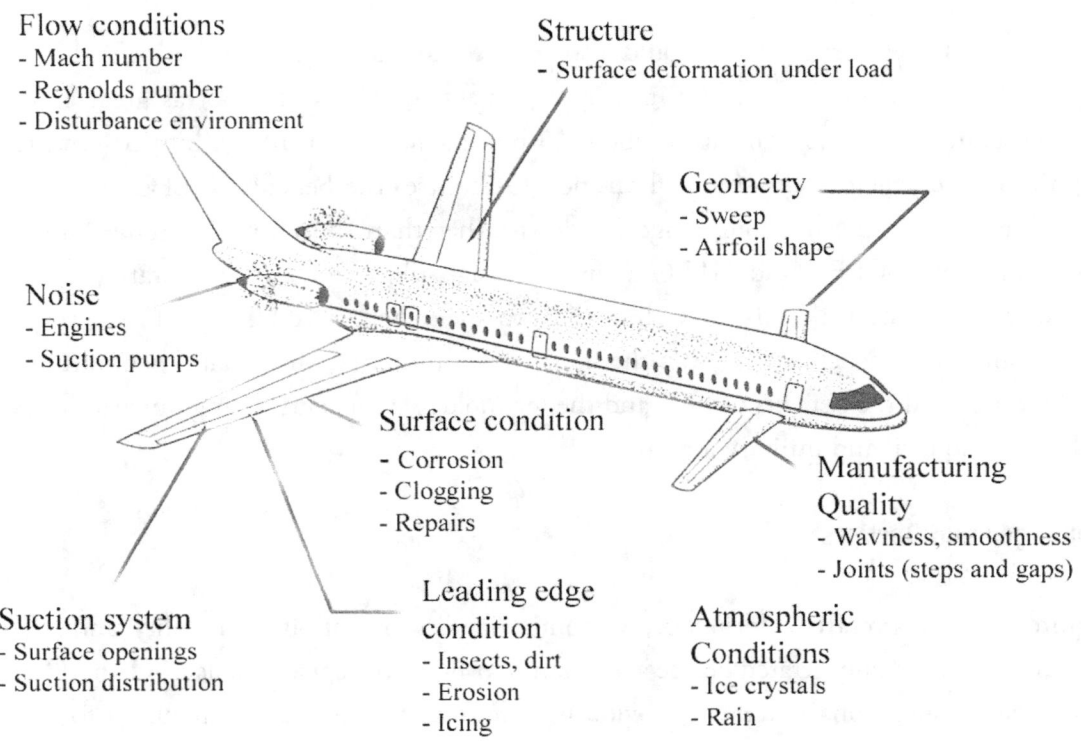

Flow conditions
- Mach number
- Reynolds number
- Disturbance environment

Structure
- Surface deformation under load

Geometry
- Sweep
- Airfoil shape

Noise
- Engines
- Suction pumps

Surface condition
- Corrosion
- Clogging
- Repairs

Manufacturing Quality
- Waviness, smoothness
- Joints (steps and gaps)

Suction system
- Surface openings
- Suction distribution

Leading edge condition
- Insects, dirt
- Erosion
- Icing

Atmospheric Conditions
- Ice crystals
- Rain

Concerns and issues addressed by laminar-flow technology.

Aircraft systems engineers must analyze the impact of losing laminar flow through mechanical or other causes, and acceptable off-design characteristics must be ensured. Aerodynamic issues also cause designers to assess the compatibility of laminar-flow concepts with conventional solutions to other aerodynamic considerations. For example, designers sometimes use vortex generators to inhibit wing flow separation for satisfactory high-speed control and buffet characteristics, but this approach would not be compatible with laminarization of wing surfaces. Management also wants to see if LFC- and HLFC-type systems can provide synergistic benefits that help such a concept "buy its way" onto the aircraft.

Issues relating to manufacturing materials with precise, economically feasible features for suction are especially critical. The layout, dimensions, and drilling of suction holes in wings, fuselages, tails, or nacelle structures are potential barrier problems that must be demonstrated with full-scale, production-type tooling and hardware.

Integrated System Challenges

Designers must also address aircraft system-level challenges for LFC and HLFC, including considerations of power systems for the suction system. Auxiliary pumps or other devices must have acceptable capacity, weight, maintainability, and economic feasibility. If engine bleed air is required, the level of bleed must not result in serious increases in the aircraft's specific fuel consumption. Assessments of ducting concepts and analysis of distributed suction levels are required to ensure adequate suction levels compatible with the physical constraints of wing structure and plumbing. Total weight of the LFC or HLFC system is usually of relatively little concern for systems envisioned, but maintenance, complexity, manufacturing feasibility, cost, and reliability are critical issues, as well as compatibility with other mechanisms located in or on the aircraft wing, such as leading-edge high-lift devices and deicing systems.

Operational Challenges

Researchers have focused on the most obvious barrier problem to laminar flow: environmental contamination of the sensitive wing leading-edge region caused by insects, rain, ice, or other debris. Concepts to minimize the effects of environmental contamination must be compatible with aircraft takeoff and landing performance, cruise performance, routine flight operations, other flight systems, and maintenance requirements. Using shielding concepts to protect the leading edge from insect debris during takeoff must be effective, not degrade the high-lift aerodynamic performance of the airplane, and either provide or be compatible with deicing capability. In-flight issues, such as the effect of encountering ice crystals during flight in clouds, must be addressed and demonstrated with actual airplane flight tests. Maintenance issues require assessments of the impact of freezing rain on suction ports and systems and of the reliability and upkeep of suction pumps and auxiliary systems. Finally, the consequences of system failures in terms of performance, mission capability (range), safety, and cost implications must be thoroughly researched.

Economic Challenges

Modern airlines are doing everything they can to conserve fuel. Throughout the history of commercial aviation, airlines have insisted on the most fuel-efficient aircraft possible and have worked with airframe and engine manufacturers to reduce fuel consumption. Today's commercial transport fleet is much more fuel-efficient than the fleet that was in operation at the time of the first Organization of Petroleum Exporting Countries (OPEC) fuel crisis in the 1970s. Changes in cruise speed, sophisticated flight planning systems, lighter composite structures, and the introduction of improved aerodynamic aircraft designs and advanced engine technology are all examples of fuel-saving technology. With paper-thin profit margins and a reluctance to invest in new technology or aircraft in the turbulent commercial marketplace of the new millennium, airlines will carefully examine all cost-benefit aspects of LFC. The lack of technical confidence in LFC or HLFC, coupled with the need for long-term, in-service experience in worldwide, all-weather conditions with such systems, has resulted in a standoff between the potentially catastrophic effects of rising fuel costs on the airline industry and the economic unknowns of unproven technology. This situation is somewhat analogous to that of the 1950s when a highly competitive industry ventured into commercial jet transport development. Thanks to the injection of new technology, a new era of aeronautics was opened and the risk takers flourished.

Langley Activities

Laminar-Flow Technology

The scope of in-house and contracted research by Langley on laminar flow technology is a rich legacy of significant Center contributions. Many outstanding technical reports, professional journal articles, formal presentations, and other media publications have ensured the dissemination of this important research to users of the technology as well as to the public. The bibliography lists many of these documents. The reader is especially referred to two superb resources that provide extensive coverage of the activities and contributions of Langley Research Center in this area. Albert L. Braslow's publication, NASA Monographs in Aerospace History No. 13, *A History of Suction-Type Laminar-Flow Control with Emphasis on Flight Research*, summarizes NASA and international contributions to LFC technology and is written for the nontechnical reader as well as the specialist. Ronald D. Joslin's NASA TP 1998-208705, *Overview of Laminar-Flow Control*, is an extensive summary directed at the technical audience. Inclusion of all activities mentioned in these references is far beyond the intended scope of this publication; however, some key points have been summarized from these documents and included herein as an introduction for the reader. The material presented here focuses on recent events in LFC and HLFC technology and emphasizes engineering activities rather than research on fundamental flow physics and boundary-layer transition mechanisms.

The National Advisory Committee for Aeronautics (NACA), NASA's predecessor, had conducted exploratory boundary-layer research in wind tunnels as early as 1926, and in 1939 researchers assessed the impact of boundary-layer suction (via slots) on wind-tunnel models. In the spring of 1941, Langley researchers conducted the first LFC flight tests ever made, installing a suction system on an experimental low-drag test panel that had been previously used for fundamental studies of boundary-layer transition (nonsuction) on the wing of a Douglas B-18 airplane. The modified test panel was fitted with suction slots and pressure tubes for a flight investigation of the transition from laminar to turbulent flow in the boundary layer.

*B-18 wing glove tests conducted by NACA (top) and close-up view of the glove panel used
in natural transition studies (bottom).*

The legendary contributions of Eastman N. Jacobs, Ira H. Abbott, and Albert E. von Doenhoff and the Low Turbulence Pressure Tunnel's (LTPT) role in the development of NACA laminar-flow airfoils are well documented in other historical works and will not be repeated here. However, the key lesson learned from applications of this family of natural laminar-flow airfoils to aircraft such as the P-51 Mustang in World War II was that large areas of laminar flow could not be achieved in daily combat operations because of "real-world" wing surface conditions. Despite this operational shortcoming, the large favorable pressure gradients used by the NACA laminar-flow airfoils resulted in superior high-speed characteristics, and they were used in several airplane designs.

Research on active suction-type LFC in Germany and Switzerland during World War II stimulated suction LFC testing in the LTPT during the late 1940s by Albert L. Braslow, Dale L. Burrows, and Fioravante Visconte. This team carried out a range of wind-tunnel experiments that culminated in the demonstration of full-chord laminar flow with LFC to very high values of Reynolds number, but the special porous bronze surface used for the tunnel experiment was not feasible for applications to aircraft.

In the 1950s, LFC research was led by the British at the Royal Aircraft Establishment and in the United States by Dr. Werner Pfenninger, who had come to the Northrop Corporation from Switzerland. Pfenninger would serve as a technical inspiration to industry, DoD (especially the Air Force), and NASA. From the 1970s until his death in 2003, he served as a close consultant to Langley researchers in LFC and HLFC studies.

The Air Force became intensely interested in LFC in the early 1960s due to looming mission requirements for a long-range subsonic transport—which would ultimately emerge as the Lockheed C-5 transport. A flight demonstration of LFC by a research aircraft was of special interest to early development activities. In particular, the Air Force desired operational and maintenance data for

The Northrop-Air Force X-21A laminar-flow control research aircraft.

application to its new transport. The Air Force and Northrop initiated the X-21 Research Airplane Program, which began flight tests in April 1963, and Al Braslow served as a NASA technical consultant during the flight-test reviews. Pioneering data were obtained in the flight program, including the effects of surface irregularities, boundary-layer turbulence induced by three-dimensional spanwise flow effects in the boundary layer (referred to as spanwise contamination), and degrading environmental effects such as ice crystals in the atmosphere. Pfenninger developed a critical breakthrough understanding of and means for preventing the spanwise contamination phenomenon.

Laminar flow had been attained over 95 percent of the intended area by the X-21 program's end. Unfortunately, the manufacturing tolerances of the X-21 wing did not meet the requirements for robust LFC. Shallow spanwise grooves existed in the wing outer surface that required filling with body putty, and the desired data on maintenance of laminar-flow systems were therefore not obtained. Many regard the shallow grooves as a particularly poor fabrication experience but did not see the grooves as evidence of any extreme difficulty in fabricating LFC wings.

Although a technical committee of industry and government engineers recommended that a new wing be made for follow-up testing, the Air Force position was that the new C-5 could not wait for the new X-21 wing and another research program. Therefore, they dropped LFC as a possible new technology for the C-5 and terminated the X-21 Program. Arguably, an even more important reason for terminating the program was that the funding and priority required by the Vietnam War severely drained the Air Force of research and development funds. Another contributing factor was the advent of high-bypass-ratio engines, which also offered significant performance and economic benefits. In 1968, Pfenninger's group at Northrop was disbanded, and he accepted a new position in the Aerodynamic Research Unit at the Boeing Commercial Aircraft Division under the direction of Adelbert L. (Del) Nagel. Although Nagel had extensive aerodynamic experience, he had a dim view of the potential of LFC. However, Nagel was very impressed with the professional knowledge of Pfenninger and he became convinced that the aeronautical community should work the concept. Nagel left Boeing in 1970 and joined the Langley staff under John V. Becker. As will be discussed, both Pfenninger and Nagel later had profound impacts on Langley's LFC research.

Following X-21 activity termination, national interest in LFC virtually disappeared from the middle-1960s to the early 1970s. One of the major factors for the lack of interest was the relatively low price of jet fuel. In the 1970s, however, that situation changed dramatically.

The Aircraft Energy EfficiencyProgram

The oil embargo imposed by members of OPEC in 1973 caused immediate chaos and concern within the commercial air transportation industry. Between 1973 and 1975 the cost of aircraft fuel tripled. Rapidly rising fuel prices and concern over future sources of fuel were rampant, and congressional attention turned to potential technologies that might help mitigate rising concerns. At the request of NASA Headquarters, the NASA field centers surveyed technologies that might help alleviate the crisis.

Del Nagel had joined the Langley staff and had been briefed by Al Braslow on the state of the art of LFC technology. Nagel prepared a memo for Langley Director Edgar M. Cortright citing the accomplishments of the X-21 Program and the huge potential benefit of LFC that urgently needed national attention. Although Cortright's senior staff was generally negative, he requested information as to Boeing's position on the topic. Nagel then contacted Boeing, which sent its Chief Aerodynamicist and others to brief Cortright, resulting in a spark of enthusiasm for NASA research in the area. Al Braslow later prepared a white paper on potential technology advances that might contribute to fuel conservation. In his opinion, the single largest potential for reductions in fuel usage would come from LFC, despite the problems that had been experienced in past research efforts. Immediate NASA response to Braslow's paper was lukewarm at best (from both management and researchers).

As Chief of the Aeronautical Systems Division, Nagel was assigned to head the Langley contingent of the NASA Aircraft Energy Efficiency (ACEE) Task Force, or Kramer Committee as it became known. Braslow and William J. (Joe) Alford, Jr., were also members of the Langley group, which delivered well-received briefings on LFC at all major domestic aircraft companies, at other NASA Centers, and at Headquarters. With the additional inputs of a number of advisory committees and coadvocacy from the Air Force, Cortright gave his support and approval to the establishment of a LFC element under the ACEE Program in 1976 led by Robert C. Leonard. Leonard's project was part of the activities of the LaRC Projects Group headed by Howard T. Wright. Other elements of the ACEE Program included the composite structures element and the energy efficient transport element. Ralph J. Muraca was assigned as Deputy Manager of the LFC element of the ACEE Program Office.

Langley 8-Foot Transonic Pressure Tunnel Tests

In 1975, Werner Pfenninger conceived a wind-tunnel experiment to determine the capability of suction LFC to control the laminar boundary layer of a large-chord, swept supercritical wing.

Objectives included an assessment of suction LFC (via slots or perforations) to laminarize flow over a supercritical region, an evaluation of the ability of transition prediction theories to predict suction requirements, a determination of the relative effectiveness of slotted and perforated suction surfaces for LFC, and an evaluation of the effects of surface conditions on laminarization. His advocacy led to tests in the Langley 8-Foot Transonic Pressure Tunnel, which required facility modifications, such as honeycomb, screens, and a sonic choke, to ensure low levels of tunnel turbulence and acoustic disturbances and avoid interference with boundary-layer flow. The test team modified the test section walls with foam liners to conform with free-air predictions of the air-flow environment. Pfenninger designed the model wing section to integrate laminar flow and a supercritical airfoil. William D. Harvey, Cuyler W. Brooks, Jr., and Charles D. Harris of the Transonic Aerodynamics Division under Percy J. (Bud) Bobbitt led the Langley research team.

The first test with a slot-suction model began in 1981 and ended in 1985; perforated-suction testing began in 1985 and ended in 1987; and the HLFC test began in the winter of 1987 and ended in 1988. Joel R. (Ray) Dagenhart, Brooks, and Harris used boundary-layer stability codes to optimize the airfoil design by analyzing suction requirements at various test conditions.

The results of the tunnel tests were impressive. The researchers achieved full-chord laminar flow on the upper and lower surfaces of the slot-suction wing for Mach numbers from 0.4 to 0.85. Laminar flow was maintained for high values of Reynolds numbers, and drag reduction for both upper and lower surfaces was about 60 percent. More importantly, the experiment demonstrated the feasibility of combined suction laminarization and supercritical aerodynamics.

The Leading-Edge Flight Test Project

In 1977, researchers at NASA's Dryden Flight Research Center joined with Langley's Al Braslow and John B. Peterson, Jr., to conduct flight tests of a modified NASA JetStar (Lockheed C-140) to determine the effectiveness of new concepts for alleviation of potential insect residue problems, which some have viewed as one of the major barriers to LFC technology. The potential impact of insect residue has remained controversial because several flight experiences (such as the X-21 Program) have shown that flight at cruise conditions eroded away insect remains. The study included several special surface coatings that researchers hoped might shed insect remains, as well as a washer-type system that would wet the wing leading edge and possibly eliminate the adherence of impacted insects. None of the special coatings used in the study were effective in preventing the accumulation of insect debris, but the washer system demonstrated favorable results if the leading edge was wet before encountering insects. Preliminary analysis of the amount of wetting agent required indicated that such a system might be feasible for large commercial transports.

In 1980, Richard D. Wagner replaced Ralph Muraca as Head of the Langley LFC element. Langley began to focus its LFC efforts in the ACEE Program on an in-flight research project called the Leading Edge Flight Experiment (LEFT) that would address LFC leading-edge system integration questions and determine the practicality of LFC systems in operational environments via simulated

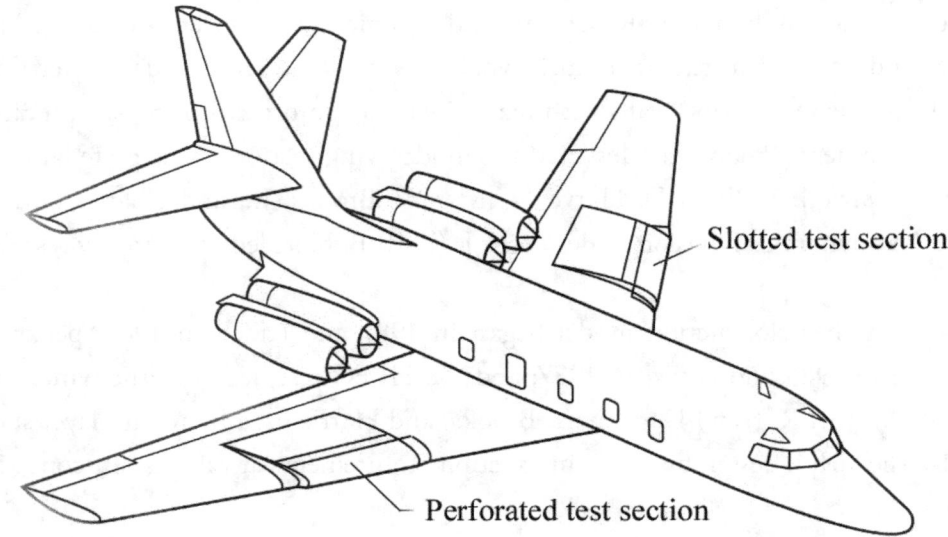

Sketch of the JetStar test panel arrangement showing the Douglas perforated panel (right wing) and the Lockheed slotted panel (left wing).

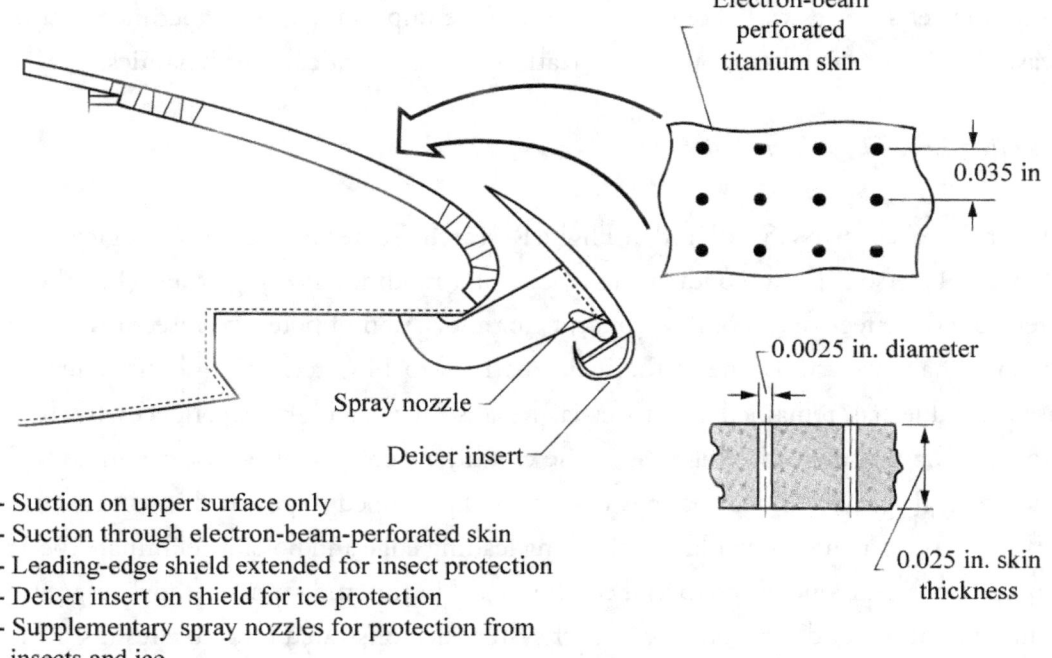

- Suction on upper surface only
- Suction through electron-beam-perforated skin
- Leading-edge shield extended for insect protection
- Deicer insert on shield for ice protection
- Supplementary spray nozzles for protection from insects and ice

The perforated Douglas Leading Edge Flight Experiment panel.

airline operations. The LFC office selected the NASA JetStar as its test vehicle, and the Douglas Aircraft Company and the Lockheed-Georgia Company were contracted to design and fabricate leading-edge test sections for the JetStar right and left wings, respectively. Douglas designed a perforated panel and Lockheed proposed the use of suction slots. Lockheed was responsible for designing modifications to the JetStar for each test panel. The actual aircraft modifications were performed at Dryden, with NASA personnel leading the efforts.

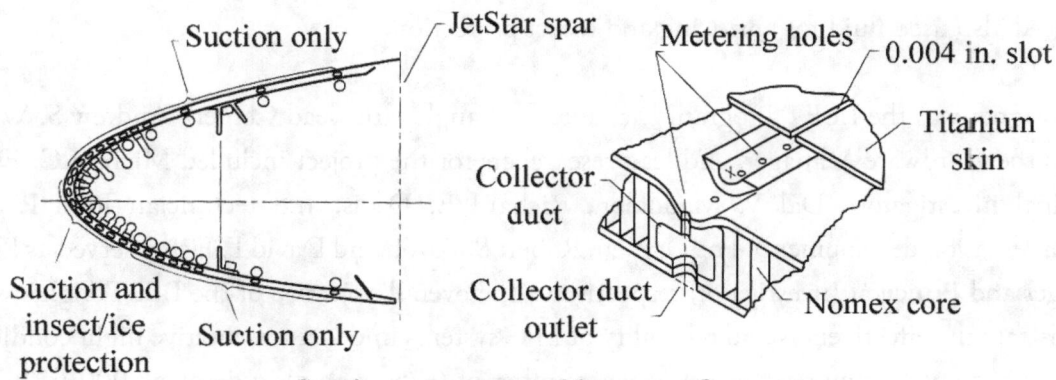

- Suction on upper and lower surface
- Suction through spanwise slots
- Liquid expelled through slots for protection from insects and icing

The slotted Lockheed Leading Edge Flight Experiment panel.

View of the Douglas test article mounted on the JetStar's right wing.

Along with the laminar-flow test panels, insect and ice protection were also incorporated in the JetStar's wings. The Douglas panel used suction through approximately 1 million electron-beam perforated holes (at 0.0025-in. diameter, smaller than a human hair) in the titanium skin to maintain laminar flow only on the upper surface of the article. In addition to a leading-edge insect shield, a propylene glycol methyl ether-water mixture spray was used for insect protection and anti-icing. The Lockheed panel used 0.004-in. wide slots (thinner than tablet paper) in a titanium surface and a design to provide laminar flow on both the upper and lower surfaces in cruise. Lockheed also used fluid for anti-icing and insect protection.

Dick Wagner was the LEFT Program Manager for Langley (the lead Center), Andrew S. Wright, Jr., was the Hardware Manager, and lead researchers for the project included Michael C. Fischer (principal investigator), Dal V. Maddalon, Richard E. Davis, and technician John P. Stack (instrumentation development). For Dryden, Robert S. Baron and David F. Fisher served as Project Manager and Principal Investigator, respectively. The overall objective of the LEFT flights was to demonstrate the effectiveness and reliability of LFC systems under representative flight conditions. The design point for the tests was a Mach number of 0.75 at an altitude of 38,000 ft, but off-design points were also flown at Mach numbers of 0.7 to 0.8. The flight evaluations by the Dryden Flight Research Center also included "simulated airline service" flights between 1983 and 1987. The primary objective of this part of the project was to demonstrate the ability of LFC systems to operate satisfactorily under conditions representative of commercial transport operations from ground queuing, taxi, takeoff, climb to cruise, cruise, descent, landing, and taxi to ramp. The aircraft would be parked outdoors, serviced routinely, flown in good and bad weather and in ice, rain, and insect-conducive conditions. Results of the simulated airline service showed that no operational problems were evident with the LFC systems, no special maintenance was required, and LFC performance was proven through the realization of laminar flow on the test article.

The LEFT Program resulted in extensive data regarding surface roughness criteria, the impact of environmental contaminants, and fabrication concepts. The Douglas panel was judged the superior concept. The electron-beam drilled panel presented no fabrication difficulties and easily met LFC criteria. The fabrication techniques for the slotted approach used by Lockheed resulted in a panel that was only marginally acceptable. However, even the most enthusiastic advocates of LFC recognized that more research would be required for the development of larger perforated panels (in terms of chord and span length) for commercial transports, and that the full-span integration of high-lift and insect protection systems would have to be addressed.

Hybrid Laminar-Flow Control

The Boeing Company did not participate in the initial phases of the ACEE Program or the LEFT Program because of commitments to near-term aircraft development programs. However, members of advanced design teams at Boeing followed the encouraging results of LEFT activities and several individuals who had considerable experience and interest in laminar-flow technology began to influence the interests of management. Del Nagel, who had previously led the Aeronautics Systems Division at Langley, had left NASA to return to Boeing, and he became an outspoken company advocate for LFC research and actively sought participation in NASA-funded studies.

Assimilating results obtained during LEFT, and cognizant of LFC technology in general, Boeing had become interested in a modified version of LFC wherein the trade-offs between system complexity for full-chord laminar flow yielded what might be a less complex concept involving partial-chord suction. Boeing's Louis B. (Bernie) Gratzer advocated a new, simpler approach to LFC. The concept, known as hybrid laminar-flow control (HLFC) used advanced airfoil design to combine some of the principles of NLF and LFC to obtain laminar flow to 50, 60, or even 70 percent of the wing chord. The HLFC concept applied active suction only to the front spar region (about the first 15 percent of the chord) to control the adverse cross flow and attachment-line effects due to wing sweep. The concept used favorable pressure gradients aft of the suction, over smooth surfaces, to promote the additional run of laminar flow to more aft-chord locations before transition occurred.

Following a joint NASA-Boeing meeting to discuss HLFC, Ray V. Hood, Jr., Head of the Langley Energy Efficient Transport (EET) element of ACEE, sponsored a Boeing exploratory study of the HLFC concept. David B. Middleton and Dennis W. Bartlett were the contract Technical Managers for Langley. This contract represented a milestone of increasing NASA-Boeing interest and collaboration in research to assess and resolve apparent issues that had to be addressed before applications of HLFC could be considered. One of the numerous issues to be resolved was the potential impact of engine noise on boundary-layer transition. Many in the aeronautical community had expressed concern that the interaction of engine noise pressure disturbances with the boundary layer might trigger flow instabilities, resulting in premature transition to turbulent conditions. The effect was thought by some to be so adverse that conventional wing-mounted engines would have to be relocated to the remote aft rear of the fuselage in order to obtain laminar flow over the wing. In 1985, Dick Wagner, Head of the Laminar-Flow Control Project Office (LFCPO), sponsored a contract under the ACEE Program for Boeing to conduct flight tests of a modified Boeing 757 aircraft evaluating the effects of engine noise on transition for a wing "glove" test article. The glove consisted of a 10-ft span smooth NLF section equipped with detailed instrumentation to measure

In-flight view of the natural laminar flow glove mounted on the Boeing 757's right wing during boundary-layer noise sensitivity evaluations.

the position of transition from laminar to turbulent conditions. The right engine, which was in close proximity to the glove, was throttled in flight at various altitudes and speeds to determine the effects of noise on transition. The results of the investigation were encouraging in that the near-field engine noise did not appear to have a significant effect on upper-surface boundary layer transition mechanisms.

Other critical research studies had to be undertaken to provide data analysis for studies of transition mechanisms to assist in the design of HLFC concepts. In one such study funded by the ACEE Program, the LFCPO at Langley, Boeing, and a team of flight researchers at Dryden conducted joint flight tests of a modified Navy F-14 Tomcat airplane at Dryden in 1986 and 1987 in a program known as the Variable-Sweep Transition Flight Experiment (VSTFE).

Dennis Bartlett was the Langley Principal Investigator for VSTFE, sharing program responsibilities with Robert (Bob) R. Meyer, Jr., Marta Bohn-Meyer, and Bianca M. Trujillo of Dryden. At Langley, Fayette (Fay) S. Collier, Jr., conducted extensive analysis and correlation of transition location predictions with boundary-layer stability theories. The objectives of the project were to

The F-14 Variable-Sweep Transition Flight Experiment airplane in flight with research gloves on both wing panels

obtain flight transition data for a range of wing-sweep angles at high subsonic speeds. Detailed data analyses provided fundamental information on the effects of sweep, Reynolds number, and wing section on critical boundary-layer transition mechanisms. The NASA research team chose an F-14 aircraft as the research vehicle for the VSTFE Program primarily because of its variable-sweep capability, Mach and Reynolds number capability, availability, and favorable wing pressure distribution.

Computational work at Langley by Edgar G. Waggoner and Richard L. Campbell, and tunnel testing in the Langley National Transonic Facility (NTF) by Pamela S. Phillips and James B. Hallissy guided the project. One of the variable sweep outer-wing panels of the F-14 was modified with a specially designed NLF glove to provide a test airfoil that produced a wide range of favorable pressure distributions for which transition locations were determined at various flight conditions and sweep angles. Under contract to the LFCPO, Boeing had also designed a candidate airfoil, but funding constraints prevented flight testing of that design. A conventional 64-series NACA airfoil glove article installed on the upper surface of the left wing was a "cleanup" for smoothing the basic F-14 wing, while the Langley-designed glove on the right wing panel provided specific pressure distributions at the design point of Mach 0.7. Data gathered in this flight program provided detailed transition information that was vital to validate predicted NLF ranges that could be experienced at relatively high sweep and Reynolds numbers. With successful test data results, optimism rapidly grew that the principles of NLF could be incorporated into the HLFC concept and that active HLFC suction could be limited to only the leading-edge region of swept wings.

In all phases of the ACEE LFC and fundamental HLFC research at Langley, the projects included significant contributions by outstanding instrumentation and flight data support. For example, John P. (Pete) Stack developed hot-film technology and instruments that provided unprecedented capability for boundary layer analysis.

In March 1987, Langley hosted a national symposium, *Research in Natural Laminar Flow and Laminar-Flow Control*, to disseminate laminar-flow technology gained during the ACEE-sponsored research projects and to disseminate additional information gathered in activities conducted within the more fundamental Research and Technology Base Program. The symposium, with over 170 NASA, industry, university, and DoD attendees, included technical sessions on advanced theory and tool development, wind-tunnel and flight research, transition measurement and detection techniques, low and high Reynolds number research, and subsonic and supersonic research. The event was one of the most important interchanges to occur within the domestic aeronautical community, which viewed it as an impressive success.

In addition to a very large number of experimentalists, flight researchers, and systems engineers, the emergence of a new talent community led by experts in CFD and transition physics began to accelerate the understanding of transition fundamentals. One of the leaders in the field was Mujeeb R. Malik, whose COSAL code became a primary tool in the design of HLFC experiments.

B757 Flight Evaluation

While NASA and Boeing geared up for increased research in HLFC, other organizations were also conducting assessment studies for future applications of LFC. For example, Lockheed-Georgia had conducted contracted studies for both NASA and the Air Force to assess potential benefits of LFC and HLFC for advanced military transports. As previously discussed, the Air Force had been the leading domestic proponent of LFC in the 1960s, as evidenced by its X-21 flight research activity. Langley found that the Air Force was interested in a cooperative program to assess the HLFC concept. Both parties subsequently signed a formal Memorandum of Understanding specifying the funding levels to be provided, which were equally shared by NASA and the Air Force. The Air Force motivation was to demonstrate HLFC for possible applications to future transports capable of taking off from the United States, flying to the Middle East fully loaded, and returning to the United States without refueling.

Both Boeing and Douglas bid on a competitive contract issued by NASA for a flight research demonstration of HLFC. After a combined team of NASA and Air Force personnel evaluated the proposals, they awarded a contract to Boeing for a flight demonstration of partial-span HLFC on its 757 prototype airplane. In 1987, NASA, the U.S. Air Force Wright Laboratory, and the Boeing Commercial Airplane Group agreed that the project would be flown out of Boeing's Seattle facilities. The 3-year, $30 million program started in November 1987 as a joint NASA-Air Force program managed by Langley and cost shared by the participants (34-percent Boeing and 66-percent government). The primary objectives of the program were to develop a database on the effectiveness of the HLFC concept applied to a medium-sized, subsonic commercial transport; to evaluate real-world performance and reliability of HLFC at flight Reynolds numbers (including off-design conditions); and to develop and validate integrated and practical high-lift, anti-ice, and HLFC systems.

Design work by Boeing on the modifications to the 757 airplane began as scheduled in November 1987. Boeing's team replaced a 22-ft span segment of the leading-edge box located outboard of the left wing engine nacelle pylon with a laser-drilled HLFC leading-edge test panel with about 22 million tiny holes. This new leading edge consisted of a perforated titanium outer skin, suction flutes under the skin, and collection ducts to allow suction to maintain a laminar boundary layer from the leading edge to the front spar. Manufacturing challenges faced in the program were among the most demanding issues, and successful solutions to numerous problems that were encountered

The prototype Boeing 757 transport used for the hybrid laminar flow control flight experiments.

remain proprietary to Boeing. The modified leading edge included a Krueger shield (somewhat similar to that conceived by Douglas in the LEFT project previously discussed) integrated for high lift and insect protection and hot-air deicing systems.

The team positioned flush-mounted pressure taps in the perforated leading edge and used tubing belts to measure the external pressure distributions over the wing box. They used hot-film sensors to determine the transition location on the wing box. For boundary-layer transition detection, they used infrared camera imaging. Also, wake-survey probes were located behind the wing trailing edge to provide data for local drag-reduction estimates. The flight test engineers monitored the state of the laminar boundary layer, the internal and external pressure distributions, and the suction system in real time onboard the aircraft during the flight test.

View of the instrumented left wing of the 757 test aircraft showing the test panel area, hot-film gauges, pressure tubing, and the leading-edge Krueger flap/insect shield.

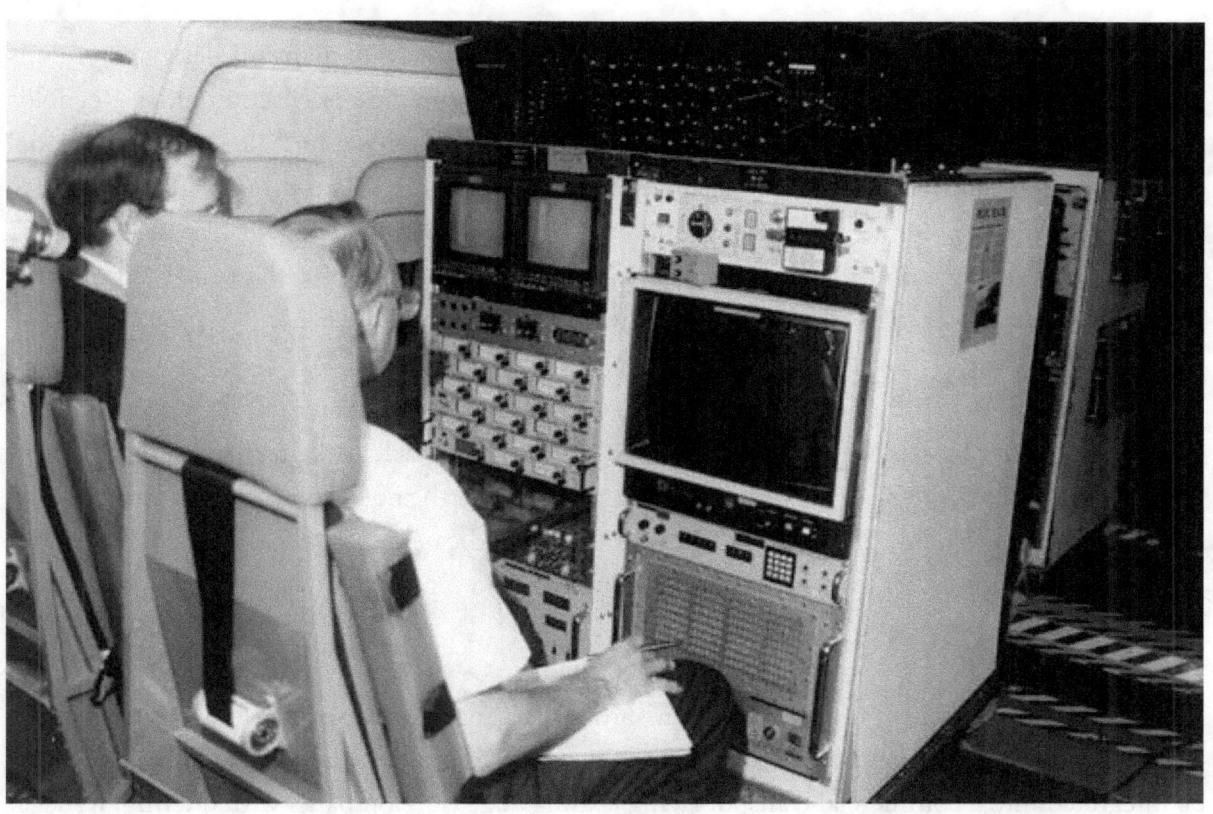

Langley researcher Dal Maddalon (right) and Boeing engineer David W. Lund (left) monitor wing transition data aboard the 757 during a typical research flight.

The Boeing 757 HLFC test bed aircraft in flight near Seattle.

The design point for the 757 flight tests was Mach 0.8, but flight tests of many off-design conditions were performed to investigate the extent of laminar flow as a function of Mach number, Reynolds number, and other parameters. Twenty research flights were conducted in 1990 and an additional ten flights were made in 1991. Additional analysis of flight data carried on through 1992. The pace of flight testing was a significant challenge successfully met by Boeing, who was also using the 757 prototype for concurrent in-flight avionics development testing for the Boeing team's entry in the Air Force Advanced Tactical Fighter Program, which later became the F-22.

The results of the 757 flight tests demonstrated that this first-generation HLFC concept was extremely effective in delaying boundary-layer transition as far back as the rear spar for the design flight condition of Mach 0.8. The data indicated that most of the hot films measured laminar flow beyond 65-percent chord. The wake-rake measurements indicated a local drag reduction of about 29 percent with the HLFC system operational, resulting in a projected 6-percent drag reduction for the aircraft.

In summary, the 757 HLFC demonstration showed that production manufacturing technology could meet the laminar-flow surface tolerance requirements and that a practical HLFC system could be integrated into a commercial transport wing leading edge (suction panel with ducting, Krueger high lift/insect shield, hot-air deicing system, suction surface purge, and suction compressor in pylon). Despite these impressive results, the flight results recorded a puzzling (although favorable) outcome—the suction rates required to achieve laminar flow to 65-percent chord were only about one third of those predicted during the initial design of the system. The disparity in design suction levels could be attributed to an overly conservative design approach, but the 757 results indicated that the suction system could be further simplified by more accurate criteria. Interestingly, Werner Pfenninger had forewarned that the suction levels would be less than the initial design based on his analysis of the application. Additional insight into the discrepancy came when Fay Collier conducted pioneering CFD analysis by developing a refined version (including curvature effects) of Mujeeb R. Malik's COSAL code, resulting in improved agreement between CFD predictions and flight results.

Following the 757 flight tests, Langley's Mission Analysis Branch under William J. Small and the Vehicle Integration Branch led by Samuel L. Dollyhigh conducted independent analyses of the benefits of HLFC for a representative advanced transport aircraft. The analyses were based on a 300-passenger transport for a 6,500-nmi mission with laminar flow over the upper wing, upper- and lower-tail surfaces, and over the vertical fin. The results indicated an impressive 13-percent savings in fuel burned for the HLFC-equipped airplane compared with a conventional transport.

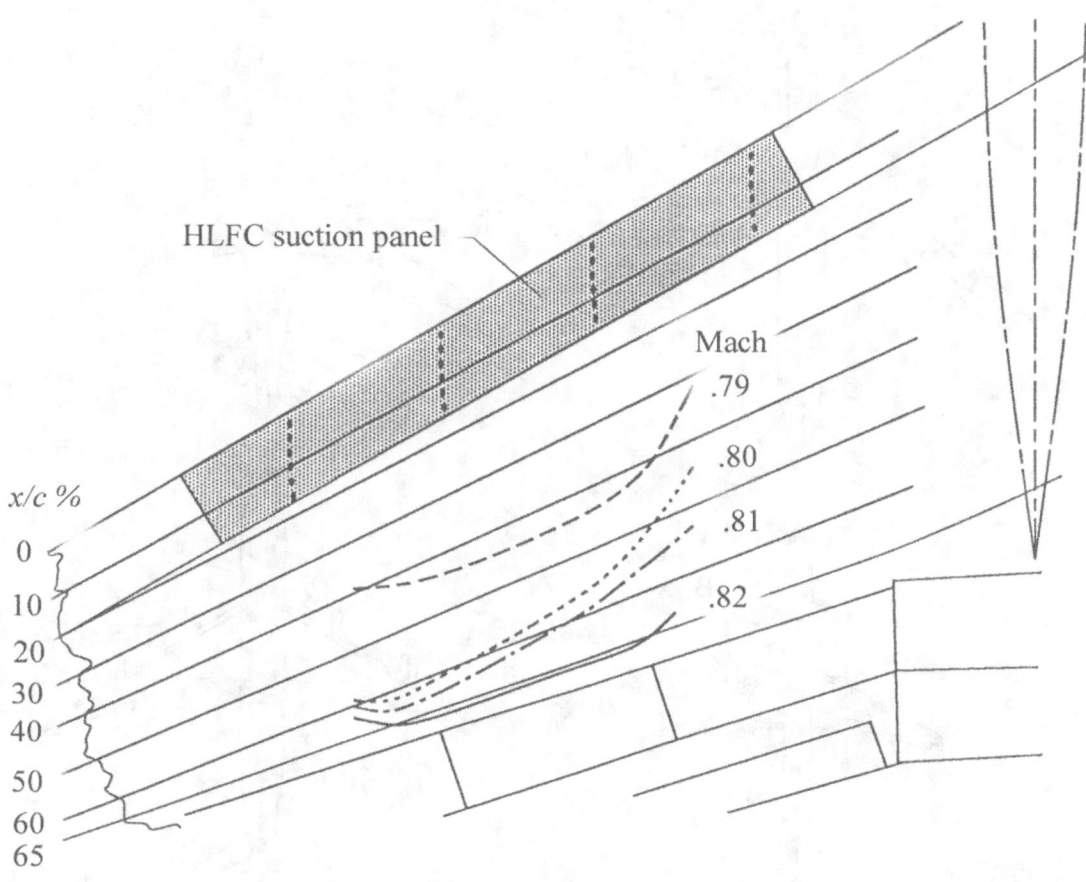

Transition contours showing the extent of laminar flow obtained on the wing upper surface of the 757 with hybrid laminar flow control.

Roy V. Harris, Jr., Director of Aeronautics, transmitted the results of these system studies to John D. Warner, Vice President of Engineering at Boeing in November 1990, to call his attention to Langley's projection of the potential benefits of HLFC and to ask Boeing to conduct its own studies for comparison. Harris was aware that Langley researchers were not exposed to highly proprietary industry methods and data for estimating the total economic impact of technology; therefore, a more refined analysis by industry would be required for a better projection of the value of HLFC. In 1991, Boeing formed a 40-person multidisciplinary assessment team that addressed all aspects of the cost-benefit trades for HLFC. The extensive Boeing study addressed aerodynamics, weight, propulsion, structures, manufacturing, safety, reliability and maintainability, marketing, finance, and other issues from an airplane manufacturer's perspective.

After extensive analysis, the Boeing team concluded that the total operating cost-benefit projections for HLFC at fuel prices of the time were positive, but not large enough to warrant the risks that

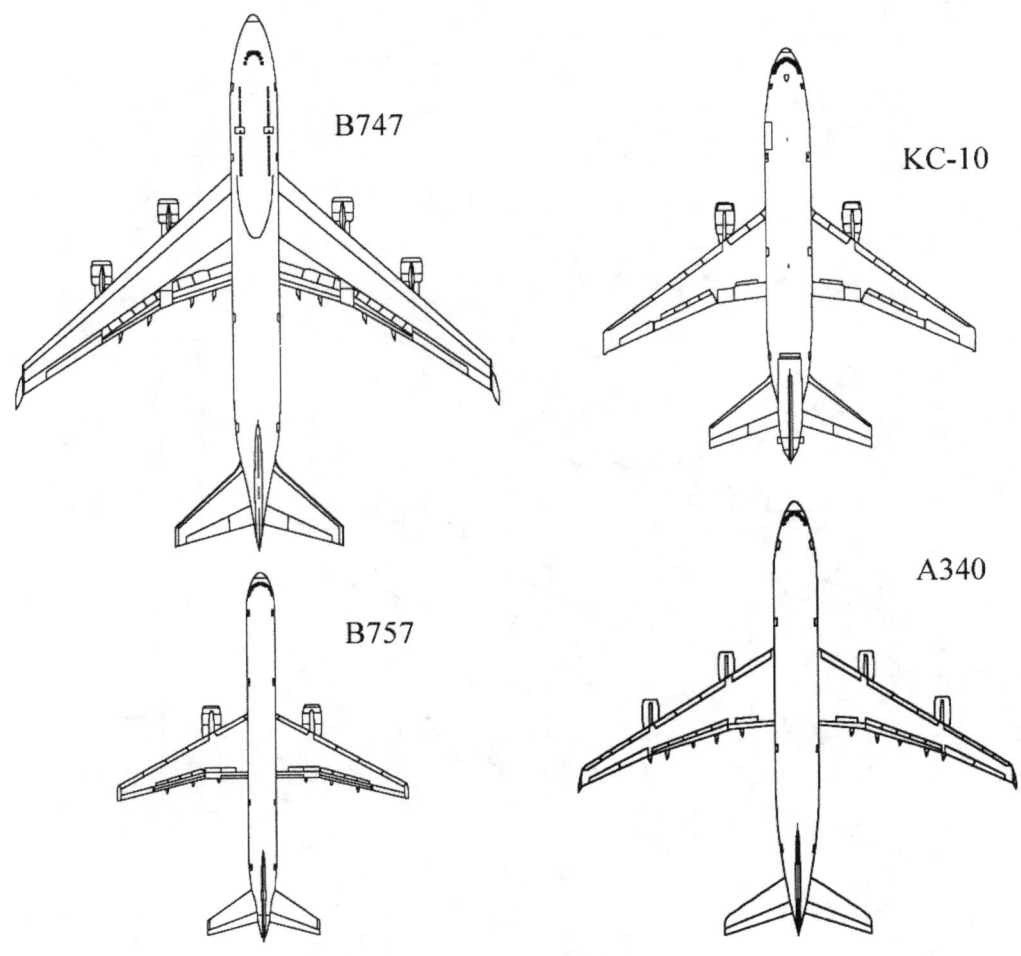

Size comparison of the 757 airplane and other civil and military transports.

still existed for applications of HLFC to commercial applications. Certifiable insect protection and deicing systems still required further research, and the use of a full-span Krueger for high lift was not an optimum solution (although the use of a Krueger for a laminar-flow airplane may well outweigh the Krueger's disadvantage). In addition, the fact that the 757 flight program did not address the inner-wing area (where transition is expected to be a more formidable challenge because of longer wing-chord lengths) represented an application barrier. Another major issue is the aerodynamic interference on the inner upper-wing surface caused by very large high-bypass-ratio engine-pylon configurations used on transport configurations. The interference problem was viewed as becoming even more critical for the larger ultra-high-bypass engines expected in the future.

Finally, the challenge of successfully applying HLFC to transports larger than the 757 remains a significant issue (recall that higher Reynolds numbers are associated with large aircraft and that high values greatly sensitize the transition process). At the termination of the 757 experiments, Boeing's immediate interest was in applying HLFC to large aircraft (of the 747 class), and considerable concern existed over the feasibility of applications to such designs. The LFCPO proposed a follow-up large-scale inboard HLFC wing experiment to NASA Headquarters in late 1990, but at that time the price of jet fuel had decreased significantly. Boeing ultimately refocused on transports smaller than the 747 and made the decision that its new transport, the B777, would not use LFC. Meanwhile, NASA's focus in aeronautics had shifted to a second-generation supersonic transport.

The loss of momentum and support for HLFC within the aircraft industry and NASA after the 757 flight program had a powerful, far-reaching negative impact on what had been a decade of rapidly accelerating development for potential future applications.

OV-1B Nacelle Experiments

With turbofan engine size rapidly growing with each generation of large transports, aerodynamic drag of the nacelle-pylon components of the propulsion system becomes significant. For a large commercial transport with wing-pylon mounted engines, an application of LFC that results in a 50-percent reduction in nacelle friction drag would be equivalent to a 2-percent reduction in total aircraft drag and cruise fuel burned. In the middle-1980s, interest began to intensify in the potential application of LFC concepts to nacelle configurations. Langley and the General Electric Company initiated a cooperative program to explore this application of NLF technology in a series of wind-tunnel, computational, and flight experiments. Initially, an NLF fairing was flown on a Cessna Citation nacelle to develop the experimental technique and establish feasibility. In the second phase of the program, the General Electric-NASA team designed and evaluated flow-through NLF nacelle concepts. The research team positioned advanced nacelle shapes below the right wing of a NASA OV-1 research aircraft. They also used a controlled noise-generating source in an under-wing pod outboard of the nacelle and a second noise source in the flow-through nacelle centerbody.

Earl C. Hastings, Jr., was Langley's Project Manager for the OV-1 Program and his team included Clifford J. Obara (aerodynamics and flow visualization), Simha S. Dodbele (aerodynamics), and James A. Schoenster (acoustics). The flight test program's scope included measurements of static pressures, fluctuating pressures caused by the noise sources, and flow visualization of the transition pattern on the nacelle using sublimating chemicals. The team collected data with the noise sources on and off and with various combinations of acoustic frequencies and sound pressure levels. During the acquisition, the right-hand aircraft engine was feathered to reduce propeller interference effects.

The NASA OV-1B in flight with an NLF nacelle under the right wing. Note feathered propeller and underwing pod containing noise source.

Prior to the flight tests, General Electric had teamed with Langley for tests of an isolated NLF nacelle as well as tests of a nacelle installed on a representative high-wing transport model in the Langley 16-Foot Transonic Tunnel. The results of the tunnel tests provided guidance for the external geometry of the nacelle to promote NLF. The pressure distribution on the outer walls of a conventional high-bypass-ratio engine nacelle is not conducive to NLF requirements. For a typical conventional nacelle, the lip of the inlet is relatively large, and the airflow rapidly accelerates to a velocity peak near the lip then decelerates over the remainder of the nacelle length. Boundary-layer transition occurs at the start of deceleration, so turbulent flow with high friction drag exists over most of the nacelle. An NLF nacelle is contoured to have a relatively small inlet radius and an accelerating flow over most of its length (about 70 percent), so transition is delayed and a relatively lower drag exists over most of the nacelle. The relatively small lip radius for the NLF design may aggravate inlet inflow distortion at off-design, crosswind, and engine-out conditions, thereby requiring careful analysis and design of the nacelle shape. General Electric designed and

Sublimating chemicals indicate laminar flow (white area) on forward region of NLF nacelle during research flight.

fabricated three fiberglass and aluminum structured nacelles for the flight test program, including instrumentation for external and internal static pressures, sound pressure levels, total pressures, and transition location.

Flight test results indicated that NLF was maintained as far aft as 50 percent of the nacelle length. No change was observed in the boundary-layer transition pattern when the noise sources were operated. These results served to broaden interest within the aeropropulsion community for potential applications to reduce nacelle drag. The unique requirements for satisfactory off-design characteristics were recognized as the major challenge for such applications of NLF. Significantly, this research project created widespread recognition as a catalyst to the propulsion community's interest in laminar-flow applications and the real-time technology transfer that occurred between Langley and its General Electric partner. This interest was soon to grow into applications of the HLFC concept to nacelles.

General Electric Nacelle

In 1990, representatives of General Electric Aircraft Engines (GEAE) visited the LFCPO to discuss mutual interest in the potential application of HLFC technology to reduce the aerodynamic drag of engine nacelles and pylons. In particular, GEAE sought Langley participation and consultation in its project with Rohr Industries, Inc., and Allied Signal Aerospace for a flight demonstration of the HLFC concept applied to the external surface of a current large turbofan engine nacelle. The project's objectives were to demonstrate laminar flow to 40 percent of the nacelle length using HLFC and to demonstrate a 1.5-percent reduction in specific fuel consumption.

Langley's expertise in HLFC technology, especially the design of effective HLFC pressure distributions for aerodynamic surfaces, instrumentation for transition measurements, and HLFC systems and flight test techniques, were of great value to the GEAE team. Fayette S. Collier, Jr. (who succeeded Dick Wagner as Head of the LFCPO in 1991), Cynthia C. Lee (Head of the Flight Research Branch), and Vernie H. Knight, Jr. (Head of the Aircraft Instrumentation Branch), led Langley's involvement in the cooperative project.

In addition to design guidance and active participation in nacelle engineering design reviews, Langley provided the project suction-system components previously used for the JetStar LEFT flight tests. This system was based on a remarkable "sonic valve" concept conceived by Langley's Emanuel (Manny) Boxer. The Langley team designed and installed an onboard, real-time data monitoring display and an instrumentation package for transition detection. The team transferred extensive NASA LEFT Program technology to GEAE and Rohr during the project.

GEAE modified a production GEAE CF6-50C2 engine nacelle installed on the starboard wing of an Airbus A300 testbed aircraft to incorporate an inboard and an outboard HLFC panel. The panels were fabricated of a perforated composite material with suction from the highlight aft to the outer barrel-fan cowl juncture. Suction was applied to the surface using circumferential flutes and was collected and ducted to a turbo compressor unit driven by engine bleed. For convenience, GEAE located the turbocompressor unit in the storage bay of the aircraft. The flow through each flute was individually metered. The GEAE-NASA team designed and fabricated the laminar-flow contour, which extended aft over the fan cowl door by using a nonperforated composite structure blended back into the original nacelle contour ahead of the thrust reverser. No provisions were made for ice-accumulation or insect-contamination avoidance systems. The team designed extensive instrumentation, including static-pressure taps that were mounted on the external surface and in the flutes, a boundary-layer rake used to measure the state of the boundary layer, and hot-film gauges used for boundary-layer transition detection. Flight engineers onboard the aircraft used surface-embedded microphones to measure noise, and they monitored the state of the boundary layer and suction system in real time. Perhaps the largest constraint to the research team was working within predetermined cowl lines.

The project's flight-test phase during July and August of 1992 included 16 flights totaling 50 flight hours. The HLFC concept was effective over the range of cruise altitude and Mach number investigated and resulted in laminar flow to as much as 43 percent of the nacelle length (the design objective), independent of altitude. Langley researchers Y. S. Wie, Collier, and Wagner believed the General Electric objective was conservative because the Langley team had designed an HLFC nacelle with a potential for 65-percent nacelle laminar flow.

Langley 8-Foot Transonic Pressure Tunnel Test

Although the Boeing 757 HLFC flight-test experiment demonstrated significant runs of laminar flow using only leading-edge suction, the fact that the amount of suction required was less than one-third of the predicted level caused uncertainty in the design tools, making the technology an unacceptable risk for the commercial market. To provide a better understanding of complex flow physics over a swept-wing geometry, generate a calibration database for the LFC design tools, and better understand the issues of suction-system design, a joint NASA-Boeing project to conduct HLFC wind-tunnel experiments on a research wing model in the Langley 8-Foot Transonic Pressure Tunnel was designed from 1993 to 1994 and conducted in 1995. Jerome T. Kegelman was the Langley Technical Leader for the project (which was funded by Elizabeth B. Plentovich of the Advanced Subsonic Technology Office). His research team included Craig L. Streett and Richard W. Wlezien, who led the computational and experimental efforts, respectively. Streett and Wlezien used a unique blend of CFD and numerical simulation of complex fluid mechanics to an unprecedented extent to guide experiment design and execution for everything from the design of the test article and measurement equipment to the analysis of the results.

Another key member of Kegelman's team was Vernon E. (Butch) Watkins, who led efforts to solve numerous potential show-stopping model and instrumentation issues resulting from tolerance requirements, vibration levels, and other test phenomena. Kennie H. Jones also contributed a testing highlight in the form of real-time data transfer from Langley's test site to Boeing's engineering offices in Seattle, Washington. All participants regarded the rapid dissemination of data for analysis and test planning as one of the most outstanding customer relations efforts by Langley. Over 200 researchers, designers, machinists, and technicians worked on this unique effort.

The test team installed a swept-wing model with a 7-ft span and 10-ft chord in the tunnel in January 1995, and tests were conducted throughout the year. They installed tunnel liners to simulate an infinite swept wing. The researchers obtained over 3,000 infrared images and 6,000 velocity profiles (hot-wire data) during the tests. Kegelman's team analyzed the influence of hole size and spacing and suction level and distribution on the transition location and correlated the results with predictions obtained from the design tools. Sufficient suction levels easily allowed

*Relative size of Langley researcher Vernie Knight illustrates the massive
size of the General Electric CF6-50 engine nacelle.*

View of the instrumented CF6-50 engine during hybrid laminar flow control nacelle flight tests.

laminar flow back to the pressure minimum. The team also made detailed surface roughness and suction level measurements.

The results of the tunnel tests, coupled with breakthrough CFD analysis by Craig Streett using a higher order modification of Mujeeb R. Malik's COSAL code, refined the suction analysis for HLFC and explained the major differences in designed and required suction levels experienced in the 757 flight tests. As a result of Langley's computational efforts, the accuracy of CFD predictions for subsonic LFC applications are considered reliable and ready for applications.

Supersonic Laminar-Flow Control

In the late 1980s, NASA initiated the HSR Program to develop technologies required for second-generation supersonic civil transports. The potential benefits of successfully applying LFC to aircraft that cruise at supersonic speeds are very attractive. Increased range, improved fuel economy, and reduced weight are among the benefits, as is the case for subsonic transports. However, the impact of weight reductions afforded by laminar flow have much larger implications for supersonic cruise aircraft, which typically have relatively low payload-to-weight ratios. Because of the greater amounts of fuel needed at a representative supersonic design cruise speed of Mach 2, even a small percentage reduction in drag could have tremendous economic benefits.

The Langley LFCPO initiated two brief exploratory flight evaluations of supersonic boundary-layer transition in 1985 and 1986 using an F-106 testbed at Langley and an F-15 at Dryden. Langley fabrication teams mounted surface cleanup gloves on both the right wing (leading-edge sweep of 60°) and the vertical tail (sweep of 55°) of the F-106. Dryden researchers installed a surface cleanup glove on the right wing of the F-15 to eliminate surface imperfections of the original wing. The glove was 4-ft wide and extended past 30-percent chord.

Advocacy for supersonic laminar-flow research within the NASA HSR Program was relatively limited, especially at NASA Headquarters. In view of the historical problems that had been encountered in subsonic LFC studies prior to the 757 flight demonstrations, and the technical difficulty of avoiding early transition on the highly-swept wings envisioned for supersonic transports, supersonic laminar-flow control (SLFC) received a low priority within the HSCT activities despite its potential benefits.

Both Boeing and McDonnell Douglas conducted benefit analyses for SLFC under contract to Dick Wagner's LFCPO in 1989. The potential mission and economic benefits of SLFC were identified in both studies, but the formidable barriers to attaining feasible SLFC on supersonic transports were obvious. Both contractors recommended supersonic flight research to evaluate the effectiveness of LFC and HLFC concepts to promote supersonic laminar flow.

Dryden used one of two F-16XL cranked-arrow delta wing prototype aircraft, on loan from the Air Force, to conduct exploratory investigations of laminar-flow technology during 1992. Dryden researchers tested a small, perforated titanium wing glove with a turbo compressor on the F-16XL (ship 1, single crew). This flight research program ended in 1996, followed by tests with NASA's two-seat F-16XL (ship 2) using a larger suction glove.

In the spring of 1992, Boeing began working with NASA Langley and Dryden, Rockwell, and McDonnell Douglas in an HSR-sponsored project for the design and testing of a supersonic HLFC glove on F-16XL ship 2. Michael C. Fischer of the LFCPO was the NASA technical principal investigator for the program, and Marta Bohn-Meyer was the Dryden flight project manager. Boeing and Rockwell were responsible for the fabrication and installation of the glove, while Boeing and Douglas assisted in the analysis of flight data.

The SLFC glove on the F-16XL covered about 75-percent of the upper-wing surface and 60-percent of the wing's leading edge. A turbocompressor in the aircraft's fuselage provided suction to draw air through more than 10 million tiny laser-drilled holes in the titanium glove via a manifold system employing 20 valves. The researchers instrumented the glove to determine the extent of laminar flow and measure other variables, such as the acoustic environment that may affect laminar flow at various flight conditions.

The flight test portion of the F-16XL SLFC program ended on November 26, 1996 after 45 test flights. The project demonstrated that laminar airflow could be achieved over a major portion of a wing at supersonic speeds using a suction system. The NASA-industry team logged about 90 hours of flight time with the unique aircraft during the 13-month flight research program, much of it at speeds of Mach 2.

Status and Outlook

The technical status of LFC technology has advanced greatly over the last 30 years. Advances in design methodology, manufacturing capabilities, and the assessment and documentation of laminar-flow phenomena across the speed range from subsonic to supersonic flight conditions have continued to mature the state of the art. A highlight of these technical activities has been the introduction of the HLFC concept and its development and evaluation through flight tests. Langley's contributions to the technology from in-house studies and experiments, as well as

Modified F-16XL Ship 2 with supersonic laminar-flow glove installed in left wing panel.

Size of holes in F-16XL test glove compared with a dime.

its contracted and cooperative efforts with other NASA Centers, industry, and DoD partners, represent some of the most significant advances to date.

Some elements of the technology still warrant further research and development, including certification issues (such as reserve fuel requirements), concepts for insect protection, and the refinement and validation of design methodology for boundary-layer stabilization. Many do not view the use of full-span Krueger-type wing leading-edge flaps for insect protection as an acceptable application because these types of flaps are normally limited to partial-span applications for structural and aerodynamic reasons, and Krueger flaps are not as effective as leading-edge slats for takeoff performance. However, the relatively low Reynolds numbers experienced on the outer wing may lead to another approach for a leading-edge device, or perhaps none at all. HLFC's lack of demonstrated success for inboard wing locations represents a serious technical challenge. Perhaps the largest concern in industry, however, is the lack of confidence in wind tunnels and CFD to work laminar-flow designs and details as opposed to building very expensive prototype aircraft. At this time, flight testing is recognized as the only reliable way to guarantee the performance of airplane HLFC applications.

In 1994, Langley Research Center reorganized and dissolved the LFCPO with its emphasis on flight demonstrations and applications-oriented research. Members of the organization were reassigned to other duties or more fundamental boundary-layer transition studies. Since the LFCPO demise, there has been no known flight development activity in the United States for LFC or HLFC.

As a final observation on the technical status of LFC, it should be noted that the leadership of the United States in LFC technology through the 1990s did not go without notice in the international community. Spurred on by successful American efforts, continuing European interest and research activities in the technology are evident today. The European community LFC efforts are being pursued by aggressive research activities for future applications to commercial transports. Notable activities have included startup of a German national program on LFC (1988); initiation of the European Laminar-Flow Investigation in 1988; various laminar-flow wing, tail, and nacelle tunnel and flight work (1986 to 1995) by France, Germany and others; and French wind-tunnel and flight research on HLFC applications using the A320 and A340 (1991 to 1995).

Unfortunately, domestic industry and airlines still regard LFC applications as an inherently high risk based on their own cost-benefit studies, which conclude that the risk outweighs the projected aerodynamic benefits of HLFC. The results of such studies are significantly impacted by the relative cost of jet fuel and its inherent dependence on world political and economic situations. The accompanying graph indicates the average domestic cost of jet fuel (not adjusted for inflation)

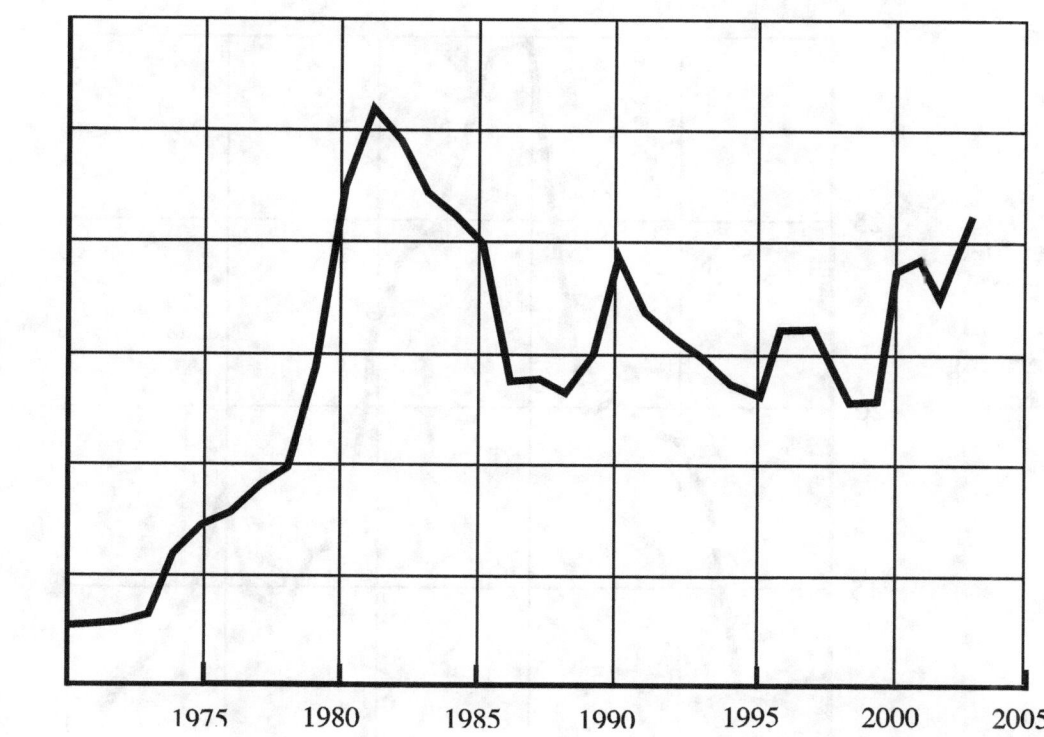

Historical trend of average cost of jet fuel.

for the last 35 years. As expected, the trends shown by the annual average cost variations follow major world events, recessions, and instabilities. Three of the major recessions of the past 30 years can, in large measure, be attributed to the steep increases in fuel prices that accompanied the 1973 OPEC oil embargo, the 1980 Iran Crisis, and the 1990 to 1991 Gulf War. During the 1990s, the average fuel price decreased and remained relatively stable compared with the traumatic increases experienced during the 1970s. At the same time, jet fuel costs as a percent of total operational costs were dramatically reduced when flight crew and personnel costs increased. For example, fuel costs represented about 30 percent of airline operating costs in 1980. By 1995, however, fuel costs had dropped to only about 10 to 12 percent of operating costs. Some have used these data to support a position that fuel costs have become a secondary player and do not warrant a drive for more sophisticated technologies to reduce fuel usage. Indeed, the decline in fuel prices of the early 1990s coupled with technical issues to deflate the momentum of the Langley HLFC research activities for subsonic transports.

The tenuous day-to-day nature of fuel costs for the airline industry are not reflected in average annual costs. Large fluctuations in fuel prices (as much as 50 cents/gal) have been experienced over time periods as short as one month. Recalling that each 1-cent rise in fuel prices increases the

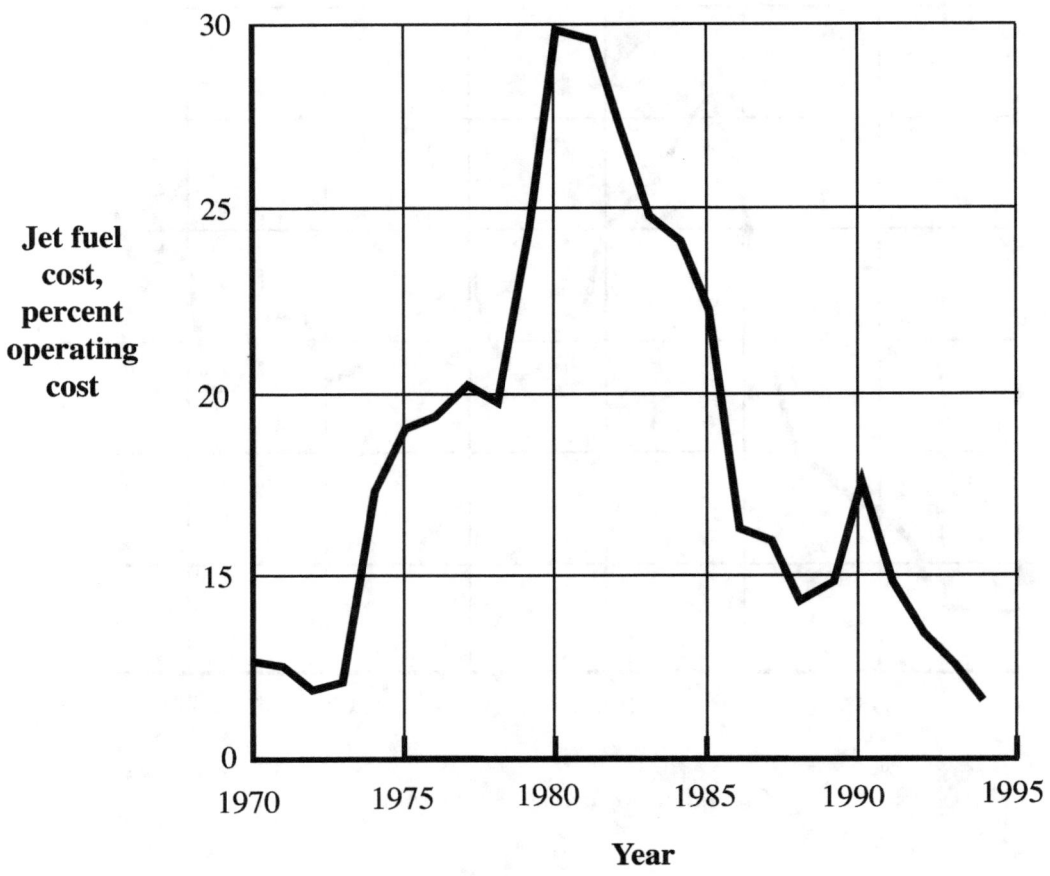

Jet fuel cost as a percent of total operating costs.

industry's annual expenses by $180 million, this level of price instability causes havoc throughout the airline industry.

In January 2004, jet fuel prices climbed to levels not seen since before the war in Iraq, prompting some airlines to raise fares and threatening to slow the industry's recovery. Prices for jet fuel rose almost 40 percent since the previous September to more than a dollar per gallon. The spot price for jet fuel reached $1.05 per gallon in New York and $1.08 in Los Angeles, according to the Department of Energy. Outlook for airline economic recovery continues to be pessimistic, with rising oil prices beginning to dominate concerns. About half the airline industry's 2003 losses were related to increasing fuel costs. Driving fuel prices up are the declining value of the U.S. dollar, OPEC pricing policies, and aggressive government purchases of oil for the country's strategic petroleum reserve. Each dollar increase in the price of oil translates into an additional 2 to 3 cents per gallon for jet fuel.

The domestic airline industry cannot afford significant energy price increases. In 1990 and 1991, almost half the airline industry filed for protection under chapter 11 of the bankruptcy code, long standing airlines went out of business, more than 100,000 employees lost their jobs, and the industry went into a financial tailspin from which it took years to recover. Today's record-setting fuel price increases (over $63 a barrel in mid-2005) take place against an even grimmer backdrop of economic chaos in the airline industry. In the period from 2001 to early 2005, the nation's airlines posted operating losses of over $22 billion. Analysts expect domestic airlines to collectively lose over $5 billion in 2005, in large part because the price of jet fuel had doubled what it was in 2003.

In summary, continued development of LFC technologies by NASA and its partners through the 1990s has removed many of the barriers to applications by airlines in the future. Work remains to be done, including advanced insect-protection concepts and validation of design methods. Systems-level studies continue to include HLFC research for future commercial long-range aircraft, raising the bar for increases in fuel efficiency and reductions in environmental impact. However, it is widely recognized that a continuation of relatively expensive flight investigations and demonstrations will be required in the future to reach adequate levels of maturity. At the present time, aerospace conglomerates in Europe are aggressively pursuing the final stages of development for this revolutionary technology while efforts in the United States have become stagnant.

Meanwhile, the domestic airline fleet continues to live in a paper-thin profit situation wherein wildly fluctuating fuel prices can result in catastrophic economic conditions. The result of this economic chaos is small profit margins, even in the best of times. Through the years airlines have earned a net profit between 1 and 2 percent, compared with an average of above 5 percent for American industry as a whole. It remains to be seen whether the continued fuel crisis (real and potential) can be tolerated, particularly if the persistent political instability of the Middle East results in new conflicts that significantly increase the price of oil.

Concept and Benefits

Performing operational short takeoff and landing (STOL) missions has been a target of innovation for aircraft designers since the beginning of heavier-than-air flight. Today, an increasing interest in using smaller airports with shorter runways for increased mobility and passenger capacity in our air transportation system is refocusing attention on aircraft concepts that provide STOL capability. Numerous aerodynamic and mechanical methods have been proposed and evaluated for such applications to fixed-wing aircraft, including the use of very low wing loading (ratio of aircraft weight to wing area), passive leading- and trailing-edge mechanical high-lift devices, boundary-layer control on leading- and trailing-edge devices, and redirection of the propeller or jet engine exhausts on trailing-edge flap systems. Using engine exhaust to augment wing lift is known as powered lift, and this approach differs from vertical takeoff and landing (VTOL) systems where power is used for direct lift. During the middle 1950s, intense research efforts on several powered-lift schemes began in Europe and the United States, resulting in dramatic increases in lift available for STOL applications. The accompanying sketch is a history of maximum lift development from the Wright brothers era.

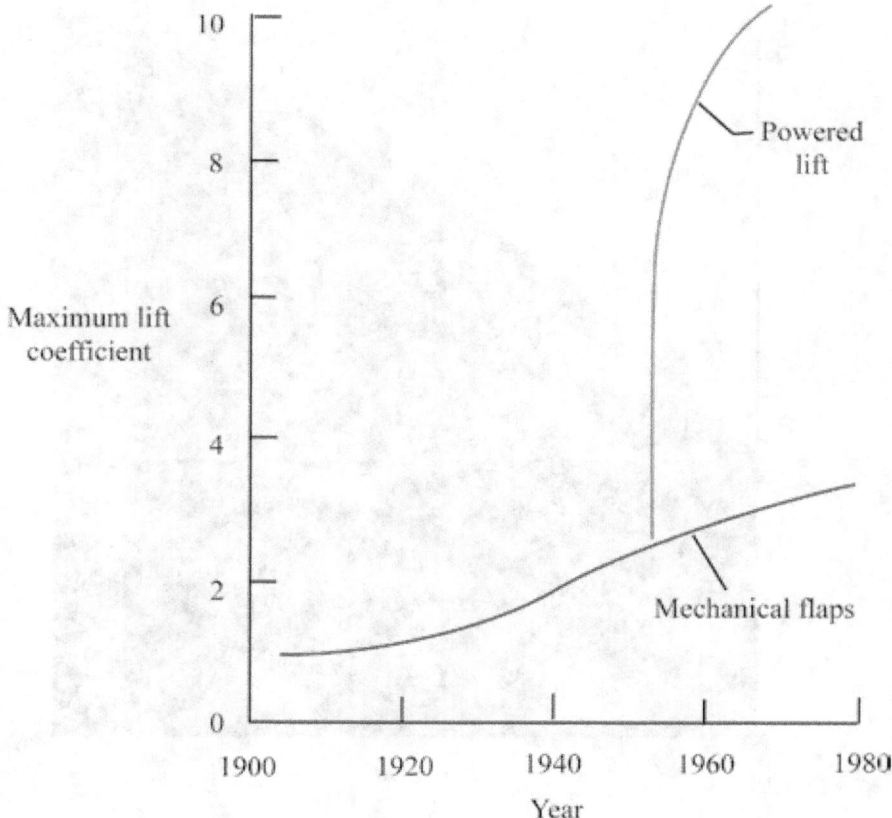

Historical development of maximum lift coefficient for aircraft.

Initially, the use of mechanical trailing-edge flaps and the refinement of these flaps led to a rapid rate of increase in maximum lift, but it later became apparent that airplanes would soon be using up most of the mechanical-flap high-lift technology made available by evolutionary advances. In the 1950s, researchers explored employing the efflux of engines to augment wing lift using the jet-flap concept to remove the limitations of conventional high-lift devices. As shown by the sketch, the magnitude of maximum lift obtained in this approach can be dramatically increased—by factors of three to four times as large as those exhibited by conventional configurations—permitting vast reductions in field length requirements and approach speeds. This revolutionary breakthrough to providing high lift led to remarkable research and development efforts.

One of the most promising powered-lift concepts is the upper surface blown (USB) flap. Before discussing this innovative concept, however, some background material on powered-lift technology is presented for the benefit of the reader.

The USB concept (or any other powered-lift concept) produces lift made up of the three components indicated in the sketch. The variation of total lift with engine thrust is presented along with the

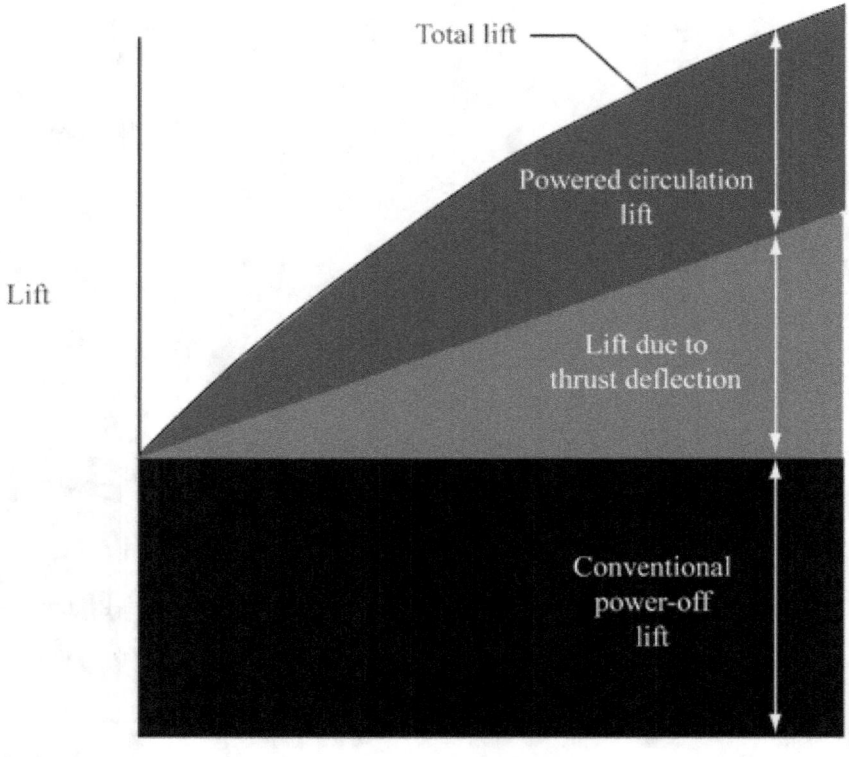

Components of lift for a powered-lift airplane.

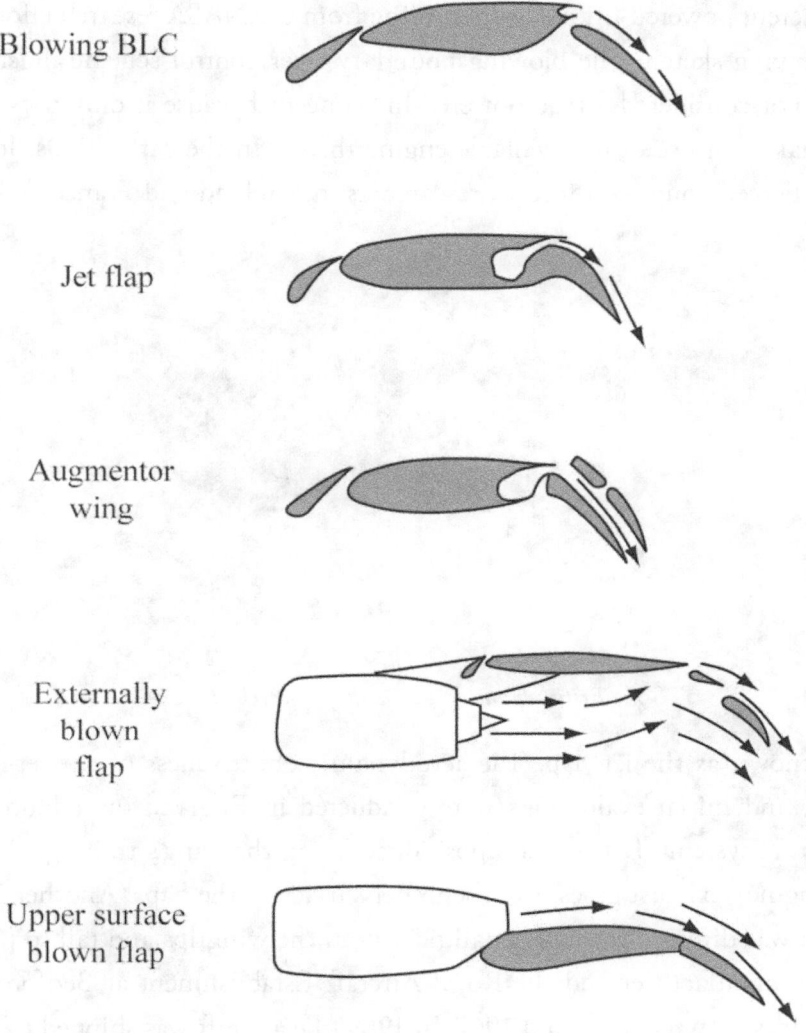

Powered-lift concepts for lift augmentation.

contributions of various powered-lift components for a fixed aircraft angle of attack and airspeed. The power-off conventional lift component is produced by the wing and its leading- and trailing-edge flap systems and usually does not vary significantly with engine thrust. The lift component due to deflection of the thrust (i.e., the vertical vector component of the thrust) and the lift component due to circulation lift (additional circulation lift induced on the wing and flap by the presence of the attached jet sheet) are strong functions of engine thrust. The specific variations of all three components with engine thrusts are strongly dependent on the geometry of the engine-flap-wing geometric arrangement.

Some of the different powered-lift concepts coming from the NASA research efforts are illustrated in the accompanying sketch. The blowing boundary-layer control scheme illustrated at the top of the figure is not considered a true powered-lift concept because it only uses engine bleed air and does not make full use of the available engine thrust. In the early 1950s, John S. Antinello of the U.S. Air Force, along with foreign researchers in England and France, led research on the

The British Hunting 126 jet-flap research airplane.

next concept, known as the jet flap. The aerodynamic effectiveness of the jet-flap concept was extremely high, and initial evaluations were conducted in England on a Hunting 126 jet-flap research aircraft. A system of 16 nozzles positioned along the wing's trailing edge directed more than half the engine's exhaust gases over the upper surface of the flaps. Another 10 percent of the engine's exhaust was directed through small nozzles in the wingtips and tail to provide control at low speeds. The manufacturer and the Royal Aircraft Establishment at Bedford made over 100 experimental flights between 1963 and 1967. In 1969, the aircraft was shipped to NASA-Ames for full-scale wind-tunnel tests. The jet flap was not developed further due to the impracticalities of the nozzle system's complex ducting and its adverse effect on engine power.

DeHavilland of Canada developed the augmentor wing concept in the late 1950s by incorporating a shroud assembly over the trailing-edge flap to create an ejector system that augmented the thrust of the nozzle by entraining additional air. In 1965, NASA and the Canadian government conducted a joint research program resulting in the C-8 augmentor wing airplane that NASA-Ames and Canada subsequently used in a joint flight research program. Both the augmentor wing and the jet flap proved very efficient aerodynamically in that they produced a very large increase in wing lift for a given amount of engine thrust. In fact, internally blown jet-flap systems remain the most efficient form of powered lift for fixed-wing aircraft today. Unfortunately, they suffer the disadvantage of requiring internal air ducting that leads to increases in the weight, cost, complexity, and maintenance of the wing structure as well as reduced internal volume for other systems.

Boeing-deHavilland augmentor wing jet short takeoff and landing research aircraft.

When it became obvious that internally blown flap airplanes would probably have unacceptable system penalties, NASA Langley started work in the 1950s on "externally blown" systems consisting of the externally blown flap (EBF) used with conventional pod-mounted engines, and the USB flap. Although such concepts may not achieve the levels of aerodynamic efficiency exhibited by the internally blown concepts, the system penalties for these concepts might be much more acceptable. Langley first carried out exploratory research on the EBF in 1956, and research on the USB flap started in 1957. Initial results appeared promising for both concepts. Langley staff conducted an extensive research program to develop technology for the EBF, but there were no indications of serious industry interest until Boeing incorporated it in its proposal for the CX-HLS (subsequently known as the C-5 transport) competition. Although Boeing did not win the C-5 contract, this show of interest by industry accelerated NASA's research on the EBF and led to a rapid buildup of the technology base required for applications of the concept. As discussed in the author's *Partners in Freedom*, the EBF concept, which John P. Campbell of Langley conceived and patented, was matured by years of intensive wind-tunnel research at Langley and subsequently applied by McDonnell Douglas to its YC-15 Advanced Medium STOL Transport (AMST) prototype in the 1970s, and then by Boeing to today's C-17 transport. This aircraft is the only U.S. production powered-lift fixed-wing airplane (excluding the vectored-thrust V/STOL AV-8).

A USB aircraft produces high lift by exhausting jet engine efflux above the wing in such a manner that it becomes attached to the wing and turns downward over the trailing-edge flap for lift augmentation in low-speed flight. In applications, a specially designed nozzle flattens the jet efflux into a thin sheet. Key to the USB concept's success is the Coanda effect, which Romanian aerodynamicist Henri Coanda (1885–1972) discovered in 1930. Coanda found that a stream of fluid (such as air or water) emerging from a nozzle tends to follow the path over a curved surface placed near the stream if the curvature of the surface or angle the surface makes with the stream is not too sharp.

The USB concept's aerodynamic performance can be equivalent to that of the EBF concept. Thrust recovery performance is usually higher for the USB because the EBF is inherently penalized by thrust loss caused by the direct impingement of engine efflux on the trailing-edge flap. This thrust loss will limit the flap setting selection flexibility. The USB also offers a critical advantage because objectionable noise levels generated during powered-lift operations are much lower for observers below the airplane. Reduced noise levels offer substantial advantages for civil aircraft applications in compliance with regulatory noise restrictions near airports, as well as military applications where noise signatures play a key role in the detectability and "stealthiness" of aircraft for certain missions.

USB aircraft: The Boeing YC-14 (upper) and the NASA Quiet Short-Haul Research Aircraft (lower).

Use of an innovative USB-type configuration for civil transports may be compatible with shorter field lengths (on the order of 3,000 ft) while retaining "good neighbor" noise characteristics. An additional operational advantage of the USB concept is the reduction of foreign object ingestion into the high-mounted engines.

NASA and industry studied the USB concept extensively from the middle 1950s to the 1980s using an extensive variety of wind-tunnel investigations, static engine tests, and piloted simulator studies that culminated with the Boeing YC-14 prototype military transport and the NASA Quiet Short-Haul Research Aircraft (QSRA) technology demonstrator. NASA conducted this research principally at Langley, which later led to other efforts in collaboration with DoD, industry, and other NASA Centers. The success of research leading to QSRA demonstrations of the USB concept's viability and advantages has matured this technology to the point where it will be a primary candidate for any future commercial or military STOL aircraft. The following discussion will highlight these past studies and examine some of the critical issues faced and solved during research at NASA Langley, resulting in a high degree of technology readiness.

Challenges and Barriers

Disciplinary Challenges

The use of propulsive lift provides STOL performance that significantly reduces field-length requirements, lowers approach speeds for improved safety, and improves the versatility of flight path management for increased runway throughput by using steep or curved approaches. One of the most important metrics for STOL is the magnitude of maximum lift produced by the aircraft configuration. An STOL airplane does not actually conduct its landing approach at maximum lift conditions, it does so at considerably reduced lift because of the various angle-of-attack and engine-out speed margins required for operational safety; however, the margins are based on specified percentages of maximum lift conditions. It is therefore imperative that the USB concept provides not only high values of maximum lift, but also high levels of propulsive efficiency. The thrust-weight ratio required for a representative STOL transport with a wing loading of about 80 lb/ft^2 for a landing field length of around 2,000 ft is about 0.5, which is approximately twice the installed thrust-weight ratio for conventional jet transports. The STOL configuration must efficiently use all of this relatively high thrust loading.

The powered-lift aerodynamic characteristics of the USB concept are critically dependent on engine nozzle-flow parameters and the geometry of the flap directly behind the engine (USB flap). In particular, the thickness of the jet efflux and the radius of curvature of the USB flap surface determine, to a large extent, the ability of the flap to turn the flow and maintain attached-flow conditions. Special features may be required for flattening and spreading of the engine exhaust to ensure the attached flow for generation of lift. These features may include a high-aspect-ratio (width much larger than height) flared nozzle exit, downward deflection of the nozzle-exit flow ("kick down") and the use of auxiliary flow control devices such as active BLC or vortex generators at the front of the USB flap.

Significant trade-offs between powered-lift efficiency and cruise efficiency will be required for USB configurations. With engines located in an unconventional upper-surface configuration, unique aerodynamic propulsion-integration challenges exist to minimize cruise drag at high subsonic speeds. Flow separation and drag during cruise must be minimized by special geometry tailoring of the USB engine nozzles and nacelles, especially interference effects with the nacelles in close proximity to the body due to engine-out concerns.

Aerodynamic loads imposed on the wing and flap structures by impinging jet flow must be predicted and accommodated by the aircraft structure. These loads take the form of relatively steady, direct normal- and axial-force loads imposed on the flap system, as well as large pressure fluctuations that can induce high vibration levels and acoustic loadings that may result in sonic fatigue. The principal sources of turbulent pressure fluctuations can be generated within the engine by combustion, the mixing region of the core or bypass exhaust jet, and in the flow impingement region by boundary layers and separated flow. If dynamic loads induced by the pressure fluctuations are significantly high, sonic fatigue failures of secondary structures can occur, and the designer must give special attention to this issue during early design.

The USB designer must meet the requirements for structural heating issues caused by the engines mounted in close proximity to the wing's upper surface. In the early 1950s when turbojet engines represented the only choice for powered-lift configurations, heating issues were especially critical. The advent of high-bypass-ratio engines having cooler bypass flow now permits heating barriers to be resolved with appropriate materials for STOL applications.

In the area of stability and control, designers must address a number of issues inherent to the USB configuration. One obvious primary concern is the critical nature of controllability when one engine becomes inoperative during flight at high-power conditions, resulting in a major loss of lift on one side of the airplane and, as a result, a large rolling moment. A basic approach to minimize roll trim problems is by locating the engines as far inboard on the wing as possible, but special control devices may have to be used to aerodynamically balance the aircraft.

The design of the horizontal tail's size and location is especially critical for USB aircraft. Longitudinal trim is a major challenge because of large nose-down pitching moments produced by powered-lift flaps at high thrust settings. Very large horizontal tails are required for trim at high-lift conditions, especially if the wing is unswept. In addition, the horizontal tail surface must be located in a high, relatively forward position to avoid the destabilizing effects of strong vortices generated at the wingtip or outboard end of the flap. Locating the tail forward removes it from the actions of the vortices and into a region of less destabilizing downwash. The very low approach speeds and high thrust engines will also require a larger than normal vertical tail and possibly a more complex rudder system.

Operational Challenges

The unique aerodynamic flow phenomena associated with powered-lift airplane configurations require an assessment of potential problem areas during actual operations. For example, the effect of flight near the ground during the landing approach is an issue. This effect is typically favorable for conventional aircraft. That is, the presence of the ground provides a cushioning, or positive, ground effect that does not result in a serious performance or handling issue at a very critical time during the mission. However, the large turning of free-stream flow due to powered-lift operations may result in unconventional ground effects, resulting in loss of favorable ground effect or significant changes in trim or stability. Thus, validated experimental and analytical predictive tools are required early in the design stage.

Arguably, the largest single obstacle to the implementation of STOL powered-lift technology for civil aircraft is the increasingly objectionable level of aircraft-generated noise for airports close to populated areas. Quiet engines are a key requirement for successful commercial aircraft applications; however, the powered-lift concepts produce additional noise that compounds the challenge. The EBF, for example, produces a very large increase in aircraft noise—beyond that for the engines—when the flaps are extended down into the jet exhausts. The benefits gained by having the exhaust flow above the wing, as in the USB concept, include substantially reduced noise levels for ground-based observers during flyovers. Sufficient STOL performance may allow all low altitude operation to remain within the airport perimeter, thereby limiting public exposure to noise.

Other operational issues for USB aircraft are representative of those faced by other powered-lift configurations, especially providing crisp, coordinated control response at low speeds and satisfactory controllability and performance if an engine becomes inoperative.

Langley Activities

Langley Research Center was recognized by the aeronautical community as an international leader in V/STOL research—especially in the conception and development of innovative powered-lift STOL configurations—from the 1950s to 1976, at which time NASA Headquarters declared NASA Ames Research Center to be lead NASA Center role for rotorcraft and V/STOL. For almost 30 years, friendly competition had existed between the two Centers in this technology area, resulting in accelerated development and maturity of the state of the art. It is generally recognized that Langley led the way in aerodynamic development of a large number of advanced STOL vehicle concepts, while Ames focused on flight and operational issues of STOL aircraft.

Langley's expertise in powered-lift STOL technology was internationally recognized, and its leaders maintained a closely coordinated in-house and contracted research program. Key individuals of this effort in the 1970s included John P. Campbell, Richard E. Kuhn, Joseph L. Johnson, Alexander D. "Dudley" Hammond, and Richard J. Margason. With unique world-class facilities that included the Langley 30- by 60-Foot (Full-Scale) Tunnel, the Langley 14- by 22-Foot (V/ STOL) Tunnel, large engine test stands, piloted simulators, and close working relationships with industry and DoD, the staff brought an immense capability and fresh innovation to the tasks at hand for powered-lift STOL research.

Early Exploratory Research in the 1950s

John M. Riebe and Edwin E. Davenport conducted initial Langley exploratory studies of blowing from nacelles on the upper surface of a wing for propulsive lift in the Langley 300 mph 7- by 10- Foot Tunnel (also known as the Low Speed 7- by 10-Foot Tunnel and subsequently replaced by the Langley V/STOL Tunnel in 1970). Thomas R. Turner, Davenport, and Riebe quickly followed this work with further systematic tests. The motivating factor for these experiments was the potential noise reduction provided by this approach for powered-lift configurations, and the studies included several innovative concepts for trimming the large diving moments produced by the blown wing flaps, such as a canard, a fuselage nose jet, and wingtip-mounted tail surfaces. Although limited in scope, results obtained from the tests, conducted in 1957, appeared promising. The aerodynamic performance of the USB concept was comparable with that of the EBF, and preliminary far-field noise studies of several different powered-lift model configurations by Langley's Domenic J. Maglieri and Harvey H. Hubbard showed the USB to be a potentially quieter concept because of the wing's shielding effect . However, because the USB arrangement involved a radical change in engine location, away from the generally accepted underslung pods, and because there was at that time no special concern with the noise problem, research on the USB flap waned. In addition, the

Model configuration used in early Langley USB research in 1957 mounted to the ceiling of the Langley 300-mph, 7- by 10-Foot Tunnel.

turbojet engines of that day had very hot exhaust temperatures and relatively heavy weight, making them impractical for providing the high thrust-to-weight ratios required for STOL applications. Interest in the USB concept at Langley was dropped after these initial studies. Research resumed in the early 1970s when it was becoming apparent that the EBF concept might have difficulty meeting increasingly stringent noise requirements for civil applications, and lightweight, high-bypass-ratio turbofan engines became available.

The 1970s

In the early 1970s, a growing national interest in short-haul transportation systems for the United States began to emerge. As part of these activities, short-field aircraft suitable for regional operations received considerable study, including assessments of the state of readiness of STOL technology and the economic and environmental impacts of such aircraft. During the decade, Langley focused extensive human and facility resources toward providing the disciplinary technology advances required to mature the short-haul STOL vision.

In September 1970, Center Director Edgar M. Cortright appointed Oran W. Nicks to the position of Deputy Director of Langley. Nicks was a hard-driving, impatient leader who demanded innovation and action in Langley's aeronautics and space programs. Although his previous NASA management positions had been in leadership of NASA lunar and planetary programs rather than aeronautics, Nicks engaged in his own research work in aerodynamics, composite materials, heat-resistant materials for re-entry, and other areas. When exposed to Langley's research efforts on the EBF concept and growing concern over noise issues, Nicks was stimulated to explore other approaches to powered lift. He became intrigued with the potential capability of upper surface blowing and strongly advocated that it would be an optimum approach for the design of STOL aircraft.

At that time, Langley operated two wind-tunnel facilities engaged in STOL research: the Langley Full-Scale Tunnel, under Marion O. McKinney, Jr., and the Langley V/STOL Tunnel (now the Langley 14- by 22-Foot Tunnel), under Alexander D. (Dudley) Hammond. Both organizations were heavily involved in the EBF concept's development. Nicks met with John Campbell and personally tasked him and his organization to quickly investigate the potential aerodynamic benefits and issues of the USB concept applied to turbofan-powered transport configurations.

The job of conducting this exploratory USB research—under the intense scrutiny of Nicks—was assigned by Campbell to Joseph L. Johnson, Jr., who formed a research team led by Arthur E. Phelps, III. Johnson, noted for his ability to quickly identify the benefits of innovative concepts by "getting 80 percent of the answer," chose to use a rapid-response approach to the task by conducting tests of a model fabricated from existing EBF model components and performing the tests in a subsonic tunnel with a 12-ft test section (formerly the Langley Free-Flight Tunnel). Johnson used this tunnel as a quick-reaction, low-cost laboratory to explore innovative concepts in a low profile, timely manner. Phelps and his team used components from a powered semispan EBF jet transport swept-wing model having a full-span plain trailing-edge flap to create a USB model with a full-span leading-edge Krueger flap and a two-engine (tip-driven turbofan simulators) podded nacelle high-bypass-ratio turbofan engine configuration. Using an auxiliary flat deflector plate attached to the nacelle exit, the team was able to thin the jet and turn the exhaust flow. The use of BLC was also explored for additional flow control. The results of this milestone investigation, conducted in 1971, showed that the high lift necessary for STOL operations could be achieved and that the performance was comparable with that of other powered-lift STOL concepts. Nicks was elated over the concept's performance and gave his influential support to further research and development activities for USB.

Following this highly successful tunnel investigation, Langley geared up for accelerated USB research efforts in tunnels and with outdoor engine test stands. At the V/STOL Tunnel, William C. Sleeman, Jr., and William C. Johnson, Jr., conducted parametric studies to define the optimum geometric and engine variables for USB performance. With superior flow quality over the 12-

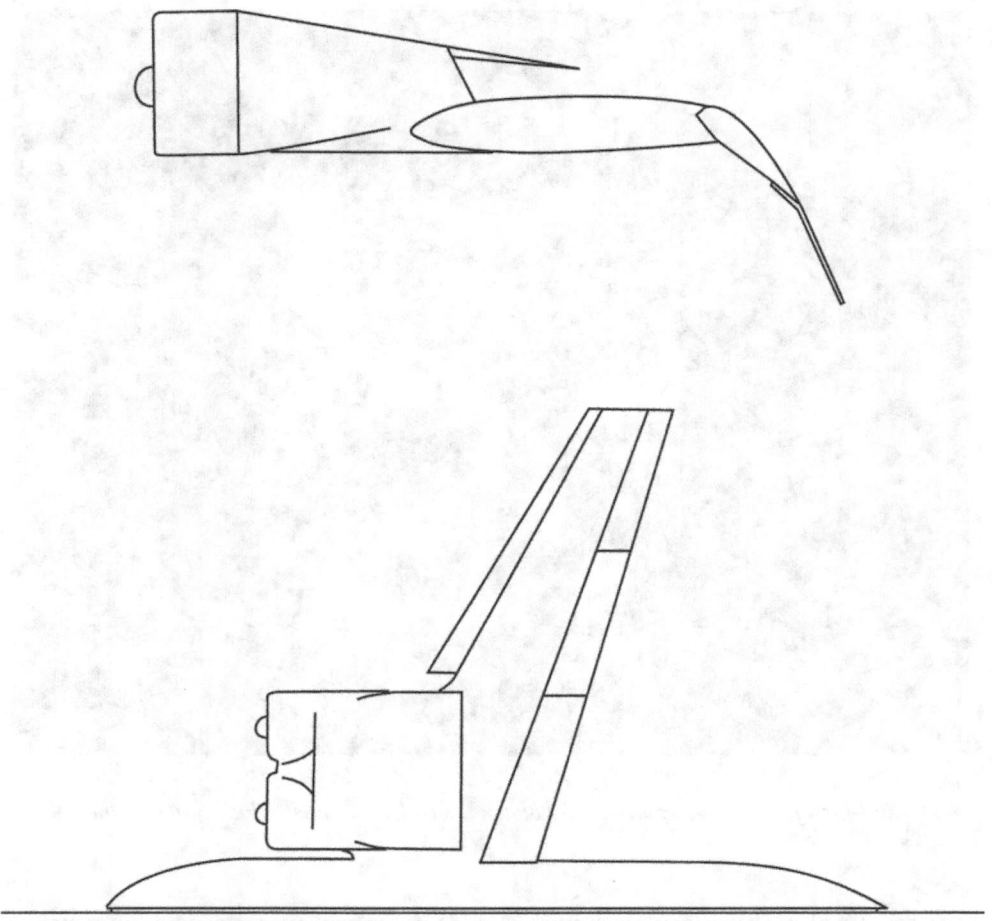

Sketch of upper-surface blown flap research model tested in 1971.

ft tunnel used by Phelps in the 1971 tests, researchers were also able to obtain fundamental information on cruise drag, such as the effects of nozzle geometry for USB configurations. Also, James L. Thomas, James L. Hassell, Jr., and Luat T. Nguyen conducted tunnel tests to determine the character of aerodynamic ground effects for USB configurations and conducted analytical studies to define the impact on aircraft flight path during approach to landings.

Meanwhile, Johnson's team changed its testing venue to the larger Full-Scale Tunnel, where they tested full-span USB models to provide additional performance information and pioneering data on the stability, control, and engine-out trim problems of USB configurations. In 1972, Phelps and Charles C. Smith, Jr., conducted tests of a four-engine USB transport model, verifying the configuration's aerodynamic STOL performance and investigating nozzle shape tailoring with contouring to direct the exhaust downward to the top of the wing for better spreading and flattening of exhaust flow over the wing and flaps. Phelps and Smith determined that a T-tail empennage configuration provided longitudinal trim and longitudinal stability; and good lateral-directional stability was obtained for the configuration at all angles of attack below stall.

Full-span four-engine upper-surface blown model in the Langley Full-Scale Tunnel.

Although the results of preliminary work with the USB concept were very encouraging, the results had been obtained at very low scale, and concern existed that higher Reynolds number testing might lead to conclusions that contradicted the small-scale results. Smith therefore conducted tests on a semispan, high-wing USB model assembled largely from components of a full-scale Cessna 210 aircraft. The investigation results indicated that the effects of Reynolds number were generally small for moderate to high-powered lift conditions. Once again, the aerodynamic efficiency of yet another USB configuration was extremely high, and noise measurements indicated large beneficial shielding effects of the wing.

In 1973, Joe Johnson formulated and led a multidisciplinary USB research program using a modified Rockwell Aero Commander configuration. The intent was to collect detailed information on subscale and large-scale models to provide data for aerodynamic performance, stability and control, effects of tail configuration, and the effects of actual turbofan engines on wing upper-surface temperatures and local acoustic loadings. In November 1973, Art Phelps conducted exploratory aerodynamic tests of an 8.8-ft span model of the configuration, providing design data for aerodynamic performance and engine-out trim strategies (leading-edge blowing and aileron deflection on the failed-engine wing panel, and a spoiler on the active engine side).

Test of a modified Cessna 210 in the Full-Scale Tunnel.

With subscale aerodynamic results of the Aero Commander USB configuration in hand, the team turned its attention to obtaining data in the areas of structures, dynamic loads, and acoustics. An outdoor engine test-stand investigation and a tunnel test of a modified full-scale Aero Commander airframe were used to attack these issues. James P. Shivers and Charles C. Smith, Jr., conducted the static test-stand investigation using an outdoor test site near the Full-Scale Tunnel. The apparatus used in the study supported a JT15D-1 turbofan engine equipped with a candidate rectangular nozzle with an aspect ratio of 6.0 (width divided by height). The engine and nozzle were oriented to direct the jet efflux onto the upper surface of a boilerplate wing-flap system, which was mounted upside down to avoid ground effects. The investigation's primary objective was to establish a configuration that would provide acceptable static turning performance over the desired range of flap-deflection angle without exceeding temperature constraints on the wing and flaps. This test was coordinated with planning for the Aero Commander wind-tunnel test, and modification of the temperature environment measured during the test-stand study ensured an acceptable condition for the tunnel tests.

Engine test-stand investigation in preparation for tunnel tests of modified Aero Commander airplane. Photo on right shows turning of flow around upper surface blown flap.

Langley researcher Charles C. Smith, Jr., and technician Charles Schrum inspect the modified Aero Commander USB research model in the Langley Full-Scale Tunnel.

In early 1974, Smith led a multidisciplinary test team to determine the aerodynamic performance, steady and laboratory aerodynamic loads, surface temperatures, and acoustic characteristics of the full-scale Aero Commander USB model. The model had a full-span, leading-edge Krueger flap using boundary-layer control and three spanwise trailing-edge flap segments: an inboard Coanda flap located behind the engine, a double-slotted midspan flap, and a drooped aileron equipped with blowing boundary-layer control. Two Pratt and Whitney JT15D turbofan engines used to power the model were equipped with rectangular nozzles. Researchers designed the internal contours of the nozzle exits so that the exhaust flow was deflected slightly downward toward the top of the wing, and deflectors were attached to the nozzles to improve the spreading and turning of the jet exhaust. The model was extensively instrumented, including pressures for aerodynamic loads, fluctuating pressure gauges, accelerometers, and thermocouples. Langley staff also made acoustic measurements to provide pioneering baseline noise data for a large-scale USB configuration having real turbofan engines. These acoustic tests included noise frequency and spectral content measurements for various flap configurations and engine thrust settings.

The large-scale test results provided detailed engineering data and matured the technology base required to reduce risk for USB applications. The aerodynamic performance of the model was especially impressive, as evidenced by a humorous experience that occurred during powered testing. At one point, the test team was in the process of evaluating the effectiveness of a horizontal T-tail for trim at high-power, low-speed conditions. The puzzled test crew detected no noticeable change in aerodynamic pitching moment when the tail was installed on the model, even with large deflections of the tail surface. After extended discussions over the unexpected result, the team surveyed the airflow in the tunnel test section behind the model and found that for high-power, low-speed conditions the aerodynamic turning of the USB concept was so powerful that it was turning the entire airflow from the wing down through the open-throat test section to the floor of the facility!

Results of static load measurements by Boyd Perry, III, indicated the expected high values of normal force for spanwise locations directly behind the engine exhaust models. Tests with one engine inoperative indicated very little lift carryover from the powered to the unpowered side of the model. Temperature and vibration characteristics measured by James A. Schoenster and Conrad M. Willis indicated that the upper surface wing-skin temperatures and accelerations on the first flap element were relatively insensitive to tunnel speed or angle of attack. Acoustic characteristics measured by John S. (Jack) Preisser confirmed the expected unsymmetrical noise radiation pattern due to wing shielding of the high-frequency engine noise and the production of low-frequency noise by jet-surface interaction. Results were in good agreement with other small- and large-scale model tests.

In addition to the foregoing studies, Langley conducted several investigations of USB configuration dynamic stability and control characteristics. Since the early 1930s, Langley had developed and continually refined a wind-tunnel free-flight model testing technique to evaluate the flying characteristics of unconventional aircraft configurations. Initially conducted in Langley's Free-Flight Tunnel, these tests were relocated to the larger 30- by 60-ft test section of the Full-Scale Tunnel

Technicians David R. Brooks and Benjamin J. Schlichenmayer prepare a free-flight model of a USB transport for tests in the Full-Scale Tunnel.

in the 1950s. The technique was used to evaluate the flying characteristics of USB configurations, including engine-out conditions during which the development and effects of asymmetric stall are difficult to predict. Lysle P. Parlett led free-flight tests in 1973 for a four-engine configuration with pod-mounted tip-driven fan engines located on top of the wing in a twin-engine (Siamese) nacelle. The results of the study showed that the longitudinal motions of the model were heavily damped and easy to control; however, the lateral motions of the model were difficult to control without artificial stabilization because of a lightly damped roll-yaw oscillation ("Dutch roll"). With artificial stabilization, the model was easy to fly. With one outboard engine inoperative, lateral trim could be restored through the use of asymmetric blowing on the wing leading edge and on the knee of the outboard flap segment.

Dissemination of Information

As Langley's staff conducted the USB research in the early 1970s, representatives of industry, DoD, and the airlines continually visited for updates on the evolving technology. National interest was being stimulated by growing interest in civil short-field transports and by an impending solicitation by DoD for a potential STOL replacement for the C-130 transport. In October 1972, NASA sponsored an STOL Technology Conference at Ames Research Center to disseminate information on the Agency's latest research results in aerodynamics, flight dynamics, loads, operational aspects, and powered-lift noise technology. The primary topics of that meeting were the rapidly maturing STOL technologies and NASA's experiences for EBF and augmentor-wing transport configurations. However, Art Phelps and Danny R. Hoad of Langley presented summaries of the initial testing that had been conducted earlier that year for the USB concept.

When the powered-lift community reconvened again at Langley for a Powered-Lift Aerodynamics and Acoustics Conference in May 1976, the major topic of discussion had shifted to USB technology and the potential noise benefits promised by the research conducted at Langley. During this time, STOL technology had received a tremendous injection of interest due to the Air Force's interest in developing prototype advanced medium transports. When Boeing selected the USB concept for its candidate transport design, interest in this innovative approach to powered-lift capability reached a new intensity.

Langley and the Boeing YC-14

Development of the Boeing YC-14 AMST is covered in great detail in the excellent American Institute for Aeronautics and Astronautics (AIAA) case history publication by John K. (Jack) Wimpress and Conrad F. Newberry (see bibliography). Wimpress received the AIAA Design Award in 1978 for the YC-14's conception, design, and development, and he was the only Boeing person to be with the YC-14 Program from its inception to its end. His publication contains an extremely interesting review of the program's technical, programmatic, and administrative aspects. The following section has summarized information from that document to indicate the interactions that occurred between Wimpress and Langley's staff, along with the contributions of Langley's research to the program.

In July 1969, the Defense Science Board produced a report urging the use of prototyping by DoD to yield better, less costly, more competitive weapon systems. Deputy Secretary of Defense David Packard was a strong advocate of the prototyping approach and, in 1971, an Air Force committee recommended six systems as candidates, including a lightweight fighter (which subsequently evolved into the F-16) and the AMST. Later that year Boeing's preparations for a response to an anticipated Air Force request for proposal (RFP) to design, build, and flight test an AMST Technology Demonstrator rapidly crystallized as the company began to develop its candidate for the competition.

Boeing had accumulated considerable expertise in powered-lift concepts, having proposed the EBF concept for its unsuccessful C-5 competitor as previously discussed and having conducted flight research with NASA using the Boeing 707 prototype (known as the 367-80) modified with sophisticated leading-edge devices and BLC on both leading- and trailing-edge flaps. Along with most of the aeronautical community, Boeing had maintained an awareness of NASA's development of various powered-lift concepts. In its RFP preparations, Boeing examined several powered-lift concepts, including boundary-layer control and, of course, the EBF. Early on, the company was convinced that a twin-engine design offered considerable advantages for the AMST from the perspectives of cost and safety. BLC would not provide the level of lift required via engine bleed air, and the use of an underwing, pod-mounted twin-engine layout for an EBF configuration would require the engines to be located very close to the fuselage to minimize rolling and yawing moments if an engine became inoperative. Boeing was concerned that large aerodynamic interference effects would occur with such an arrangement, particularly at cruise conditions. Thus, Boeing was searching for a new concept that would permit the deflection of jet flow behind a twin-engine arrangement.

Boeing had analyzed the previously discussed exploratory upper-surface blowing tests published a decade earlier by Turner and was interested if NASA had since conducted additional research on the concept. While attending a conference on STOL aerodynamics at Langley in November 1971, Jack Wimpress was informed of the previously discussed semispan USB research being conducted by Joe Johnson and Art Phelps at the request of Oran Nicks in the 12-ft tunnel. An examination of the preliminary results by Wimpress revealed that the magnitude of lift generated was as high as had ever been seen for any powered-lift system. Johnson provided Wimpress with a set of preliminary data, which he enthusiastically took to the Boeing team working on the AMST's presolicitation design. Wimpress was very excited about the Langley data because they were the key enablers for a twin-engine STOL configuration layout. In particular, with the engines on top of the wing, they could be placed close to the centerline of the airplane without causing large aerodynamic inference with the fuselage. Boeing immediately started to build wind-tunnel models to verify the NASA data with geometric and engine parameters more closely representing configurations that Boeing was actually considering. By the end of 1971, Boeing was hard at work in several wind tunnels assessing and refining the twin-engine configuration.

When the Air Force RFP for the AMST prototypes was released in January 1972, it called for the very impressive capability of operations into and out of a 2,000-ft semiprepared field at the midpoint of a 500 nmi mission while carrying a 27,000-lb payload both ways. By comparison, the C-130 series in operation at that time required field lengths almost twice as long to lift a 27,000 lb payload. In February, Wimpress visited Dick Kuhn, Dudley Hammond, Jim Hassell, and others at Langley to brief them on Boeing's progress for the USB configuration, and he briefed Langley's high-speed performance researchers regarding approaches to minimizing drag of the over-the-wing engines.

Following the submittal of its proposal in March, Boeing conducted many wind-tunnel and engine test-stand investigations to refine its proposed design and to identify and solve potential problems. In November, Wimpress again visited Langley for an update on NASA's USB research activities. Joe Johnson and Dudley Hammond both reported on testing being conducted in their organizations and showed Wimpress experimental data that verified the high-lift performance that Boeing had submitted in its proposal.

On November 10, 1972, the Air Force selected Boeing and McDonnell Douglas as contractors to work on the AMST prototypes. At the request of the Air Force, Langley's Dick Kuhn had participated in the evaluation process for the AMST competition. Following the contract award, Boeing launched an aggressive development program to actually design the airplane. Considerable efforts were required for the development of an acceptable USB nozzle, and a major technical surprise occurred when Boeing discovered that the forward flow over the airplane during low-speed operations had a degrading effect on the USB flap, reducing the jet spreading and causing separation ahead of the flap trailing edge. This phenomenon had not been noted in earlier NASA or Boeing wind-tunnel testing. Results from those earlier tests had led to the conclusion that forward speed effects would not significantly impact the flow-turning capability of the nozzle. Boeing added vortex generators to the YC-14 configuration to re-energize the flow and promote attachment on the USB flap during STOL operations. The vortex generators were extended only when the USB flap was deployed beyond 30° and were retracted against the wing surface during cruise.

Boeing adopted a supercritical airfoil for the wing of the YC-14 based on internal aerodynamic research following the 747's design. Initially, senior aerodynamicists at the company were reluctant to accept such a radical airfoil shape. After reviewing ongoing supercritical wing research at Langley led by Richard T. Whitcomb, they were impressed by the performance of a supercritical airfoil applied to a Navy T-2C aircraft in a research program by Langley (see *Concept to Reality* for additional information on Whitcomb's supercritical wing activities). Confidence in the design methodology for the new family of airfoils was provided by close correlation of wind-tunnel predictions and actual flight results obtained with the T-2C. With the NASA data in hand, Boeing proceeded to implement the supercritical technology for the YC-14 and for its subsequent civil commercial transports, including the 777.

During the development process, Boeing was faced with determining the size of the horizontal tail and its placement on the configuration. The initial proposal airplane had a horizontal tail mounted on the end of a long extended body atop a vertical tail with relatively high sweep. However, as the design evolved it became apparent that the proposal configuration would not adequately accommodate the large nose-down pitching moments of the powered-lift system or ground effects. Boeing examined the parametric design information on longitudinal stability and trim that Langley tests had produced in the Full-Scale Tunnel and the V/STOL Tunnel, indicating that it was very desirable to place the horizontal tail in a position that was more forward and higher than the position that Boeing had used for the proposal configuration. These Langley data provided critical guidelines in the tail configuration's revision for the YC-14's final version.

By December 1975, Langley had negotiated with Boeing to obtain full-scale data on a USB high-lift system. Boeing conducted full-scale powered ground tests of a complete YC-14 wing-flap-fuselage segment at its Tulalip test facility to evaluate the effectiveness and noise levels of its powered system. During the tests, sound levels and pressure distributions were measured by Boeing over the USB flap and the fuselage next to the flap. These data were made available to Langley under the special research contract. Langley's interest was stimulated in part by the fact that the engine nozzle of the YC-14 design incorporated a D-nozzle (a semielliptical exit shape), which differed from the high-aspect-ratio rectangular nozzles that had been used at Langley in the full-scale Aero Commander tests previously discussed. With the full-scale YC-14 data in hand, Langley proceeded with a test program to determine the adequacy of subscale models to predict such information, including the development of scaling relationships required for the various technologies involved.

Under the leadership of Jim Hassell, 0.25-scale model static ground tests of the Boeing YC-14 powered lift system were conducted at the outdoor test site near the Full-Scale Tunnel for correlation with full-scale test results. The model used a JT-15D turbofan engine to represent the CF6-50D engine used on the YC-14. The tests included evaluations of static turning performance, static surface pressure and temperature distributions, fluctuating loads, and physical accelerations of portions of the wing, flaps, and fuselage. Results were obtained for the landing flap configuration over a range of fan pressure ratio for various ground heights and vortex generator modifications.

The YC-14 prototype's first flight occurred on August 9, 1976. YC-14 and YC-15 airplane capabilities were evaluated in a flight test program at Edwards Air Force Base in early November 1976. By the end of April 1977, the very successful YC-14 Program had exceeded all its projected goals in terms of flight hours, test conditions accomplished, and data accumulated. The performance goals were met in terms of maneuvering, field length, and touchdown dispersion. Following the flight test program, Boeing demonstrated the YC-14 to U.S. forces in Europe, including an appearance at the

James L. Hassell inspects the 0.25-scale model of the YC-14 USB arrangement.
Note vortex generators deployed on USB flap.

Paris Air Show in June. The airplane impressed the crowds at the air show, performing maneuvers formally considered impossible for a medium-sized transport. After the European tour, the YC-14 arrived for a demonstration at Langley Air Force Base on June 18, 1977, where its outstanding STOL capability and crisp maneuvers stunned not only the Air Force observers but many of the NASA-Langley researchers who had participated in USB studies that helped contribute to the design and success of this remarkable airplane.

The YC-14 flight test program ended on August 8, 1977, exactly 1 year after it began. Unfortunately, the anticipated mission of the AMST did not meet with Air Force funding priorities at the end of the flight evaluations (the B-1B bomber was by then the top Air Force priority), and the AMST Program ended. In 1981, the Air Force became interested in another transport, one having less STOL capability but more strategic airlift capability than the AMST YC-14 and YC-15 airplanes. That airplane was ultimately developed to become today's C-17 transport. The two YC-14 prototype aircraft were placed in storage at the Davis Monthan Air Force Base, and one was later moved to the Pima Air Museum in Tucson, Arizona, where it is displayed next to one of the YC-15 aircraft.

One of the two Boeing YC-14 prototype aircraft.

The Boeing YC-14 demonstrates its low-speed maneuverability.

The Quiet Short-Haul Research Airplane

In 1972, Langley advocates had led a growing NASA interest in developing a contract for a high-performance STOL research airplane to be called the Quiet Experimental STOL Transport (QUESTOL). NASA envisioned a USB research airplane that would push the boundaries of technology further than the requirements of the AMST Program. Langley had examined the possibility of using a modified twin-engine high-wing B-66 Air Force bomber for the configuration. However, this study was subsequently terminated by a new focus for a USB demonstrator aircraft. In January 1974, NASA began a program known as the QSRA Project. The objective of the program was to develop and demonstrate the capabilities of a low-cost, versatile, quiet jet research aircraft with next-generation STOL performance. The powered-lift system selected for the QSRA was a four-engine USB concept designed to yield high lift and very low sideline noise levels. In addition to demonstrating outstanding aerodynamic lift efficiency, the aircraft would be used in conducting terminal-area research, including operations from field lengths ranging from 1,500 ft to 4,000 ft with sideline noise levels of only 90EPNdB on landing approach.

The QSRA aircraft design was based on a modified deHavilland C-8A Buffalo airframe, and the program was led by Ames with Boeing as the primary contractor. Although the research airplane operated only in the low-speed regime, the wing and nacelles were designed by the team to be conceptually representative of a high-speed commercial transport aircraft. Development of the airplane included tests in the Ames 40- by 80-Foot Tunnel and extensive research on the Ames Flight Simulator for Advanced Aircraft. First flight of the airplane took place on July 6, 1978. The QSRA was tested extensively at Ames, including joint operations with the U.S. Navy in 1980 from the *U.S.S. Kitty Hawk* aircraft carrier. The QSRA conducted 37 touch-and-go landings and 16 full-stop takeoffs and landings on the carrier without the use of an arresting gear during landing or a catapult during takeoff. Noise levels of 90EPNdB were obtained at sideline distances of 500 ft, the lowest ever obtained for any jet STOL design. The acoustic signature of the airplane was highly directional, being projected in a small, 35° cone in front of the aircraft. Most observers were comfortable standing next to the QSRA without ear protection during powered operation.

The low-speed maneuverability of the QSRA was demonstrated by a turn radius of only 660 ft at an airspeed of 87 kts with the critical engine failed. To illustrate the maneuver capability, assume civil operations from a terminal with parallel runways. The QSRA could take off, climb to safe altitude, perform a climbing curved departure, and depart with a 180° change in heading while operating within the airspace over the center of the terminal and within the boundary between the two parallel runways. Terminal area operations of this type would alleviate the dangers of aircraft proximity and the nuisance of noise as problems for surrounding communities. The aircraft also

The NASA-Ames Quiet Short-Haul Airplane lands on the carrier Kitty Hawk without arresting gear in 1980.

demonstrated high descent paths as large as 17° during approach (flaring out to 6° near touchdown) compared with the 1.5° flown by conventional airliners.

With the 1976 transfer of rotorcraft and V/STOL technology from Langley to the Ames Research Center decreed by NASA Headquarters, Langley researchers played a minor role in QSRA activities, and all STOL research at Langley essentially was terminated. However, the revolutionary capabilities indicated by the early research at Langley had been vividly demonstrated by the research airplane including a breathtaking performance at the Paris International Air Show in 1983.

Foreign Applications

The Soviet technical community characteristically followed development of the AMST prototype aircraft and ultimately produced the An-72 aircraft (North Atlantic Treaty Organization code named Coaler) whose features very closely resembled the YC-14. The first prototype flew on December 22, 1977, and the aircraft entered service in 1979. A few years later, a derivative version known as the An-74 appeared at the Paris Air Show.

In 1975, the advisory board to the Japanese government's Science and Technology Agency (STA) suggested that an STOL experimental aircraft be developed. Consequently, the Japanese National

Aerospace Laboratory (NAL) embarked upon the Quiet STOL project in 1977. For this research, a Kawasaki C-1 transport aircraft was heavily modified into an STOL aircraft with modifications being completed in 1985. Designated ASKA, the modified aircraft was fitted with four Japanese-developed turbofan engines on the wing's upper surface and used a USB concept similar to the YC-14. Flight testing occurred between October 1985 and March 1989.

Status and Outlook

The remarkable technical ingenuity and progress made by NASA and industry on the USB concept during a very short time in the 1970s stands as one of the most significant accomplishments of the NASA aeronautics program. The conception of an extremely efficient powered-lift principle, the development of design data, the discovery of potential problems and development of solutions, and the flight-demonstrated capabilities of research vehicles involved NASA research capability at its best. Fundamental data were available when industry and DoD needed it. The advantages of the USB compared with other powered-lift concepts—especially in reduced noise—continue as a viable reason for considering this concept a primary candidate for future military and civil aircraft requiring short-field capability.

As a result of NASA's pioneering work, it is generally accepted that the technology challenges to develop medium and large transport aircraft having STOL capabilities have been met. Day-to-day demonstrations of the military C-17 EBF airplane have provided tremendous confidence in the engineering community and the military users of powered-lift technology. At the 2002 International Powered-Lift Conference in Williamsburg, Virginia, the scope of papers presented by the engineering community emphasized the need for reexamining the potential benefits of powered lift for the civil transportation system rather than disciplinary technologies that have already been demonstrated.

Unfortunately, from a commercial transport perspective, there is still a very limited current requirement for medium or large STOL airplanes in the U.S. transportation system. Airport space and runways have not been expensive enough to warrant compromising the capability of the conventional commercial transport to any extent to achieve very short field lengths. City hub and spoke operations continue to be the emphasis of today's air transportation system, although traffic delays, congestion, and unavailability of more flexible travel schedules continue to be a hindrance to the traveler. The issue of noise generated by powered-lift aircraft continues to surface as a challenge to the application of the technology. The USB concept provides an approach to attack this issue, and recent experiences with advanced propulsion system concepts, such as extremely high-aspect-ratio nozzles for stealth benefits on military aircraft, might be pursued for advanced

USB applications. Another interesting concept currently under study is the use of many small "distributed" propulsion units that spread out the induced circulation lift across the wing span. It is expected that many small engines would generate noise at high frequencies, which typically mix with entrained flow much faster and might further reduce the noise for powered-lift systems.

A currently emerging NASA interest in developing the technology required for futuristic runway-independent fixed-wing transport aircraft addresses the benefits that can accrue by the operation of STOL aircraft. If, as predicted by the FAA, commercial air traffic continues to increase in the new millennium, a fresh look at the ability of STOL technology to relieve limitations of the future transportation system will be in order. Thanks to research conducted by NASA and its partners over three decades ago, the technology appears to be ready for applications.

Concept and Benefits

The flexible characteristics of aircraft structures can result in dramatic, sometimes catastrophic, behavior of civil and military aircraft. When the inherent structural flexibility of an airplane interplays with the aerodynamic, gravitational, and inertial forces and moments acting on it, steady or dynamic deflections or oscillatory motions of aircraft components can result. Such interactions can cause reduced structural life of airframe components, undesirable coupling with control systems, severe reductions in ride quality, and even abrupt and violent structural failures. The three most important aeroelastic phenomena for aircraft designers are loads (static and dynamic), flutter, and buffet. The subject of loads is concerned with the a structural airframe's ability to accommodate external loads encountered during the flight envelope, with emphasis on the airplane's performance, stability, control, and structural integrity. Flutter is a dynamic aeroelastic

Photographs of B-52 airplane on the ground and in flight graphically show the structural flexibility of the wing.

phenomenon that involves the interactions of a structure's elastic and inertia characteristics with the aerodynamic forces produced by the airflow over the vehicle. It is a self-excited oscillation of the aircraft structure involving energy absorbed from the airstream. When an aircraft's elastic structure is disturbed at speeds below flutter speed, the resulting oscillatory motions decay. However, when the structure is disturbed at speeds above flutter speed, the oscillatory motions will abruptly increase in amplitude and can rapidly lead to catastrophic structural failure. In some instances, flutter oscillations are limited to just a single airplane component such as the wing, while in other instances the oscillations may be considerably more complex, involving coupling of natural structural modes of wing, fuselage, and empennage motions. Buffet is a randomly varying structural response often triggered by intense and chaotic aerodynamic forcing functions associated with stalled or separated flow conditions. Fluctuating pressures present during buffet conditions can cause highly undesirable responses from wings, fuselages, pod-mounted engine nacelles, and empennages. Dynamic loads experienced during buffet can lead to pilot fatigue or structural fatigue, resulting in serious reductions in the anticipated structural life of airframe components.

The traditional solution to these aeroelastic issues has been primarily to stiffen the airframe structure, thereby either eliminating undesirable excitation of structural characteristics or ensuring that the undesirable phenomena occur only at conditions beyond the flight envelope. Unfortunately, this "passive" approach involves adding additional structure to stiffen that which is already sufficiently strong to carry normal flight loads. These weight penalties adversely affect manufacturing and acquisition costs, mission performance, and add to operational costs throughout the life of the airplane.

The state of the art in aeroelasticity has steadily advanced to the point that, by the 1960s, many fundamental physical phenomena, predictive methodologies, and processes for the resolution of problems had been identified for conventional airplanes. Researchers began turning attention to the use of "active" controls technology (ACT) to favorably modify the aeroelastic response characteristics of aircraft to permit structural weight reduction, optimal maneuvering performance, and multimission capability. As the name implies, ACT uses aircraft control surfaces that are linked to a computer and sensors in a manner to automatically and immediately limit any unwanted motions or aerodynamic loads on the aircraft structure.

The potential benefits of active control of aeroelastic response are significant. For example, if the stiffening requirements for wings can be reduced, the weight reduction could be absorbed in additional passengers and payload revenues for commercial transports. If an active flutter suppression system was incorporated in the design of an advanced configuration, such as a highly swept supersonic transport, the substantial weight savings translates into increased range or payload, with a significant reduction in airplane direct operating costs. Active control of tail buffet for highly maneuverable fighter aircraft could result in weight savings, increased structural service life, and reduced maintenance and cost.

The transition of technology for effective control of aeroelastic response from laboratory experiment to extensive fleet applications has involved a few success stories, but the general application to aircraft to date is relatively limited. For example, gust load alleviation using active control laws on commercial transports has only been implemented on aircraft such as the Lockheed L-1011 and the Airbus A320, and active flutter suppression has not achieved operational status on any civil aircraft. In the early 1970s, the first practical demonstration of active flutter suppression was carried out by the U.S. Air Force in its Load Alleviation and Mode Stabilization (LAMS) Program. A Boeing B-52 bomber was equipped with an active flutter suppression system that was demonstrated during flight tests to increase the airplane flutter speed by at least 10 kts. As will be discussed, another success story has been the development and application of an active system by McDonnell Douglas (now Boeing) to suppress an unacceptable limit-cycle structural oscillation exhibited by preproduction F/A-18 aircraft with certain external store loadings. The system was subsequently incorporated into the production control system for fleet F/A-18 aircraft.

Challenges and Barriers

Disciplinary Challenges

The primary disciplinary challenges for active control of aeroelastic response are requirements for highly accurate predictions of critical aerodynamic and structural phenomena at test flight conditions of interest; reliable prediction and analysis methods for the aeroelastic interactions that occur; and the design of robust, redundant control systems that are tolerant to parametric uncertainties. The consequences of inadequate or invalid design methods are not acceptable, and concepts such as active flutter suppression are viewed as inherently high risk.

Active control of aeroelastic response is dependent on the specific configuration and flight condition under analysis. For example, design of an active flutter suppression system for a commercial transport, with a typical high-aspect-ratio wing, may focus on the interactions between the first few structural vibration modes of the wing, usually relatively simple combinations of bending and twisting of the structure. However, flutter mechanisms of a low-aspect-ratio fighter design will be more complicated because the structural vibration modes are more complex, including interactions of basic wing motions with the motion of a variety of different pylon-mounted external stores.

Paramount to all these design challenges is the difficulty of precisely predicting the complex, steady and unsteady, aerodynamic characteristics present at the high subsonic or transonic conditions where the onset of flutter becomes most critical.

Operational and Economic Challenges

Operational issues pose special challenges for active control systems. Clearly, systems used for flutter suppression must be reliable and designed to be totally failsafe, with requirements similar to those for automatic control systems used in the automatic stabilization and control of highly unstable airplanes. The design of such systems clearly calls for redundancy, and multiple systems will be required for safety. Other concerns for active controls include manufacturing and maintenance costs and additional complexity. The active control system will have to be as reliable as the structure that it replaces.

As is the case for any advanced technology, the use of active control for tailoring aeroelastic response will be successful only if the potential benefits are feasible from a cost/benefit perspective. The requirements for robust, redundant sensors, maintenance schedules and costs, environmental protection, and certification compliance and costs must be favorably addressed before widespread application of the technology can occur.

Langley Activities

Langley Research Center is the lead NASA Center for research in structures and materials, and it is internationally recognized as a world leader in aeroelastic research. This reputation stems from a rich legacy of contributions in technology, cooperative research, consultation and problem solving, and unique experimental facilities. Langley's efforts in structures, materials, and loads as an NACA laboratory prior to 1958 are well documented and will not be discussed herein. Rather, several major contributions and research activities since the early 1960s are highlighted as examples of the critical role Langley researchers and facilities played in advancing the state of the art in active and passive control of aeroelastic response. Although Langley's research accomplishments have included both fixed-wing and rotary-wing vehicles, the discussion is limited to research on fixed-wing aircraft. Two areas of activities are overviewed: contributions and ongoing aeroelastic research over the last 35 years by the staff and facilities of the Langley 16-Foot Transonic Dynamics Tunnel (TDT); and the NASA ACT Program of the late 1970s. The discussion draws extensively on the excellent summaries of Boyd P. Perry, III, and Ray V. Hood, Jr. (see bibliography).

The 16-Foot Transonic Dynamics Tunnel

Without question, the centerpiece facility for Langley's research in aeroelasticity is the Langley TDT. Converted from an existing 19-Foot Pressure Tunnel, with operations commencing in 1960, it is the only facility in the world capable of studying a full range of aeroelastic phenomena at transonic speeds. Research in aeroelasticity in the TDT ranges from flutter clearance studies of new vehicles using aeroelastic models to the development and assessment of new concepts to control aeroelastic response, and to the acquisition of unsteady pressures on wind-tunnel models for providing experimental data to validate unsteady theories. Analytical methods are developed and validated to solve the aeroelastic problems of fixed- and rotary-wing vehicles, including the control of instabilities, loads, vibration, and adverse structural response.

The TDT is a closed-circuit continuous-flow wind tunnel capable of testing with either air or R-134a refrigerant as the test medium over a Mach number range from 0 to 1.2. The R-134a gas is very attractive for use in wind-tunnel studies of aeroelastic phenomena because, as compared with air, it has a low speed of sound and high density. Since the first test was conducted in 1960, the tunnel's testing capabilities have been continuously expanded by introducing a number of new features, such as airstream oscillators, sophisticated data acquisition systems, a variety of model mounting and suspension systems (including a two-cable suspension system for full-span "free-flying" flutter models), and excellent model-monitoring visual systems. Many very significant contributions of the tunnel and its staff to military and civil aircraft programs are summarized in NASA SP-2000-4519 *Partners in Freedom* and NASA SP-2003-4529 *Concept to Reality* (see bibliography).

The Langley 16-Foot Transonic Dynamics Tunnel.

After almost 45 years of operations and over 500 tests, the TDT staff has led the way as experimental aeroelasticity reached relative maturity. This progress has involved an intense coupling of experimental and computational research including the use of advanced CFD and advanced control theory. However, the research required to mature and extend the use of active controls for aeroelastic response has not diminished. Understanding and predicting the effects of transonic aerodynamic phenomena such as shock waves, flow separation, viscosity, and the interactions of these complex aerodynamic features with active control systems in controlling aeroelastic response are still major challenges.

Flutter Suppression

Delta Wing Flutter Suppression Study

The first practical demonstration of an active flutter suppression system was accomplished during the early 1970s in TDT tests. This effort was fueled by flutter concerns of large SST aircraft. Most studies had shown that large SST configurations, such as those of interest in the United States, had relatively severe flutter problems, requiring the addition of thousands of pounds of structural weight to provide the stiffness needed to ensure that flutter occurred well outside the operating envelope. Conceptually, an active flutter suppression system would require the addition of perhaps only a few hundred pounds of added weight as opposed to the thousands of pounds of a passive system. Because there was little information available on the design, implementation, and operation of active flutter suppression systems, a delta wing flutter suppression study was undertaken. The research team was lead by Langley's Maynard C. Sandford with critical support provided by a number of other Langley researchers (especially Irving Abel and David C. Grey) and from Boeing-Wichita under contract.

In the benchmark 1971 experiments, a simplified 1/17-scale semispan model representative of the Boeing SST (2707-300) wing configuration was mounted to the wall of the TDT with a rigid sidewall-mounting block used to simulate the fuselage faring. The model was equipped with both leading- and trailing-edge control surfaces as well as high-fineness-ratio bodies on the wing lower surface that were used to simulate the mass properties of engine nacelles. The model also incorporated advanced, miniaturized hydraulic actuators to move the active control surfaces. The development of these actuators was a significant advance in the state of the art for model construction and surface actuation at that time.

The flutter suppression systems implemented were based on the aerodynamic energy concept developed by Elihu Nissim, who worked at Langley as a postdoctoral research fellow. Three control laws were studied, including Nissim's basic method and two variations developed by Langley researchers. Two of the systems used both leading- and trailing-edge control surfaces, whereas the third system used only the trailing-edge surface. The control laws were implemented on an analog computer located in the wind-tunnel control room. Response of the model was sensed by two accelerometers located at the same outboard station as were the control surfaces. Model response signals were routed to the analog computer for processing through the control laws. The processed signals were then routed to servo valves that provided hydraulic power to the actuators and caused them to move in such a way that the aerodynamic forces generated by the motion of the surfaces added damping to the wing, thus preventing flutter from occurring.

When the "open-loop" flutter characteristics of the model with the suppression control system off were compared with results obtained with the suppression control activated, it was found that all three active control systems demonstrated significant increases in the dynamic pressure for flutter onset at transonic speeds (Mach = 0.9). The increase in flutter dynamic pressure ranged from 11 to 30 percent for the systems.

In addition to dramatically demonstrating the potential of active control systems to extend flutter speeds, this investigation made major contributions to the fundamental understanding of aerodynamic prediction methods for complex transonic flows and the understanding of inertial coupling between the control surfaces and the main wing. In retrospect, this particular investigation is widely regarded as a landmark study and a major contribution to subsequent advances made in active control of aircraft aeroelastic responses.

Wing-Store Flutter Suppression

YF-17 Program

High-performance military attack aircraft typically carry vast arrays of air-to-air or air-to-ground weapons on wing-mounted pylons. The substantial weight and aerodynamic characteristics of these external stores can dramatically change the aeroelastic characteristics and structural response characteristics of wings, resulting in unacceptable flutter constraints or airplane motions. Each of the literally hundreds of different combinations of external store configurations is a new dynamic system with its own set of flutter characteristics. In particular, the onset of flutter can occur at lower airspeeds than the baseline aircraft, thus restricting the aircraft's operating envelope and limiting military operations. It appeared that active control techniques might be applicable to the wing-store flutter problem. Initial U.S. efforts to develop active wing-pylon-store flutter suppression systems were begun by the military with the F-4 airplane in the early 1970s. Langley has participated extensively in domestic and international research programs to advance and validate store-induced flutter design methods, including assessments of active control systems for wing-store flutter suppression.

In 1977, Langley researchers began a cooperative program with Northrop and the Air Force Flight Dynamics Laboratory (AFFDL) to conduct long-term wind-tunnel investigations in the Langley TDT of several concepts for wing-store flutter suppression. The program's focus was a 30-percent scale semispan, aeroelastic model of the YF-17 Lightweight Fighter prototype consisting of a wing-fuselage and horizontal tail. The model was mounted on a special support system that provided rigid-body pitch and plunge freedoms. In addition to having powered leading- and trailing-edge control surfaces, the model was equipped with three different external store configurations that produced widely different flutter characteristics (flutter frequency, coupling of structural modes, and relative violence of the flutter mode).

Moses G. Farmer led the cooperative test team during the TDT test program. During initial testing in 1977, results demonstrated that active flutter suppression could be achieved for the significantly different aeroelastic characteristics of the wing-store configurations tested and that, for the first time, the use of only leading-edge control surfaces could achieve suppression. In a second series of TDT tests conducted in 1979, more sophisticated multiple control loops were conceived and assessed for further expansion of the flutter-free envelope. Concentrating on the store configuration with the most violent flutter mode, researchers developed an innovative "flutter stopper" electromechanical internal system to rapidly change the distribution of store mass in such a manner as to decouple the critical elastic modes. This unique system proved a valuable tool in suppressing flutter of the model during the tunnel test.

Through the auspices of an Air Force data exchange agreement with certain European nations, an international assessment of control laws developed by individual organizations was conducted using the YF-17 model in the TDT. European participants included British Aerospace and the Royal Aeronautical Establishment (RAE) from the United Kingdom, the Office National d'Etudes et de Recherches Aerospatiale (ONERA) from France, and Messerschmitt-Bolkow-Blohm GmbH (MBB) from Germany. Substantial information was mutually shared on the effects of suppression system design, including the number of sensors and control surfaces used. Some concepts produced extremely effective flutter suppression, including one that was tested to a dynamic pressure 70 percent above the passive flutter dynamic pressure. Many notable firsts were achieved in this international program, including demonstrations of the ability to switch between flutter-suppression control surfaces above flutter speeds without undesirable transients, and the validation of design procedures and techniques.

Members of the international active store flutter suppression team pose with the YF-17 model in the Transonic Dynamics Tunnel.

The initial collaborative wing-store flutter suppression activities in the TDT had been based on the use of analog controllers, but the advances and application of digital controllers in an adaptive manner was the next target of researchers. During 1981, some control laws previously implemented on an analog computer were converted to a digital computer and retested. In 1982, another phase of investigations of digital controllers was conducted with the objective of demonstrating adaptive flutter suppression. In this approach, the controller was required to discriminate between flutter modes and select the appropriate control law with changes in flight condition. The tests were highly successful and proceeded to the point of demonstrating the release of a wingtip-mounted store designed to transform the model configuration from a stable condition to a violent flutter condition. The adaptive controller rapidly recognized the unstable behavior, implemented a new control law, and stabilized the model in a fraction of a second.

The highly successful YF-17 store flutter suppression program extended through seven different entries in the TDT and is known as a critical NASA accomplishment in the field of aeroelasticity.

F-16 Program

The General Dynamics YF-16, winner of the Air Force's Lightweight Fighter Program, was initially conceived as a highly agile, lightweight fighter with emphasis on close-in air-to-air combat maneuverability. As the Air Force developed the airplane into today's F-16, mission requirements for the aircraft changed to emphasize the air-to-ground mission, thereby leading to an extensive application of external stores to the configuration. Early in the airplane's development program, the Air Force requested Langley to support the flutter clearance requirements for flight testing by conducting traditional flutter tests in the Langley TDT. Subsequently, over 18 different TDT test entries were conducted for the F-16 to cover flutter characteristics of the basic airplane and the airplane with external stores. With the very large number of potential external stores and aeroelastic characteristics to be encountered with the F-16, a cooperative program on active flutter suppression was established between Langley, the Air Force Wright Aeronautical Laboratories (AFWAL), and General Dynamics. Langley's lead researchers for the program were Moses G. Farmer, Raymond G. Kvaternik, Jerome T. Foughner, Frank W. Cazier, and Michael H. Durham. Using a 0.25-scale flutter model of the F-16, the team investigated a range of potential control concepts from analog-type systems to digital adaptive systems.

Tests of a single wing-store configuration were led in 1979 by Foughner to investigate the suppression of flutter for that specific configuration. Study results demonstrated that an antisymmetric flutter mode could be suppressed with an active control system, and detailed research data and analyses of the details of mechanizing such systems produced vital information for further developments. Test program highlights included a demonstrated ability to switch control laws on above the

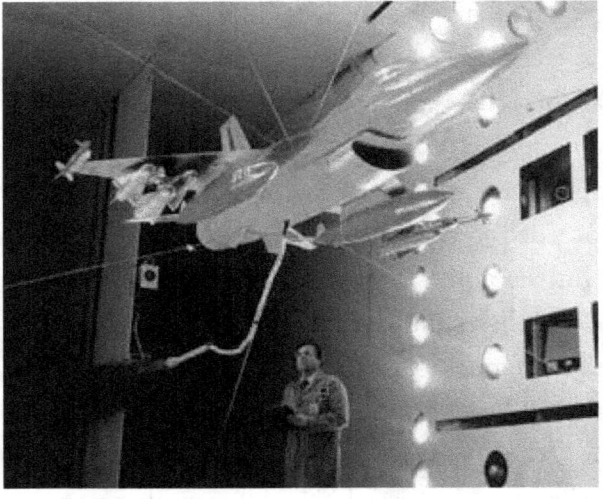

Jerome T. Foughner with F-16 model in Transonic Dynamics Tunnel during flutter tests with external stores.

unaugmented flutter condition without undesirable transients and demonstrated flutter suppression to a dynamic pressure 100 percent above the unaugmented flutter dynamic pressure.

Further testing of the F-16 model by Cazier in 1981 introduced a second store configuration with successful demonstrations of flutter suppression. Additional information on the dynamic response requirements for the control system was determined, and assessments of the effectiveness of individual control surfaces to suppress flutter were made.

Based on YF-17 test program successes and the F-16 demonstrations, the joint F-16 research team pursued the goal of developing and demonstrating a totally digital, adaptive suppression system. The work tasks included developing a suppression system for three different external store configurations, demonstrating a 30-percent improvement in flutter speed for each configuration, and demonstrating the suppression of flutter following the separation of a store from the wing. Conducted in 1986 by Cazier and Moses Farmer, these TDT tests contributed critical technology to the development of adaptive digital suppression controls. By demonstrating the feasibility of a digital adaptive system that required no prior knowledge of the wing-store configuration, coupled with successful simulated launching of missiles from a free-flying model at conditions below and above the unaugmented flutter boundary, this highly successful cooperative research project has been recognized as a benchmark event for adaptive control technology development.

In 1999, the Air Force designed and tested a prototype active flutter suppression system (AFSS) designed to suppress the F-16's tendency to oscillate when flying at high speeds while carrying certain combinations of fuel tanks and different types of weapons. The motion is known as limit cycle oscillation, or LCO. Although the oscillations are not serious enough to damage an F-16, they can affect a pilot's ability to precisely fulfill his mission, such as accurately launching a missile during air-to-air combat. Referred to as being "like driving a car with an out-of-balance tire," the phenomenon is caused by the antisymmetric flutter mode discussed earlier. Due to this mode's excitation, the pilot experiences a side-to-side rolling motion in the cockpit. The F-16 active flutter suppression system was designed by Lockheed Martin Tactical Aircraft Systems and uses ailerons for flutter suppression. The flight tests included flights with the F-16 configured in five different store loadings, including heavy air-to-ground weapons under the wings, AIM-9 Sidewinder missiles attached on the wingtips, and wing-mounted fuel tanks. The program successfully suppressed LCO at the desired speeds and altitudes for each combination of loadings. The test team flew 21 flights with the system totaling more than 48 flying hours.

Despite the improved characteristics experienced in the flight test evaluations, the AFSS has not been implemented in operational F-16s at this time.

F/A-18 Application

A remarkable accomplishment in the suppression of wing-store limit cycle oscillations occurred in the early 1980s during the development and deployment of the F/A-18 aircraft by McDonnell Douglas. As an attack aircraft for the U.S. Navy, the F/A-18 is required to carry a wide variety of air-to-ground stores up to transonic speeds. During pre-production flight tests of certain external store configurations, the aircraft exhibited an unacceptable LCO of about 5.6 Hz during flutter testing. The resulting lateral accelerations at the pilot's station greatly exceeded allowable levels, with values of as much as 1.0 g peak to peak experienced in the cockpit. The oscillations typically occurred only at altitudes less than 12,000 feet and at speeds greater than Mach 0.8.

The specific store loading susceptible to LCO involved wing store combinations which included high pitch inertia stores on the outboard wing pylons along with wingtip-mounted AIM-9 missiles. The fundamental structural contributor to the LCO mechanism was an outboard pylon/store antisymmetric pitch structural mode. Extensive flight testing demonstrated that the oscillations were not due to classical flutter and were not reinforced by coupling with the flight control system. Testing further demonstrated that the wing oscillation amplitude was not sufficient to cause a structural integrity or fatigue problem. However, the accelerations at the pilots station were of sufficient magnitude to create a very uncomfortable ride and thereby degrade pilot performance. A flight test program was initiated to solve the LCO problem, with an initial focus on potential mechanical passive solutions. The testing results indicated that a practical wing reconfiguration could not be found to satisfactorily reduce the oscillations over the flight envelope.

McDonnell Douglas engineers were aware of the research being conducted at Langley on wing/store flutter for the YF-17 and the F-16, and the promising results of that research gave the company additional confidence that an active system might provide a feasible solution to the F/A-18 problem. The program was in a situation requiring a rapid and reliable solution which led to development of a solution from flight testing. One of the major issues encountered by the McDonnell Douglas team was the fact that the LCO phenomenon had not been predicted by flutter analyses conducted at that time. The LCO had been experienced almost 200 knots below the speed predicted by linear flutter analysis.

An Active Oscillation Suppression (AOS) system was subsequently developed under the programmatic constraint of only using the existing F/A-18 flight control system components and interfaces. After extensive analysis and flight test evaluations, an effective AOS system was developed and implemented using feedback from an existing lateral accelerometer to actuate the aircraft's ailerons via the flight control computer. During other flight testing, it was found that

McDonnell Douglas successfully developed and applied an active control system to suppress wing/store limit cycle oscillations in production F/A-18 aircraft

certain outboard pylon store configurations would not require engagement of the AOS. Thus, logic to interrogate the type of store configurations carried was implemented within the AOS system. Since the LCO was not regarded as a classical flutter problem and therefore not a safety of flight consideration, the use of a single string AOS concept did not violate any redundancy requirements. However, if more serious flutter is encountered for future aircraft requiring active flutter suppression, redundancy will have to be considered from a flight safety perspective.

Following the highly successful development of the AOS system by McDonnell Douglas, the system was applied on all production F/A-18A/B/C/D aircraft and has been extremely effective for over 20 years. Although NASA was not an active participant in the development of the solution to the F/A-18 problem, the precursor research that had been conducted by Langley played an important role in establishing confidence and risk reduction in this critical activity.

Decoupler Pylon

In concluding the discussion of wing-store flutter suppression, it is appropriate to mention an innovative "quasi-passive" concept conceived by Langley's Wilmer H. (Bill) Reed, III, in the early 1980s. As an alternative to conventional passive methods of incorporating additional structure to increase stiffness or to use advanced active control methods, Reed devised a spring-mount system for external stores called a decoupler pylon, which isolates or decouples the external store's pitching vibratory motions from those of the wing, thereby increasing the flutter speed. The concept was substantiated by analysis, demonstrated in TDT tests, and validated at full-scale conditions through flight test using an F-16 airplane. Langley participants in studies of this revolutionary concept, in addition to Reed, were Frank W. Cazier, Jr., Moses G. Farmer, and Harry L. Runyon, Jr.

Reed's decoupler pylon concept was relatively simple yet very effective. It consisted of soft-spring and damper components that, in combination, isolated the wing from the pitch inertia effects of the external store. A low frequency, automatically controlled alignment system was provided to keep the softly supported store properly positioned relative to the wing during maneuvers. The decoupler pylon could be made robust in that a variety of different stores could be mounted on the same decoupler pylon without changing the overall wing flutter characteristics.

Following some carefully crafted analytical studies, and after highly successful wind-tunnel studies in the TDT using YF-17 and F-16 flutter models demonstrated increases in flutter speeds of over 100 percent could be obtained by using the decoupler pylon versus the same store mounted on a conventional pylon, a flight demonstration program was initiated. The plan was to design, build, and flight test a decoupler pylon on an F-16 airplane. A pair of decoupler pylons was designed, fabricated, and ground tested by General Dynamics (now Lockheed Martin) under contract to NASA. The flight tests were conducted on an F-16 from the Joint Task Force at Edwards Air Force Base, California. Langley's Frank W. Cazier, Jr., served as Project Manager for the flight test in a joint activity with NASA Dryden Flight Research Center, the Air Force, and General Dynamics.

The chosen test configuration was an asymmetric loading of AIM-9J wingtip missiles, a GBU-8 bomb near midspan, and a half-full 370-gallon fuel tank inboard. That particular configuration exhibited the well-defined limited-amplitude antisymmetric flutter when the bomb is carried on a standard F-16 pylon. Analyses and wind-tunnel tests indicated that mounting the bombs on decoupler pylons in place of standard pylons would appreciably increase the airplane's flutter speed. The flight test objectives were to demonstrate an improvement flutter speed of at least 30 percent over the conventional pylon flutter boundary, assess the requirements for the alignment system, and demonstrate that store separation from the decoupler pylon was satisfactory.

Flight tests for the F-16 in 1985 with a standard pylon-store configuration were first conducted at an altitude of 10,000 ft and for Mach numbers above 0.7. This configuration experienced the antisymmetric LCO discussed previously. Pilots described the oscillation as a "continual pounding oscillation that was of sufficient amplitude to cause visual blurring of the cockpit displays." With the decoupler pylon, the LCO that had been experienced with the standard pylons was suppressed throughout the flight envelope tested. The investigation expanded in scope to develop methods to reduce friction in the pylon mechanisms, as well as assessments of the effectiveness of the alignment system. During one flight test, a GBU-8 was ejected, demonstrating that weapons separation from the decoupler pylon was satisfactory. Flight tests, including maneuvers, demonstrated an increase in flutter speed of 37 percent over the standard F-16 pylon configuration.

As was the case for the F-16 AFSS, technical success in the decoupler project did not lead to applications to the F-16 fleet.

Load Alleviation

The reduction of loads imposed on aircraft by maneuvers, gusts, and turbulence has been explored extensively by Langley researchers during in-house studies and cooperative programs with their partners from industry and DoD. The range of airplane configurations studied has included general aviation airplanes, commercial transports, military transports, and bombers. Although some activities were based on passive control that did not include elements of active control systems, they are included herein for background and completeness.

C-5A Active Lift Distribution Control System

Lockheed-Georgia was awarded an October 1965 Air Force contract for a C-5A heavy transport with a specified wing fatigue life of 30,000 flying hours. The first flight of the new transport was in June 1968 and, unfortunately, static fatigue testing of a wing test specimen revealed wing structural cracks in July 1969. Although a structural modification program was immediately begun to reinforce the critical wing stations, structural modifications turned out not to be an acceptable long-term solution to the problem. After the C-5A had been in service for several years, a wing tear-down inspection on one aircraft revealed cracks in the structure that projected to a fatigue life of less than 8,000 flying hours, approximately one quarter of the desired life. Lockheed proceeded to explore solutions, including an active aileron system to alleviate gust loads on the wing, local structural modifications to improve fatigue life, and redistribution of fuel within the wing to reduce bending moments. Active ailerons were retrofitted to C-5As during 1975 to 1977 as part of an active lift distribution control system (ALDCS) that increased fatigue life by symmetric

C-5 model mounted in the Transonic Dynamics Tunnel.

deflections of the ailerons in response to gusts and maneuvers. The concept also used an automatic elevator deflection to null out pitching moments caused by the aileron deflections. Additional redesign of the center wing and wing box sections was also incorporated in the modification program, and by 1987 all surviving C-5As had been modified. Also, in 1982 the decision was made to have Lockheed build 50 C-5Bs, which incorporated the wing improvements of the C-5A.

Wind-tunnel tests conduced in the Langley TDT during 1973 were a key component of the successful development of the ALDCS. A full-span cable-mounted 1/22-scale C-5A model was tested to experimentally verify the effectiveness of the ALDCS system in reducing loads. Langley's Charles L. Ruhlin and Maynard C. Sandford led the NASA-Lockheed team that conducted this first-ever scaled model study of an ALDCS.

The C-5 test was very successful. The results showed that the ALDCS was very effective in reducing both wing dynamic bending and torsion loads. Bending moments at the frequency of the wing first bending mode were reduced by more than 50 percent across the wing span. Although the reduction for torsion loads was less, it was still substantial. Later correlation of results from airplane flight tests and the aeroelastic wind-tunnel model tests were in very good agreement for the critical low frequency bending mode. Once again, this study validated the use of active control technology to reduce aircraft aeroelastic response and further demonstrated the valid application of aeroelastic wind-tunnel models for developing active control technology.

Passive Gust Load Alleviation

The ride quality for passengers in light general aviation airplanes in turbulent weather is characteristically rough and uncomfortable. Particularly offensive are the large up-and-down heave motions encountered because of relatively light wing loadings of such aircraft. Researchers at Langley have investigated the human response to typical accelerations encountered in flight, and have identified the critical frequencies that lead to highly undesirable effects on humans, including airsickness. An extensive investigation into the subject of ride quality was led by Langley's D. William (Bill) Conner during the 1970s.

Although these highly undesirable passenger accelerations could theoretically be alleviated by an automatic control system using appropriate sensors, computers, and rapid-actuation controls, the complexity, costs, and maintenance of such systems are beyond the capabilities of typical airplane owners. Despite the long-term interest of designers in reducing the effects of turbulence on ride quality, and the continuing dissatisfaction of public passengers with undesirable accelerations due to turbulence, no current general aviation aircraft are equipped with gust-alleviation systems. As part of a long-term research in aircraft response to gusts, Langley Research Center has investigated several concepts for gust alleviation for this class of aircraft.

In the late 1940s Langley's W. Hewitt Phillips was exposed to an earlier French gust-alleviation concept by René Hirsch wherein the horizontal tail surfaces were connected by pushrods to flaps on the wing. On encountering an upward gust, the tail surfaces would deflect up, moving the wing flaps up and thereby offsetting the effects of the gust. The system had been analyzed and designed to minimize adverse interactions on other airplane characteristics, such as pitching moments. (Phillips later traveled to France in 1975, met Hirsch, and inspected some of the aircraft that he had designed.) Intrigued by the possibility of achieving gust alleviation with automatic controls rather than the complex aeromechanical interconnects of Hirsch's design, Phillips began studies of airplane response characteristics to sinusoidal gusts and the character of control inputs required to alleviate accelerations. After studying several systems, he arrived at the idea of using a gust-sensing vane mounted on a boom ahead of the nose to operate flaps on the wing through a hydraulic servomechanism.

Following analytical studies, a flight demonstration project was conceived to demonstrate gust alleviation in flight. A Navy C-45 twin-engine airplane was modified to include a nose boom to hold an angle-of-attack vane; the wing flaps, which normally deflected only downward, were modified for deflections in both up and down directions; the elevator was split into three sections with two sections being linked to the flaps for gust alleviation; and small segments of the wing flaps near the fuselage were driven separately from the rest of the flap system so that they could be used in either the same or opposite directions as the rest of the flaps. Following these NACA tests at Langley, jet transports were introduced into commercial service and the higher wing loadings, higher cruise altitudes, and the use of weather radar to avoid storms resulted in less likelihood that passengers might become airsick. Also, the problem of active gust alleviation was made more difficult because the structural flexibility of jet transports placed structural frequencies closer to the range of interest for gust alleviation. Thus, the interest and momentum for gust-alleviation systems waned.

Following a visit to France and meeting with Hirsch in 1975, Phillips revisited the aeromechanical approach to gust alleviation and initiated a Langley study of the concept. Eric C. Stewart, L. Tracey Redd, and Robert. V. Doggett, Jr., led analytical and experimental studies of a 1/6-scale model of a typical general aviation airplane equipped with an aeromechanical gust alleviation system.

Nose boom with angle of attack vane on C-45 transport used for gust alleviation research.

The project was designed as a cooperative venture between NASA, Cessna, and the Massachusetts Institute of Technology (MIT). The gust alleviation system consisted of two auxiliary aerodynamic surfaces that deflected the wing flaps through mechanical linkages to maintain nearly constant airplane lift when a gust was encountered. The dynamic model represented a four-place, high-wing, single-engine light airplane, and was rod mounted in the Langley TDT for tests. The effects of flaps with different spans, two sizes of auxiliary aerodynamic surfaces, single and double-hinged flaps, and a flap-elevator interconnect were studied. Investigation results showed that the gust-alleviation system reduced the model's root-mean-square normal acceleration response by 30 percent in comparison with the response in the flaps-locked condition. Despite these promising results, the aeromechanical concept was not pursued and has not been applied to production aircraft.

About 10 years later, Langley briefly pursued a concept for an active, computer-based gust-alleviation system for general aviation aircraft. Teamed with Cessna and the University of Kansas, Langley researchers conducted analytical studies of the application of computer-driven controls with a view toward flight demonstrations using a Cessna C-402 twin-engine research airplane. The analysis included the use of advanced modern control theory to develop the control architecture. Unfortunately, the response characteristics required of the control actuators could not be accommodated within the budget and time allotted for the project, and the activity was terminated.

Combined Aeroelastic Control Concepts

Although some studies examined the effectiveness of a single active control concept, others emphasized more than one: for example, the simultaneous application of active flutter suppression and active load control. Some of these latter studies are described in this section.

B-52 Control Configured Vehicles Program

The B-52 Control Configured Vehicle (CCV) Program was the first in a number of studies addressing multiple applications of active controls. It was a natural follow-up to work of the 1960s in applying flight controls systems to attenuate the structural response (especially cockpit accelerations) of large military airplanes such as the B-52E and the XB-70.

During the early 1970s, AFFDL sponsored the B-52 CCV Program at The Boeing Company to demonstrate the benefits of applying advanced flight control technology to a large flexible airplane. The effort was initiated in July 1971 and was completed in 1974. A highly modified Boeing NB-52E bomber was used to investigate four active control concepts: ride control, flutter mode control, maneuver load control, and augmented stability. The existing elevators and rudder of the B-52 were not sufficient to implement the control systems, so it was necessary to add additional control surfaces consisting of three-segment flaperons, outboard ailerons, and horizontal and vertical canards. On August 2, 1973, the B-52 CCV test aircraft made aviation history by flying 10 kts faster than its flutter speed. Although the flight tests were halted at this point, there was no indication of a decrease in damping in the structural vibration mode important to flutter, so the actual flutter speed was considerably higher. This event was the first time that an aircraft had been flight tested above its flutter speed relying solely on an active flutter control system to augment the structural damping.

At Langley, an investigation sponsored by AFFDL with Boeing and NASA participation was conducted for correlation with flight results. The objective was to demonstrate that wind-tunnel models and testing techniques could be used to design and assess active control concepts. An existing 1/30-scale, full-span, free-flying B-52 aeroelastic wind-tunnel model was modified and tested in the TDT. Although capability to study all four active control concepts was incorporated into the model, only active vertical ride control (VRC) and active flutter suppression (AFS) were actually tested during three separate wind-tunnel tests in 1973 and 1974. The Langley Project Managers for the wind-tunnel studies were Jean Gilman, Jr., and L. Tracy Redd.

The airplane VRC system was designed to reduce the gust-induced vertical acceleration at the pilot's station by at least 30 percent. This system processed vertical acceleration signals sensed at the pilot's station through a computer implemented control law to drive horizontal canards. The performance of the model's VRC closely matched the performance of the full-scale airplane system, resulting in a dramatic reduction in vertical accelerations at the cockpit location

The AFS consisted of feedback loops using signals from accelerometers mounted on the model's external fuel tanks (fed back to the aileron control surfaces) and from accelerometer signals located near the midwing (fed back to the flap segments). Wind-tunnel tests results demonstrated that, with the AFS on, the damping in the flutter mode showed a large improvement over that displayed with the AFS off, verifying the full-scale flight results and indicating the potential for a significant increase in flutter speed.

Follow-up AFS tests with yet another modification to the B-52 model were conducted by Robert V. Doggett, Jr., Rodney H. Ricketts, and Maynard Sandford in 1978. For this study, the model was converted from a free-flying model to a sting-mounted model. In this case, the digital-computer-implemented control laws had to simultaneously deal with two distinct flutter modes, one involving antisymmetric wing motion and the other involving symmetric wing motion. Because the control laws were implemented on three separate computers, it was possible to evaluate the effects of system failures on the effectiveness of the AFS. This study provided the first successful demonstration of multimode, digital active flutter suppression, including considerations of redundancy management.

From a research viewpoint, the most significant result of B-52 CCV experiments in the TDT was validation that dynamically scaled, actively controlled wind-tunnel models could be used to study and demonstrate advanced active control concepts. Based on the proven success of this pioneering effort, wind-tunnel models in the TDT are now used routinely to increase the confidence level in active control concepts by providing data to verify analytical models and methods used in design and to eliminate the risks and lower the costs associated with flight testing such concepts.

The Aircraft Energy Efficiency Active Controls Technology Program

In 1976, NASA initiated its ACEE Program in response to the dramatic increase in fuel prices that began in the early 1970s. The program included several elements of technology in aerodynamics and active controls with an emphasis on concepts that traded cruise speed for increased fuel efficiency. A major part of ACEE activities was the EET Program. Langley's leaders in the active controls element of the EET Program were Ray V. Hood (Program Manager) and David B. Middleton (Deputy Program Manager). A detailed program summary and bibliography of the EET activities has been prepared by Middleton, Bartlett, and Hood (see bibliography). One element of the EET Program included in-house research activities and cost-shared contracts with Boeing, Douglas, and Lockheed-California for the analysis, preliminary design, testing, and in-depth assessments of selected advanced concepts for ACT for improved mission efficiencies. Because higher aspect-ratio wings quickly became a focal point for aerodynamic efficiency, control of aeroelastic responses became a vital segment of the program.

Active wing flutter suppression concepts were pursued that increased the damping of wing structural modes important to flutter to the extent that the flutter placard speed was increased beyond the airplane's expected maximum operating speed without adding any structural weight. In addition, maneuver load control concepts that reduced wing-bending moments during maneuvering flight were conceived, as well as active gust load alleviation systems that reduced structural loads during encounters with vertical gusts. Collectively, these two load alleviation systems comprised an active control function called wing-load alleviation.

Douglas pursued the design and assessment of active systems for flutter suppression and load alleviation on a derivative of the DC-10 configuration that had an increased wing span. Wind-tunnel testing to determine dynamic wing loads was conducted in industry tunnels, and control laws derived by Douglas using conventional methods increased flutter speed by up to 19 percent and significantly decreased wing-bending accelerations. Within this coordinated effort, Langley supplied alternate control laws based on advanced design methods. Both NASA control system designs increased flutter speeds by more than 25 percent.

The Lockheed L-1011-500 was the first commercial transport to use active load control.

Lockheed studies involved extending the wing span of its existing Lockheed L-1011 transport configuration and providing a load alleviation system using symmetric operation of outboard ailerons at high speeds. Outboard ailerons on most conventional transports are designed to be inoperative at high speeds because of adverse aeroelastic issues, and inboard ailerons are used for roll control. The load alleviation system for the L-1011 redistributed the wing lift and thus eliminated the need for significant structural redesign and increase in structural weight to support the extended wing span. This configuration was ultimately implemented by Lockheed with company funds and flight tested on Lockheed's L-1011 research airplane, demonstrating a 3-percent fuel savings. Based on these very favorable results, Lockheed immediately pursued FAA certification of the active control system and later incorporated the system in its derivative long-range Advanced TriStar L-1011-500 transport in 1980, representing the first significant application of active controls to a modern wide-body transport.

The ACEE Program's EET element greatly accelerated the state of the art in active control of aeroelastic response, and the resulting application by Lockheed to the L-1011 was a major event in the acceptability and certification of such systems.

Drones for Aerodynamic and Structural Testing Program

In keeping with its mission for conducting high-risk research, Langley conceived and initiated a flight test project known as Drones for Aerodynamic and Structural Testing (DAST) in the early 1970s to validate analysis and synthesis methods for active control of aeroelastic response and analysis techniques for aerodynamic loads prediction. Flight tests provided the opportunity to simulate characteristics that could not be accurately simulated or properly accounted for in wind-tunnel tests, such as maneuvering flight. Because of the inherent risks in flight testing advanced active control concepts, an unmanned, remotely controlled Teledyne-Ryan BQM-34 Firebee II was chosen as the test vehicle, with the flight test to be conducted at NASA's Dryden Flight Research Center. Langley's Harold N. Murrow was the Project Manager and headed a virtual "who's who" team of Langley aeroelasticians, aerodynamic and structural analysts, and control theory specialists. Some key Langley researchers were Irving Abel, William M. Adams, Jr., Clinton V. Eckstrom, Jerry R. Newsom, Boyd Perry, III, Maynard C. Sandford, and Vivak Mukhopadhyay. An equally competent team was assembled at Dryden to conduct the flight tests.

NASA F-8 research airplane with supercritical wing used as basis for design of ARW-1 wing.

The plan was to fit the Firebee with two aeroelastic research wings (ARW). Both wings were to be representative of advanced subsonic transonic transport configurations. ARW-1 was to have the same planform as the research wing that had been used in Dryden flight demonstrations of the supercritical airfoil section on the NASA F-8 research airplane. The ARW-1 test evaluated two active flutter suppression systems that had been carefully selected from a number of proposals. The objective was to demonstrate in transonic flight at least a 20-percent increase in flutter velocity. Wind-tunnel tests in the TDT were conducted using a simplified model of the ARW-1 to add confidence that the proper choices had been made.

The ARW-2 wing was an even more ambitious activity, including three active control systems: flutter suppression, gust load alleviation, and maneuver load alleviation. The ARW-2 had a higher aspect ratio than ARW-1. The wing configuration was chosen to represent a design derived during a NASA-contracted Boeing study of EET configurations. Fabrication of the ARW-2 began while the ARW-1 portion of the program was still in progress. Part of the ARW-2 plan was to test one of the flight test wing panels in the TDT as opposed to building a separate simplified model, as was done for ARW-1. TDT testing of the ARW-2 wing by Maynard Sandford began in 1978.

The flight-test approach involved launching the test drone from a wing-mounted pylon on NASA's B-52B launch aircraft, conducting the active control experiments, then recovering the test vehicle by deploying an onboard parachute that was "air-snatched" by an Air Force helicopter/aircrew during descent. During the free-flight portion of the experiment, a NASA pilot controlled the drone from a remote ground-based cockpit while researchers monitored flight data transmitted via telemetry. In case the telemetry link between the drone and the ground was lost, the Firebee could also be flown to the recovery site using a backup control system in a NASA F-104 chase airplane.

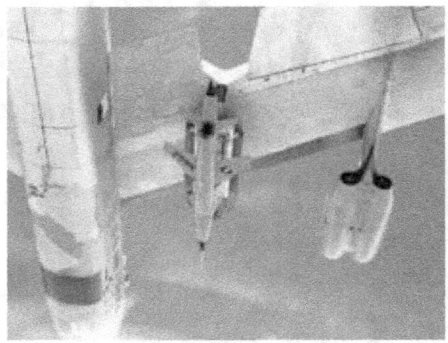

Drone with standard Firebee wing mated to B-52 in 1977 captive flight at NASA Dryden.

Drone during flight with ARW-1 research wing on June 12, 1980, before catastrophic flutter occurred.

Research flights for the DAST program at Dryden were conducted from 1977 to 1983. Initial fight tests were conducted with the Firebee fitted with an instrumented standard wing (also called the "Blue Streak" wing) to (1) develop test procedures and experience to be used during assessments of the flutter-suppression concepts for the ARW-1, and (2) to obtain wing data on surface pressures and bending moments using strain gauge instrumentation. The wing had been designed for a predicted flutter speed of Mach 0.95 at an altitude of 25,000 ft.

Unfortunately, the DAST project was fraught with operational problems, so only a few flights were completed successfully. Research studies of ARW-1 were halted unceremoniously when the test vehicle crashed on June 12, 1980. A programming error in implementing the active flutter suppression control law went undetected, despite careful review by all participants. This error resulted in the system gain being only one-fourth the desired value, and the wing fluttered unexpectedly at flight conditions where it should have been well safe from flutter. This catastrophic flutter resulted in the breakup of the wings and subsequent crash of the test vehicle.

The ARW-1 wing was rebuilt after the crash and again prepared for testing with the control law error corrected. On June 1, 1983, the ARW-1's misfortune continued when, following launch from a Navy DC-130 airplane routinely used to launch military drones such as the Firebee, the recovery parachute system malfunctioned and the parachute inadvertently disconnected from the drone, resulting in a second crash.

Following this second crash, the DAST project was terminated for several reasons. The program's initially planned 5-year lifetime had elapsed, and a combination of reduced funding and resource demands for other emerging high-priority unmanned airplane projects at Dryden made additional flight tests unlikely. However, the planned TDT testing of the ARW-2 wing was completed prior to the program's final termination.

Some view the DAST project as a technical disappointment because the program's original objectives were not attained. However, all the program's inherent research and active control law development considerably advanced the overall state of the art in applying active control techniques to favorably modify aeroelastic response. Perhaps the program's most important legacy was the dramatic experience with the challenges and difficulty of achieving some of these advanced concepts in practice.

Active Flexible Wing Program

In the early 1980s, engineers at Rockwell International Corporation conceived and studied analytically an active control concept that became known as the active flexible wing (AFW). Rockwell's early work was so promising that a cooperative research program involving Rockwell, the U.S. AFWAL, and NASA Langley was initiated in 1985 to further develop the concept and demonstrate it in tests in the TDT.

In the AFW concept, an active roll control system was used to optimize the airplane's rolling response while minimizing maneuver loads. This was achieved by taking advantage of inherent flexibility characteristics of the wings in a carefully controlled manner in conjunction with actuating leading- and trailing-edge control surfaces. The system monitored both flight conditions and wing structural deformations. Using this information, the system selects the best control surfaces to produce the desired rolling motion and commands those surfaces to deflect accordingly. An active roll control system offers the potential for significant savings in structural weight. For example, because the system works effectively at angles of attack above the control-surface reversal condition, it would eliminate the need for the "rolling horizontal tail" and render unnecessary the structural weight required by the rolling tail. If the AFW incorporates other active control applications, such as active flutter suppression, gust load alleviation, and maneuver load control, additional weight savings are possible. Rockwell predicted that by taking full advantage of the AFW concept, a weight savings of at least 15 percent of takeoff gross weight was possible for advanced fighter configurations.

Testing of the AFW concept in the TDT was conducted between 1986 and 1991. Langley's leading researchers for the AFW investigations included Boyd Perry, III, Carol D. Wieseman, Jennifer Heeg, Jessica A. Woods-Vedeler, Anthony S. Pototzky, Sherwood T. Hoadley, Vivak Mukhopadhyay, Maynard C. Sandford, Stanley R. Cole, William M. Adams, Jr., Carey S. Buttrill, Jacob A. Houck, and Martin R. Wazak.

The AFW TDT study used an aeroelastically scaled, 1/6-scale, full-span wind-tunnel model of an advanced fighter concept that was fabricated by Rockwell and tested during four different tunnel entries. The model featured eight separate active control surfaces with two leading and two trailing edges on each side of the wing. As per the name, the wing of the AFW model was designed to be extremely flexible and lightweight. The model test set up included a novel single-degree-of-freedom internal bearing arrangement, which permitted the model to roll freely about the wind-tunnel sting mount. Extensive instrumentation and sensors were also implemented in the model, including accelerometers, strain gauges, and a roll-rate gyro. Because the flutter speed of the basic configuration was too high, it was reduced by the addition of a specially designed wing tip mounted store. A remotely controlled weight within the store could be rapidly moved to raise the flutter speed should violent flutter be encountered unexpectedly.

The investigation included two distinct research parts. In the first part, the Air Force, Langley, and Rockwell coordinated efforts to demonstrate the effectiveness of the basic AFW concept during TDT tests in 1986 and 1987. In the first of these tests, a data base of static forces and moments produced by control surface deflections was determined. These data were required to provide accurate values of the control surface effectiveness needed to design the active roll control system. After several active roll control laws were synthesized by using this data base, the different control was implemented on the wind-tunnel model system and each successfully evaluated during the second wind-tunnel test. All the digital-computer implemented control laws performed well, with the experimental results being in good agreement with theoretical predictions. The test results clearly showed that the AFW concept worked as advertised and, therefore, offers a viable means of improving the maneuver and roll control characteristics of advanced fighter type airplanes.

The second part of the AFW study was considerably more complex than the first. The objective was to demonstrate multi-input/multi-output (MIMO) single function and multifunction digital control of aeroelastic response. Three active control capabilities were incorporated into the wind-tunnel model system: active flutter suppression, the roll rate tracking system (RRTS), and rolling maneuver load alleviation (RMLA). The RRTS was designed to limit loads only when loads reach a predetermined level. The RMLA was designed to reduce loads during rolling maneuvers up to 90 degrees in amplitude. The control laws were implemented on a digital computer. Single function MIMO studies were conducted for each control system. Multifunction studies were conducted for active flutter suppression in combination with each of the two roll control systems

Key accomplishments of this sophisticated investigation included successful demonstrations of single- and multiple-mode flutter suppression, load alleviation and load control during rapid roll maneuvers, and MIMO active-control demonstrations above the open-loop flutter boundary. Rolling maneuvers representative of goals defined by military specifications were performed, and wing loads were controlled at dynamic pressures 24 percent above the open-loop flutter condition. In addition to significantly advancing active controls technology, this study also provided significant advances in the wind-tunnel test methodology needed to evaluate active control of aeroelastic response.

The Benchmark Active Controls Technology Project

The analysis and accurate prediction of aeroelastic phenomena is one of the most difficult challenges facing aerospace engineers. Not only are the phenomena affected by complex interactions of aerodynamic and structural forces, but they often are most troublesome in nonlinear flight regimes, such as transonic speeds. The addition of active controls to the technology poses even new challenges to aeroelasticians. In the late 1980s, Langley initiated the Benchmark Models Program (BMP), with goals of providing high quality experimental data that could be used to the evaluate the accuracy of advanced CFD codes applicable to aeroelastic analysis and to study the effects of new aerodynamic concepts on aeroelastic phenomena. The basic idea was to conduct relatively simple experimental studies where it would be possible to isolate the effects of key parameters, such as airfoil shape. Although active control technology was not included in the initial program plan, such studies were added after the program was initiated.

The BMP Program was a collaborative effort among several working groups of the Structural Dynamics Division and was supported by the entire Langley infrastructure. The Configuration Aeroelasticity Branch, the Unsteady Aerodynamics Branch, and the Aeroservoelasticity Branch all participated in the research activities, which were based on about two tests in the TDT per year over the program's 5-year duration. TDT researchers Robert M. Bennett, Clinton V. Eckstrom, Jose A. Rivera, Jr., Bryan E. Dansberry, Moses G. Farmer, Michael H. Durham, David A. Seidel, and Walter A. Silva collaborated in early benchmark studies. Researchers David M. Schuster, Robert C. Scott, and Sherwood T. Hoadley joined the team as the program evolved.

The program used a basic benchmark active controls technology (BACT) model, which was a rigid semispan configuration that had an NACA 0012 airfoil section. The unswept rectangular-planform model could be mounted on either rigid or flexible supports. The relatively simple, flexible support system provided for pitch and plunge motion of the model, the two most important motions to aeroelastic response. This system greatly simplified the structural aspects of the experiment and allowed the focus to be on aerodynamics and active controls. The model was well instrumented, with a number of pressure transducers to determine aerodynamic pressures and accelerometers to measure model motion. The model had a remotely controlled trailing-edge aerodynamic control surface that could be positioned either statically or dynamically. Remotely controlled upper and lower surface aerodynamic spoilers were also provided. The trailing-edge control and the spoilers were driven by miniature hydraulic actuators similar to those developed during the delta wing flutter suppression study.

The BACT model offered the opportunity to conduct a number of pioneering active control studies. Many of these are very technical and can be fully appreciated only by those well versed in controls theory, whereas others are relatively easy to understand. A couple of the latter studies will be cited here. Although it had been shown previously by another investigator that statically deflected spoilers were effective in increasing flutter speeds, BACT model tests represented the first time that actively controlled spoilers were effectively used as flutter suppressors. The second example was application of artificial intelligence (neural network) concepts to active flutter suppression. Artificial intelligence systems learn based on experiences and, depending on the application, may actually improve themselves as they are used or gain experience. This effort was part of the Adaptive Neural Control of Aeroelastic Response Program, which was a joint effort between NASA Langley and McDonnell Douglas Corporation (now part of The Boeing Company). A number of control systems, both adaptive and nonadaptive, were developed using neural network concepts implemented on the BACT model and successfully demonstrated in TDT tests.

The BACT model provided an opportunity not only to learn more about the characteristics of different aeroelastic phenomena, but also to evaluate very advanced active control techniques during an experiment that is easily managed as compared with many active controls studies conducted heretofore. Although the model system might be relatively simple, the phenomena being studied were not.

Piezoelectric Aeroelastic Response Tailoring Investigation

Before discussing the details of the Piezoelectric Aeroelastic Response Tailoring Investigation (PARTI) some introductory comments are in order. Previous active control studies to favorably change aeroelastic response of airplanes had focused on the use of traditional aerodynamic control surfaces to effect the changes in excitation forces needed to accomplish the desired performance improvements. As advances were made in structural and other technologies, it became apparent that the use of "structural actuators" might be viable alternatives to "aerodynamic control surface" actuators. Piezoelectric materials appeared to offer much promise. When electric voltages are applied to these materials, internal strains develop that cause the material to change shape. By controlling the applied voltages to piezoelectric actuators either embedded in or mounted on a structure, it is possible to deform the structure in a desirable manner.

Inspired by graduate student Robert C. Scott's (later a TDT staff member) thesis in 1990, Jennifer Heeg designed and implemented an exploratory wind-tunnel experiment to assess the use of piezoelectric actuators in active flutter suppression. Following the detailed development of a candidate control law, a wind-tunnel experiment of a simple, free to pitch and plunge, aeroelastic wing model was conducted in the Flutter Research and Experiment Device (FRED), which was a small open-circuit wind tunnel with a 6- by 6-in. test section. The experiments, which included open-loop and closed-loop flutter testing, demonstrated that the use of piezoelectric control could increase flutter speed of the test wing by about 20 percent. Almost simultaneously, Heeg expanded her study to include active control of buffeting response. A modified version of the model was used for additional tests in FRED. This study resulted in the first successful application in the United States of active controls to attenuating buffeting response.

The favorable results of Heeg's early work and of studies performed elsewhere led Langley to establish a cooperative research program with MIT. The program's purposes were to further evaluate the ability of distributed strain actuators to control aeroelastic response and to demonstrate selected concepts on a research model wing to be tested in the Langley TDT. The principle used for control in the investigation involved the use of piezoelectric actuators. The piezoelectric actuator concept consists of a series of electrical strain-gauge patches (potentially hundreds per wing) wired for a low-current, high-voltage electrical charge. Wing response measurements, either static or dynamic, are fed back through control laws that output voltages to these actuators, either individually or in selected combinations. These voltages produce internal actuator strains that cause the wing to deform either statically or dynamically in a desired manner.

The PARTI project used an aeroelastic semispan model with 72 distributed piezoelectric actuator patches on the upper and lower surfaces of the wing. (An actual airplane application may require hundreds of actuator patches.) Various groups of actuator patches were oriented to facilitate bending and torsional responses of the model. In addition to the piezoelectric actuators, the model had a trailing-edge aerodynamic control surface driven by an electric motor located in the wing root. Extensive research activities were allocated to the development of instrumentation, control law development, and experimental demonstrations of flutter suppression.

During the first TDT entry in early 1994, the open-loop characteristics of the model were determined, including supercritical (below flutter) response, basic flutter characteristics of the model, and time-dependent response functions for each important piezoelectric sensor group. These data provided the foundation for the Langley-MIT research team to construct mathematical models of candidate control laws and validate analysis techniques prior to additional wind-tunnel testing.

Objectives of the second TDT entry in late 1994 included an assessment and demonstration of the capability of piezoelectric actuators to suppress flutter and to reduce aeroelastic response caused by tunnel turbulence. Several control laws, based on different design techniques, were implemented to assess input-output control effectiveness for various sensor and actuator groups. For the most successful control law, an increase in flutter dynamic pressure of 12 percent was demonstrated, and the peak value of strain measured by the instrumentation was significantly reduced for dynamic pressures below flutter.

The PARTI project successfully completed its primary objective of demonstrating flutter suppression and aeroelastic response control by using distributed piezoelectric actuators on a large-scale aeroelastic wind-tunnel model. Key Langley researchers for PARTI included Anna-Maria R. McGowan, Jennifer Heeg, Donald F. Keller, and Renee C. Lake.

Control of Aeroelastic Response of Vertical Tails

During the 1970s, the operational doctrine of U.S. military air forces began to focus on highly maneuverable fighter tactics. Extensive advancements in aerodynamics, propulsion, and structures—coupled with effective digital flight controls that provided "carefree" maneuvering—resulted in significant operations at high angles of attack. Many recently developed advanced U.S. fighter configurations have used vortex-control techniques for enhanced lift during strenuous maneuvers, as well as twin-tail configurations to provide satisfactory stability and control during these conditions. A number of these configurations, including the F-14, F-15, F-18, and F-22, have experienced problematic buffeting loads and oscillatory stresses to the vertical tails at high angles of attack (above about 25 degrees) because the tails were immersed in high-intensity turbulence and chaotic airflow caused by phenomena, such as stalled wing wakes or vortex "bursting." The resulting randomly varying structural response of the tails caused by the applied buffet loads severely degrades the fatigue life of these components. Tail buffet loads have necessitated structural modifications for some airplanes, or even mandated maneuver limitations for others. In addition to structural modifications, special and costly inspections are required to check for damage due to buffet loads. Analysis based on available usage history of two aircraft configurations suggests that the tail surface fatigue life could be doubled if the tail stresses could be reduced by only 10 percent.

In the case of the F/A-18, an aggressive problem-solving exercise by industry, DoD, and NASA over a period of years had resulted in a passive approach to the fin buffet issue. Specifically combined modifications consisting of structural cleats at the bottom of the vertical tails and small fences on the wing leading-edge extension (LEX) were incorporated on operational aircraft to meet fatigue requirements. The effects of different LEX lengths on tail buffet loads were examined on an F/A-18 model in the Langley 30- by 60-Foot Full-Scale Tunnel by researcher Gautam H. Shah. McDonnell Douglas (now Boeing) also examined other passive techniques to increase fatigue life of its aircraft. Because these types of passive techniques do not solve the buffeting problem for all flight conditions, an active tail buffet alleviation study was initiated. With active control techniques offering so much promise for solving other aeroelastic problems, it was only natural that research would be initiated to reduce the buffeting response of vertical tails. Except for some work in France and the aforementioned efforts of Heeg, little research had been conducted previously on the active control of buffeting response.

Langley and its DoD, industry, and international partners have conducted extensive research on the fundamental aeroelastic phenomena associated with tail buffet and have conducted several studies to assess and demonstrate active control to reduce the loads and stresses encountered. Led by Robert W. Moses, a series of wind-tunnel tests have been performed in the TDT since 1995 to develop and mature active control concepts. The initial activity, known as the Actively Controlled Response of Buffet Affected Tails (ACROBAT) project, focused on the F/A-18 configuration that had experienced significant operational tail buffet loads due to vortex bursting at high angles of attack.

A 1/6-scale, sting-mounted model of the F/A-18 served as the ACROBAT study workhorse. Objectives of the project were to apply active controls technology using various force producers, such as aerodynamic control surfaces and piezoelectric structural actuators, to alleviate buffeting for twin vertical tails; and to determine detailed unsteady aerodynamic data at high angles of attack with the buffet alleviation controls on and off. A variety of vertical tail surfaces was fabricated for the tests, including rigid (nonflexible) as well as flexible surfaces. Extensive instrumentation, including strain gauges and accelerometers, was used to obtain steady and unsteady characteristics during the tunnel tests. The investigated angle-of-attack range varied from 20 to 40 degrees. Early results of the ACROBAT studies indicated that control systems using either the rudders or piezoelectric actuators worked best for suppressing the buffeting loads and for angles of attack up to about 30 degrees, both approaches were equally effective in buffet alleviation. Exhibiting a strong interest in applying the rudder and piezoelectric actuators to reduce tail buffet loads, Daimler Benz Aerospace of Germany participated in the tests through a set of international agreements in aeroelasticity research.

Through an interagency agreement, NASA joined forces with the Air Force Research Laboratory (AFRL) to develop buffet scaling techniques by comparing the ACROBAT unsteady pressure data with full-scale, low-speed pressure measurements on an F/A-18 aircraft tested in the 80- by 120-Foot test section of the National Full-Scale Aerodynamics Complex (NFAC) Facility at NASA Ames. The scaling technique was later demonstrated by Moses and Shah through comparisons with unsteady pressures measured on a vertical tail of the NASA High Angle of Attack Research

NASA's F/A-18 High Angle of Attack Research Vehicle uses smoke injected into vortex flow to illustrate vortex breakdown position for angles of attack of 20 degrees (top) and 30 degrees (bottom).

Vehicle F/A-18 aircraft at NASA Dryden while the airplane was flying at high angle-of-attack conditions. In addition to the scaling technique, the spatial correlation of the buffet, a random process, was demonstrated by Moses for the ACROBAT pressure data and comparisons with limited aircraft data. This information subsequently proved vital to modeling unsteady buffet pressures on the F-22 configuration for evaluating active control system models or minor changes to the tail structures and materials.

Building upon the successful ACROBAT Program, the collaborative F/A-18 tail buffet suppression studies were later expanded to include participation by Australia and Canada (operational users of the F/A-18). The research program was coordinated by AFRL and was conducted under the auspices of The Technical Cooperation Program (TTCP). The collaborative program involved tests of a full-scale F/A-18 empennage, including assessing the use of commercially available patch piezoceramic actuators to provide buffet alleviation. This ground test program used the International Follow-On Structural Testing Program (IFOSTP) facility located at the Australian Defence Sciences & Technology Organisation. The purpose of this collaborative program was to investigate the feasibility of piezoceramic actuators to withstand and control severe buffet loads applied to the F/A-18 vertical tails. Open- and closed-loop tests of the concept's effectiveness were completed successfully during ground tests in 1997 and 1998, respectively. This highly successful cooperative program has served as a pathfinder for future buffet loads alleviation research.

In 1998, another test entry of the F/A-18 model in the TDT involved a project known as Scaling Influences Derived from Experimentally-Known Impact of Controls (SIDEKIC). In this study, Bob Moses and his team cooperated with the Australian Aeronautical and Maritime Research Laboratory (AMRL) to correlate data during mutual investigations of the F/A-18 configuration. Because the F/A-18 is also flown by Australian military forces, mutual sharing of data and technology on tail buffet alleviation was especially valuable to the participants. New vertical tails were fabricated for the TDT F/A-18 model, and an effort was made to match the arrangement of piezoelectric actuators used during full-scale airplane ground tests at AMRL. One of the model's vertical tails used both an active rudder and active piezoelectric actuators for controlling responses over specific frequency ranges. This approach to providing buffet alleviation was referred to as a "blended" system because two different actuator technologies were combined by Bob Moses. Several other control schemes were evaluated during these tests, including one contributed by Boeing.

The F/A-18 research program's contributions and other studies of the F-15 configuration resulted in extensive studies using flow visualization, flow velocity measurements, pressure transducers, and response gauges. The state of the art for predicting buffet loads and fatigue life has rapidly matured and has been updated with tests of additional configurations. In 1999, Langley and AFRL conducted a cooperative TDT investigation of vertical tail buffeting characteristics of an early model of the F-22 fighter. Led by Bob Moses, the investigation used a 13.3-percent-scale model of the F-22 equipped with various types of instrumentation and sting-mounted in the TDT for testing at low Mach numbers (up to 0.12) and high angles of attack. A variety of measurements, including flow visualization techniques, was used to identify key features of the buffet-inducing flows. Model configuration variables such as wing leading-edge flap deflection were also assessed, and the general results obtained for the F-22 model were compared with the F/A-18 results for correlation and general conclusions. A rudder on the starboard-side vertical tail was actively controlled using feedback of buffet-induced accelerations near the tip of that tail. This approach proved quite effective in reducing buffet-induced responses.

F-22 model mounted in Transonic Dynamics Tunnel for tail buffet studies.

Highly successful demonstrations of the blended control system in the TDT, under the SIDEKIC Program presented earlier, led to full-scale actuator development, including systems-level considerations of cost and operational environmental conditions for electronic components. To validate the latest technologies in piezoceramic actuators and piezo drive amplifiers on an F/A-18, another international ground test program was formed in 2002 under the auspices of the TTCP. A series of ground tests were conducted in the Australian IFOSTP facility, as before; however, this test concentrated the piezo actuators near the vertical tail tip to control buffet-induced responses there and near the rudder to reduce vibratory response in the bending mode. Completed in 2004, this ground test program successfully demonstrated the feasibility of the "blended" control system to alleviate buffet loads as designed for an aircraft.

The success of international collaboration has peaked interest in the next generation of vertical tail active buffet suppression systems and the capability to predict systems performance. This interest was especially intense in 2001, when an early version of the Lockheed Martin X-35 Joint Strike Fighter aircraft experienced high tail buffet loads when attempting to fly at high angles of attack. Bob Moses was contacted by Lockheed Martin for consultation and assistance in the development of an in-house capability to design for tail buffet. Together, this team implemented an aggressive wind-tunnel test and tool development program that benefited from Langley's experience in model instrumentation, data acquisition and analysis, and predictive tool development. Within

15 months of the initial consultation, this team had scaled and implemented wind-tunnel pressure measurements into design methods not only to predict the buffet loads on existing designs, but also to redesign the tails to mitigate buffet-induced fatigue. Plans are underway to implement similar capabilities at Boeing to augment its current buffet loads design capabilities.

Status and Outlook

The challenges inherent in active control of aeroelastic responses have been the target of research at Langley Research Center for over 35 years. Progress in defining the complex transonic aerodynamic flow fields of importance has increased tremendously, as has the ability of CFD methodology to predict these phenomena. Experimental demonstrations in the TDT and in flight have been impressive and provided confidence in the ability of technology to alleviate aeroelastic problems using active control techniques.

Nonetheless, there has been very little application of active control for fixed-wing aircraft in the civil or military sectors. Significant widespread application barriers remain, especially issues regarding the additional complexity and cost of active controls. As yet, the cost-benefit consideration has not been in favor of such systems. More importantly, the critical safety-related margins comfortably enjoyed today for aeroelastic issues such as flutter are the result of years of experience in worldwide operational scenarios.

Using active controls for control of aeroelastic response within the U.S. commercial transport industry has not significantly advanced beyond Lockheed's early application to the L-1011 configuration in the 1970s. Meanwhile, the European Airbus Industrie Consortium has explored numerous areas using active controls for drag reduction, active center-of-gravity control, active-load control, variable-camber control, and active sideslip control. Airbus has subsequently applied the early principles derived from the Lockheed efforts by designing a wing load alleviation system into its A-320 transport from its early design, thereby reducing wing weight and improving passenger ride quality in turbulence by actively controlling wing bending moments. The A-320 entered commercial operations in 1988. Military applications of the technology have now progressed to in-depth assessments and flight evaluations for control of vertical tail buffet concerns at high angles of attack and for limite-cycle flutter alleviation for wing/store combinations. The successful application of active controls by McDonnell Douglas to production versions of the F/A-18 prior to the F/A-18E/F represents a milestone in the technology.

The Joined Wing: Diamond in the Sky

Concept and Benefits

One of the most attractive aircraft design areas for innovators has been the challenge of optimizing trade-offs among aerodynamic efficiency, structural effectiveness, and aircraft weight. Although requirements for aerodynamic performance may stimulate the designer to consider wings with very high aspect ratios, the attendant structural weight penalties and requirements for strength and rigidity for such configurations limit the geometric approaches that may be used for a feasible design. For conventional configurations, which use cantilevered-wing arrangements, the loads that must be safely accommodated by the wing-fuselage structure include critical bending moments induced by the aerodynamic and weight loads on the wing panels. Such loads always play a critical role in the aerodynamic and structural integration of new aircraft. Since the advent of heavier-than-air flight, the aeronautical community has continually investigated unconventional and innovative schemes to optimize these trades.

One approach used by designers has been to lay out configurations that use tandem fore-and-aft wings that are joined to form a diamond-type shape when viewed from above and from the front or rear. Depending on the specific geometry involved, potential reductions in structural weight or improved aerodynamic characteristics may be generated. Early designs included a glider, designed by Reinhold Platz in Europe in 1920, and a rudimentary multijoined-wing airplane built by Ben Brown of the University of Kansas in 1932. A more recent joined-wing configuration is the "box plane" concept designed by Luis R. Miranda of the Lockheed-Georgia Corporation in the early 1970s. The box plane concept has been proposed by Lockheed Martin for potential applications for commercial transports, freighters and military tankers.

Also in the 1970s, Julian Wolkovitch of ACA Industries advanced a joined-wing concept wherein the root of the rear wing was intentionally designed to be at a higher elevation than the front wing. With this arrangement, the fore-and-aft wings form a truss structure that relieves some of the loading from the front wing and significantly stiffens the structure. This joined-wing concept is obviously a highly integrated approach to aerodynamic and structural design.

For aircraft applications, the principal benefit of this particular joined-wing configuration is that the rear wing acts as a strut brace to support some of the wing bending moments. This loading feature can be exploited as a reduction in wing weight or as an increase in wing span (aspect ratio), or a combination of both. A secondary benefit of the joined-wing configuration is that the nonplanar arrangement of lifting surfaces can theoretically result in lower induced drag for a given span and weight.

Boeing concept for a joined-wing flight demonstrator.

In addition to these fundamental considerations, the joined-wing configuration offers other potential benefits that are unique to its unconventional geometry. For example, because of the wings' diamond-shaped arrangement when viewed from above, the lifting surfaces can be used to support various types of radar antennas to provide a 360-degree azimuth coverage with little or no aerodynamic penalty. Equipped with wing conformal electronically scanned array radars, a joined-wing research aircraft could offer a substantial increase in radar capability and improved range and endurance. The multiple lifting surfaces result in a compact configuration, requiring less deck space for shipboard military naval applications. In another potential military application, the relatively stiff outer wing of a joined-wing tanker (with a forward/rear wing joint at about 70 percent of the semispan) could accommodate refueling booms on fairing pods at the two outer-wing joints. This capability would enable simultaneous air-to-air refueling of two aircraft, which is not currently possible with today's tanker configurations.

An interesting potential application of the joined-wing configuration would be for advanced aircraft designed for aerial applications, such as crop treatment and seeding, or for fire fighting. In these potentially hazardous missions, structural robustness and crashworthiness can be more important than aerodynamic efficiency or structural weight. The rigidity and structural strength afforded by the joined-wing geometric arrangement offers the promise of significantly enhanced safety and reduction of fatalities. In another civil application, the use of the joined-wing layout with its inherent rigidity might significantly increase the flutter speed encountered by conventional high-altitude sensor vehicles, such as those used to monitor earth environmental and resource characteristics. These vehicles conventionally have been configured with very high-aspect-ratio wings that can result in undesirably low flutter speeds.

Yet another potential application of the joined-wing concept involves the design of supersonic aircraft configurations with relatively low sonic boom levels. The intensity of sonic booms is a strong function of vehicle length, and a joined-wing configuration has a greater "effective length" because of the elevated rear wing junction to the vertical fin. Additionally, current concepts for engine nacelles that reduce takeoff and landing noise have rather long silencers extending aft from the wing trailing edge. These nacelles provide a natural location for the wing-tail joint and may provide some bending or torsional moment relief to the wing.

Challenges and Barriers

The joined-wing concept has faced many challenges and barriers from technical considerations in the areas of structures, aerodynamics, and stability and control. NASA, industry, DoD, and universities have addressed many of these issues with analytical and experimental studies.

The greatest structural benefit of the joined-wing configuration occurs when the front- and rear-wing joints are all fixed cantilever connections. Unfortunately, this arrangement results in a structure that is more difficult to analyze (referred to as statically indeterminate) and can result in counterintuitive characteristics. Another major challenge results from the fact that typical joined-wing configurations are designed with the root of the rear wing above the front wing, so the rear wing is loaded in combined bending and compression. The rear wing, which acts as a compression strut, must be designed with enough stiffness not to buckle. Typical low-fidelity structural weight-estimation tools used during early conceptual and preliminary design are not capable of determining realistic loads, moments, stresses, or weight of a joined-wing structure.

The necessity for more sophisticated structural design methods and capability—early in the vehicle conceptual development—is a powerful economic barrier for companies that might otherwise consider a joined-wing configuration. Because of the lack of detailed design experience with such an unconventional structure, the potential advantage of lower structural weight is regarded as a significant technical risk. Companies have been reluctant to make the investment in design tools and training, and they have neither sufficient funding nor schedule margin to allow longer design evolution/iteration to occur in the detailed design. Any nontraditional structural arrangement will encounter similar barriers when the groups performing detailed structural design within the companies are faced with such a radical departure from established methods and procedures.

In the area of aerodynamics, the most dominant challenge to the joined-wing configuration is the minimization or elimination of separated flow at wing and fuselage junctures and aerodynamic component interference effects across the flight envelope, including cruise, takeoff, and landing. With the added component juncture formed by the wing joint, the joined wing provides added challenges to the aerodynamicist. If the configuration experiences unacceptable juncture-flow characteristics (particularly at high subsonic cruise conditions), overall drag levels may be significantly higher than those of conventional transports.

The joined-wing configuration may also exhibit unique challenges in the critical area of propulsion integration. For some applications, engine nacelles may have to be located on lateral stubs on the fuselage near the configuration's center. Aerodynamic interference effects from the forward wing/fuselage components (particularly for high angles of attack or sideslip) may result in unsatisfactory engine inlet flow characteristics or inefficient propulsion performance at cruise. In addition, the engine efflux may cause interference effects on the aft wing or vertical tail.

Finally, inadequate design of the rear wing or vertical tail juncture may cause flow separation, which can result in a significant increase in drag and a large impact on stability and control. In addition, the overall consideration of trimmed lift for operational conditions across the envelope must be analyzed and the vehicle configured to ensure satisfactory characteristics. The relatively short moment arm of the aft wing control surfaces of most diamond-wing type joined wing aircraft aggravates the classical problem of longitudinal trim or lift trades at low-speed landing conditions. For a stable aircraft, the short-coupled rear wing may have to produce excessive download to trim the pitching moments experienced during various phases of flight, resulting in a significant loss of lift. Other approaches to joined-wing configurations, such as an auxiliary aft-mounted tail surface, might be employed to alleviate unacceptable levels of lift loss due to trim.

Many joined-wing configuration wind-tunnel models have exhibited a nosedown ("pitch down") characteristic at moderate angles of attack below wing stall, thereby limiting the maximum lift of the configuration to less than desirable values. The phenomenon is attributed to stalling of the front wing, resulting in loss of lift on the forward wing and a reduction in downwash onto the rear wing, which increases the pitch-down contribution of the rear wing. Although this effect is favorable as a natural stall-prevention mechanism for the airplane, it can severely limit the magnitude of attainable lift. Thus, longitudinal stability of the joined-wing design requires a careful integration of individual wing stall characteristics.

Operational challenges specific to joined-wing configurations are relatively unknown because of the lack of applications and flight experiences with aircraft other than personal sport vehicles. Issues such as icing characteristics, detailed handling quality assessments, and other real world issues have not been assessed at the current time.

Langley Activities

NASA's participation in research on joined-wing aircraft has involved Langley Research Center, Ames Research Center, and Dryden Flight Research Center. The following discussion highlights critical activities at the participating Centers, with an emphasis on activities that have occurred at Langley. More detailed information on activities at Ames and Dryden is provided in references listed in the bibliography.

Exploratory Study of Aerial Applications Aircraft

In 1979, Julian Wolkovitch approached Joseph L. Johnson, Assistant Head of the Dynamic Stability Branch, with a request for a cooperative wind-tunnel test of an advanced joined-wing general aviation airplane designed for aerial applications. The configuration, which the legendary Elbert L. (Burt) Rutan had designed, featured a tractor-propeller-driven, joined-wing layout with the rear wing joined at the mid-span location of the forward wing, which had winglets. The pilot was located in the 18-percent thick vertical tail of the vehicle. Wolkovitch had crash resistance in mind as a primary design objective when he first pursued the joined wing as a sport glider in 1974, and he and Rutan believed that the proposed agricultural plane design would offer significant safety improvement over conventional designs.

Because of its interest in providing data for advanced configurations, NASA fabricated a scale model of the design and conducted a cooperative test in a 12-foot low-speed subsonic tunnel at Langley. Lead engineer for Langley during the exploratory tests was E. Richard White. The tests were regarded as exploratory and limited because of the low Reynolds number of the test conditions, and all participants had expected premature flow separation on the wings and junctures due to lack of simulated flight conditions. Nonetheless, it was felt that any aerodynamic data on stability and control characteristics of this remarkable configuration would be of great interest to the engineering community.

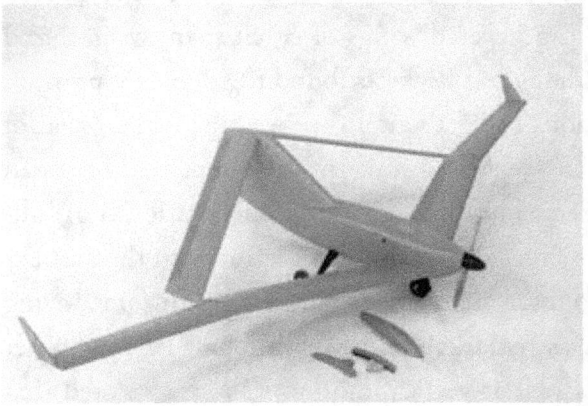 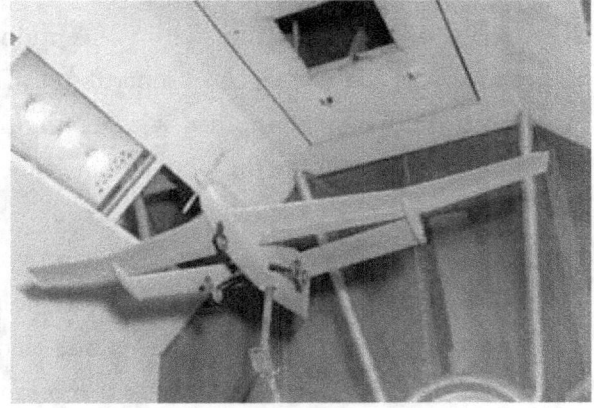

Advanced agricultural airplane model tested at Langley.

The results of the test verified the expected flow separation regions, especially at the wing-joint locations at moderate and high angles of attack. Of more concern, however, was the impact of flow separation at the rear wing-vertical tail juncture, which resulted in a loss of directional stability contributed by the thick, short-span vertical tail. Although the design was not subsequently pursued for a commercial product, this early test identified a number of performance, stability, and control issues that have resurfaced as challenges throughout later studies of joined-wing vehicles.

High Altitude Vehicle Flutter

Aircraft flying above 100,000 ft must operate near the drag-divergence Mach number while generating high lift coefficients. For such flight conditions, thin supercritical airfoils are desirable. Cantilever wings employing these thin airfoils tend to be heavy or excessively flexible. For joined wings, however, reducing thickness-chord ratio gives only small penalties in structural weight and rigidity. The net effect is that the joined wing can potentially increase the altitude and payload capabilities of very high altitude aircraft. A key consideration of this benefit is the joined wing's impact on potentially catastrophic flutter.

In 1984, Langley's Michael H. Durham and Rodney H. Ricketts of the Aeroelasticity Branch teamed for an analytical and experimental study of the joined-wing configuration's benefits on flutter characteristics of very high-aspect-ratio (21.6 and 42) vehicles. In the investigation, they studied two types of joined-wing models in the Langley TDT at Mach numbers of 0.4 and 0.6. Durham and Ricketts investigated semispan wall-mounted models of conventional and joined-wing designs, as well as full-span flutter models, on the unique free flying cable-mount system used for flutter testing in the TDT. Results obtained with the sidewall-mounted models compared characteristics of joined wings with conventional cantilevered wings of equal span, weight, and projected area. For each Mach number tested, Durham and Ricketts found the dynamic pressure for onset of flutter for the joined-wing configurations to be about 1.6 times higher than that of cantilever wings, verifying the joined wing's expected benefits. Testing the cable-mounted full-span models provided more excitement and some unexpected results. The lower aspect-ratio (21.5) full-span joined-wing model experienced an aerodynamic instability and was destroyed in the ensuing out-of-control motions. In addition, the cable-mounted high-aspect-ratio full-span joined-wing model exhibited a symmetric flutter mode that was remarkably unconventional. In this flutter mode, the model displayed fore-and-aft motion as well as vertical motion. Observers noted that the model appeared to be performing a "butterfly stroke" similar to a swimmer. Durham studied the motion and developed an approach for analysis that correlated well with the experimental results for both flutter speed and mode. He subsequently disseminated the investigation's results at specialists meetings.

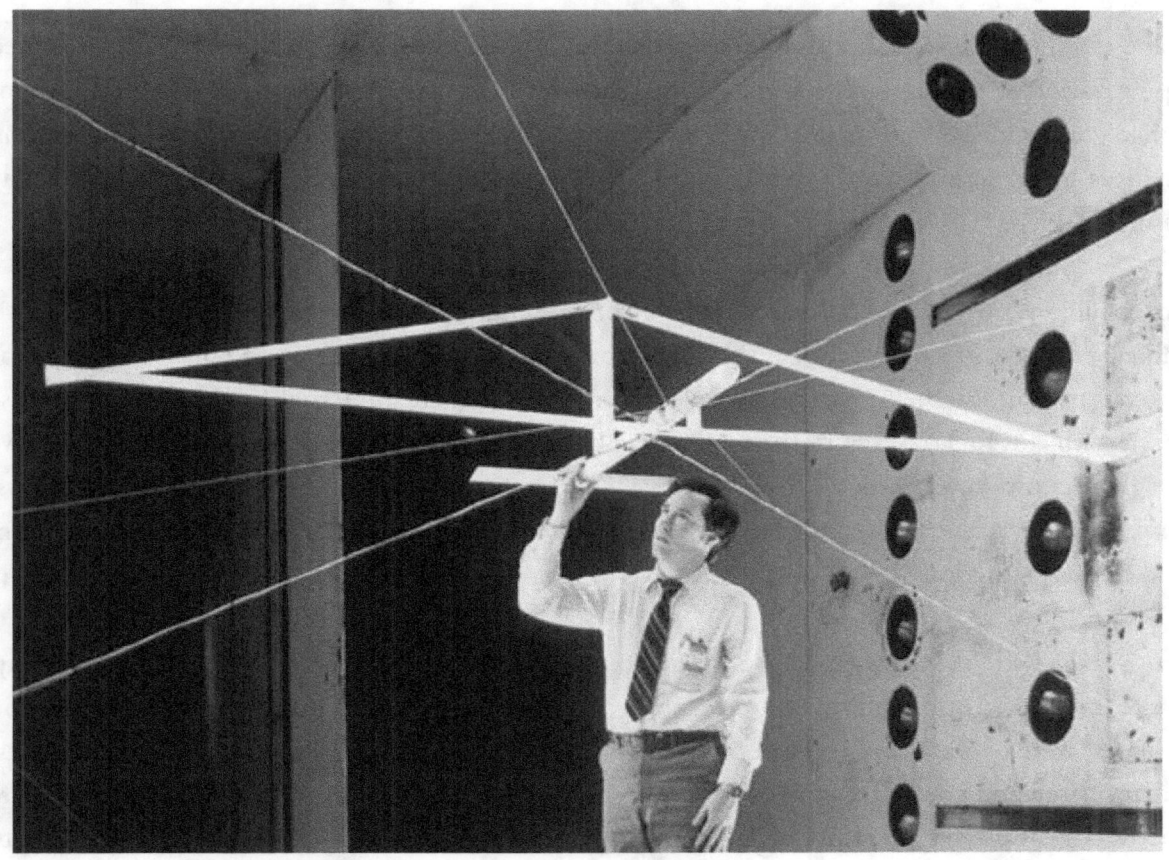

Researcher Mike Durham with flutter model of joined-wing high-aspect-ratio configuration.

Joined-Wing Studies at NASA Ames Research Center

While Langley was engaged in assessing the benefits of joined-wing vehicles for civil and military applications, similar efforts were underway at NASA Ames Research Center, including investigations of civil transport applications. Although not directly coordinated with Langley, this work mentions these studies for completeness and perspective on the scope of studies at Ames.

As researchers at Ames began studying the joined-wing concept, they recognized that more sophisticated design and analysis tools would be required to properly assess performance trends that are dependent on structural weight and trimmed-drag prediction. In 1986, work began on a combined structural and aerodynamic analysis code that would be appropriate for conceptual design. Stephen C. Smith at NASA Ames and Ilan M. Kroo and John W. Gallman at Stanford University collaborated on this work. They based the aerodynamic model on a vortex-lattice representation of the configuration and included a coupled optimization routine to find optimum twist distribution and tail incidence to minimize induced drag and achieve pitch trim with fixed static stability. The structural model was based on a finite beam-element method with a coupled optimization to determine the minimum structural weight with maximum-stress and minimum-gauge constraints. These tools allowed parametric studies of the effects of various configuration changes on structural weight and cruise drag. Smith, Kroo, and Gallman published the study results in 1987.

Ames subsequently hired Gallman, and he incorporated these models into a full mission-synthesis model that performed a complete vehicle optimization subject to real world constraints, such as takeoff and landing field length, engine-out climb requirements, internal fuel volume and cruise range with IFR fuel reserves, static stability and trim over allowable center of gravity range, positive weight on nose wheel, structural loads and weights in compliance with FAR 25, and many others. Improvements to the analysis models included maximum trimmed lift capability, buckling margin, and flutter prediction.

In parallel with the conceptual design efforts, Ames supported Julian Wolkovitch's company, ACA Industries, in designing and developing a manned flight demonstrator aircraft to develop a representative joined-wing structural arrangement and demonstrate satisfactory flying qualities. SBIR phase I and phase II awards funded this effort. A wind-tunnel test was conducted to measure the aerodynamic characteristics of a joined-wing research aircraft (JWRA), which was designed to use the fuselage and engines of the existing NASA AD-1 research aircraft. The AD-1 had completed a very successful piloted flight program to demonstrate oblique-wing technology. The JWRA was designed to have removable outer-wing panels to represent three different configurations with the interwing joint at different fractions of the wing span. A 1/6-scale model of all three configurations of the JWRA was tested in the Ames 12-Foot Pressure Tunnel to measure aerodynamic performance, stability, and control characteristics. These test results indicate that the JWRA had very good aerodynamic performance and acceptable stability and control throughout its flight envelope. Although the wind-tunnel results showed satisfactory performance, stability, and control, with no adverse interference drag using well-designed fairings at the wing-tail joint, the funds available for research were exhausted before the flight demonstrator vehicle could be fabricated.

Ames design study results of commercial civil transports indicated that, for the specific mission application chosen, the joined wing had a few percent higher direct operating cost. However, they also showed that several adverse characteristics of the design could probably be mitigated with further design. Chief among these was the larger wing size required because of poor trimmed maximum lift, a consequence of high tail downloads required to trim. Alternative high-lift systems that produce less pitching moment and longer fuselage layouts may have improved the trimmed lift enough to make the joined wing competitive with conventional configurations. At the same

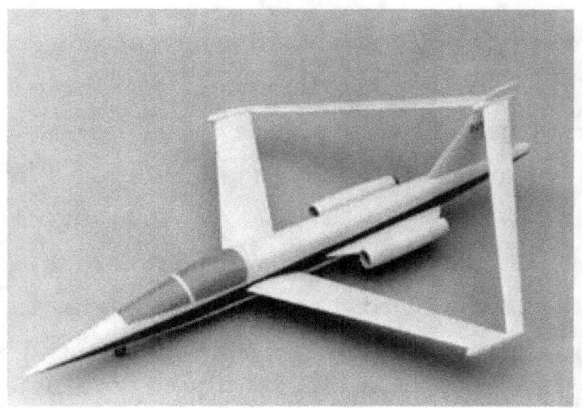

Model of the Ames Joined-Wing Research Aircraft concept shows two of the three wing arrangements.

The 1/6-scale model of the Joined-Wing Research Aircraft in the Ames 12-Foot Pressure Tunnel.

time, tailored composite tail structures may have increased stiffness and buckling margin with less weight penalty, again improving the joined-wing performance relative to the conventional airplane arrangement. Alternatively, exploiting the wing strut bracing's structural benefit while retaining the efficient trimming capability of a conventional horizontal tail may be an even more efficient configuration. Each potential design fix was regarded as beyond the scope of the Ames studies, which were concluded in 1993. Such approaches, however, could potentially make the joined wing attractive and successful. The Ames experience shows that the joined wing, more than most other vehicle concepts, requires a well-established multidisciplinary design approach throughout the vehicle development process, from conceptual and preliminary design through detailed design.

Participation in Boeing's EX Program

The safety of U.S. Navy carrier battle groups depends strongly on an early warning of incoming aircraft and missiles launched by beyond-the-horizon enemies. For over 30 years the responsibility for providing early warning has been assigned to the Navy E-2C Hawkeye aircraft, which uses a 24-ft rotodome atop the vehicle to enclose its radar antenna. Anticipating the need for a more capable replacement surveillance aircraft as the E-2C reaches the end of its lifetime in the fleet, the Boeing Defense and Space Group's Military Airplane Division embarked on studies of a radical new joined-wing surveillance aircraft design in response to a new Navy program known as the Electronics Experimental (EX) Program in 1990. The EX Program achieved Milestone 0 definition in 1992, but the Navy did not pursue the program because of defense funding reductions.

The Boeing EX aircraft concept incorporated advanced active-aperture radar arrays in each joined-wing segment to create an ideal arrangement for the radar arrays and a more aerodynamically effective design than the conventional E-2C. The joined-wing EX concept was only about 80 percent the size of the larger E-2C, yet it incorporated four 31.5-ft wing-mounted radar apertures, compared with the single 22-ft aperture carried by the E-2C.

In the early 1990s, the Navy E-2C Program Office approached NASA Langley researchers for discussions of a cooperative study of the EX configuration in the Langley 16-Foot Transonic Tunnel. In accordance with NASA's mission to explore advanced configurations of interest, Division Chief William P. Henderson and Branch Head Bobby L. Berrier agreed to Langley participation in the project, and researchers Richard J. Re, Jeffery A. Yetter, and Timmy T. Kariya served as key Langley engineers on the Boeing-NASA team. In July 1993, the team tested a model of the EX design to evaluate longitudinal and lateral aerodynamic characteristics and the effectiveness of various control surfaces. Measurements were also made to determine the effects of the wings and fuselage on engine inlet fan-face total pressure distortions at angles of attack and sideslip. The test program's results showed that the initial EX configuration exhibited several regions of separated flow for all values of Mach number investigated, including cruise conditions.

Artist's concept of the Boeing EX joined-wing aircraft.

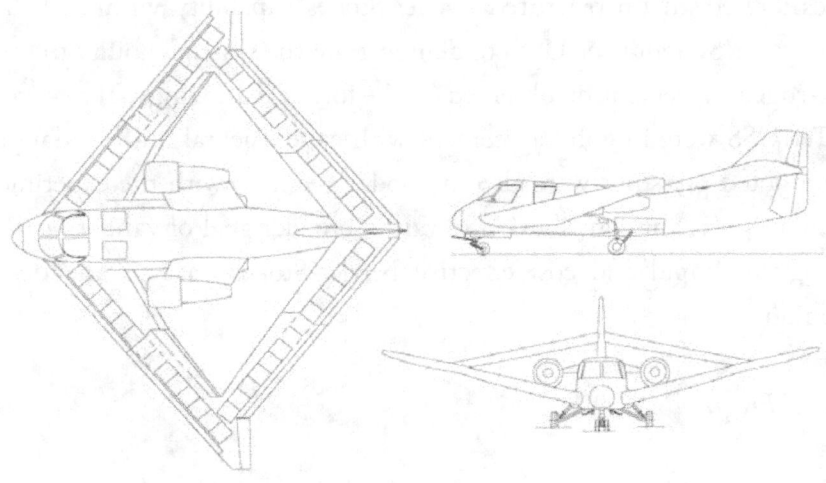

Three-view sketch of the Boeing EX configuration.

Guided by the results of this first tunnel entry, Boeing modified the configuration's wings, and a second entry in the tunnel occurred during October 1998. E. Ann Bare led Langley's participation and was assisted by Wesley L. Goodman. Early test results indicated that undesirable flow separation still existed on the modified configuration. Langley's Steven E. Krist and Boeing provided additional analysis and guidance by conducting CFD analyses. One of the configuration's more challenging flow separation areas was the juncture of the aft-wing root and the vertical tail. Aerodynamic drag caused by massive separation in this area resulted in large performance penalties for the configuration. Responding in an extremely timely fashion, Krist quickly analyzed the flow field at the critical junction area using the OVERFLOW code and designed a leading-edge modification ("bump") for the vertical tail that minimized the separation phenomenon. Technicians quickly fabricated the tail modification for the model and provided quick turn around for testing of the modification. Test results for the revised model showed that the new tail configuration dramatically reduced drag. Krist's valuable contribution to the joint investigation was widely recognized and appreciated by all members of the Boeing-NASA team.

In addition to the pioneering information provided on the aerodynamic characteristics of joined-wing configurations, and the EX in particular, the test entries in the 16-Foot Transonic Tunnel and the interactions of the Langley and Boeing staffs provided the foundation for a follow-up NASA RevCon project to be discussed in a later section.

Other Langley CFD efforts were also directed at the unconventional joined-wing EX configuration operating at transonic, separated-flow conditions. Neal T. Frink, Shahyar Pirzadeh, and Paresh Parikh calibrated an unstructured Navier-Stokes capability within NASA's Tetrahedral Unstructured Software System (TetrUSS) to demonstrate the system's ability to predict the shock-induced trailing-edge flow separation observed on the fore and aft wings. The surface-flow patterns obtained with TetrUSS were in good agreement with experimental oil-flow data obtained in the tunnel tests. Computed pressures were also in good agreement with the experimental data. This study represented a significant contribution toward a broader goal of validating a next-generation CFD methodology for rapid and cost effective Navier-Stokes analysis and design of complex aerodynamic configurations.

The NASA RevCon Program

As previously discussed within the topic of the blended wing body concept, in 1997 Darrel R. Tenney, Director of the Airframe Systems Program Office, and Joseph R. Chambers, Chief of the Aeronautics Systems Analysis Division, formulated and proposed a new research program based on the selection of precompetitive advanced configurations that would be designed, evaluated, fabricated, and test flown using remotely piloted vehicle technology at Dryden. The program, known as RevCon, would be based on a 4-year life cycle of support for concepts selected. Initial reactions to the proposed program from NASA Headquarters and Dryden were favorable, and following intercenter discussions with the additional participation of Ames and Glenn, a formal NASA RevCon Program was initiated in 2000 that was to be led by Dryden. Robert E. McKinley led the RevCon activities at Langley under the RACRSS element of Airframe Systems.

In June 2000, NASA's Office of AeroSpace Technology selected nine aeronautical concepts in its initial RevCon Program, including a teamed effort by Langley (team lead) with partners from Dryden, Boeing (Phantom Works), Naval Air Systems Command (NAVAIR), and AFRL for the design, development, fabrication, and flight testing of a joined-wing integrated structures demonstrator. The Air Force involvement in the program came about due to rapidly growing interest in surveillance unmanned air vehicles (UAVs). The project would receive approximately

Jeff Yetter inspects the EX model in the Langley 16-Foot Transonic Tunnel in 1993.

$300,000 from NASA for phase I research, and the industry-DoD partners were expected to commit similar levels of funding. Objectives of the 4-plus-year project were to (1) enable the integration of large radar apertures into smaller aircraft for improved detection range and resolution, (2) reduce drag and weight for improved aircraft speed and endurance, and (3) reduce system costs. Flight experiments would be conducted with a full-scale piloted research aircraft using a modified U.S. Navy S-3 Viking fuselage with new joined wings.

In an 8-month phase I activity, the team explored demonstration alternatives, conducted risk-reduction experiments and analyses, and planned phase II details and costs. The primary research and technology objectives of the Joined-Wing Flight Demonstrator (JWFD) Project fell into three broad categories: (1) aerodynamics, flight controls and flight characteristics; (2) multifunctional structures, and (3) wing-integrated RF apertures. During phase II, the demonstrator aircraft would be fabricated and flight tested. Within the RevCon Program, flight testing would focus on aircraft performance, flying qualities, flight-envelope expansion, and validation of structural behavior. Following the RevCon phase II flight test activities at Dryden, plans included U.S. Navy flight testing at Patuxent River, Maryland, to evaluate carrier suitability and the radar aperture performance.

Langley's Program Manager for the teamed phase I effort was Jeff Yetter, manager of the Advances through Cooperative Efforts (ACE) Program of the Aerospace Vehicles Systems Technology Office. The research Integrated Product Team (IPT) leaders at Langley were Phillip B. Bogert (structures), Steve Krist (aerodynamics), and James W. Johnson (electromagnetics). The phase I and phase

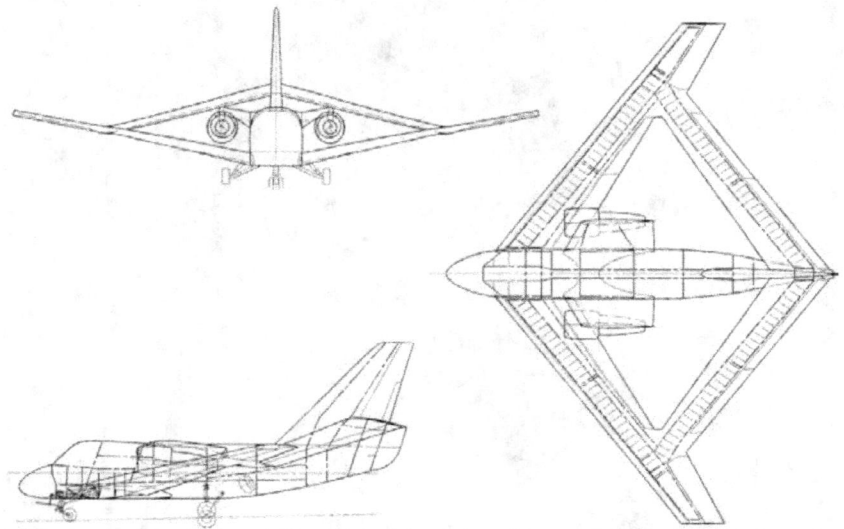

Three-view sketch of the Joined-Wing Flight Demonstrator configuration.

II plans identified the use of several unique Langley facilities, including tentative entries in the Langley 16-Foot Transonic Tunnel, the Langley 14- by 22-Foot Low Speed Tunnel, the Langley 20-Foot Vertical Spin Tunnel, the Langley Electromagnetics Test Facilities, the Langley Structures and Materials Laboratory, and the Langley Nondestructive Test Laboratory.

The project would make extensive use of existing Navy flight vehicle hardware and new joined-wing hardware. The 35,000-pound (takeoff gross weight) JWFD would be assembled from new joined wings adapted to an existing Navy S-3 aircraft fuselage. The forward and aft wings of the JWFD would contain integrated phased array antennas. The forward fuselage, aft fuselage, and vertical tail would be modified to accept the new wings. The S-3's existing wing would be terminated outboard of the fuselage sides and new wing stubs would be added to accommodate the pylon/engine installations. The engines would be TF-34 turbofans (existing S-3 engines) provided from the Navy inventory.

The phase I aerodynamic design of the JWFD expanded upon knowledge gained from Boeing-Navy-NASA studies of the earlier Boeing EX configuration. The JWFD's forward and aft wings were essentially identical to those for the EX, with supercritical airfoil sections and slightly different sweep and dihedral angles. The JWFD's wing span was increased from that of the EX in order to provide adequate aileron area for desired roll control authority. This change, together with the minor sweep change, increased total span from 63 ft on the EX to 72 ft on the JWFD. The JWFD planform, like that of the EX, permitted the integration of 31.5-ft conformal apertures into each of the four wings.

Model of the Joined-Wing Flight Demonstrator in the Langley 14- by 22-Foot Tunnel.

The JWFD was designed to have 13 flight control surfaces consisting of inboard and outboard trailing-edge flaps on the forward wing, inboard trailing-edge flaps on the aft wing, upper and lower split trailing-edge flaps on the outer-aft wing, trailing-edge flaps on the wing tips, and the rudder. Leading-edge flaps were provided on the forward wing for high lift during takeoff and landing. This robust suite of flight controls made the JWFD an excellent platform for further control system development and optimization of handling qualities for joined-wing aircraft.

The project started its phase I risk reduction and phase II planning activities on September 1, 2000, rapidly advancing the definition of the JWFD. In aerodynamics activities, an exploratory low-speed test of the JWFD configuration was immediately formulated, a model prepared, and tests conducted in the Langley 14- by 22-Foot Tunnel during February 2001 to determine performance of the high-lift system, static longitudinal and lateral-directional stability, control effectiveness, inlet flow qualities, and ground effects. Langley's JWFD test leaders were Richard J. Re and Harry L. Morgan. The configurations tested, including takeoff, approach, landing, and patrol configurations, determined the effects of various control surface deflections (individually as well as in combinations) on stability and control. The leading- and trailing-edge flaps of the forward wing were evaluated for high-lift capability, as were the wing-tip ailerons. On the aft wing, the effects of elevator, outboard flap, and speed-brake deflections were investigated. Data were also obtained for maximum rudder deflection. Runs were conducted with, and without, the landing gear and gear doors extended, and a limited number of flow visualization runs were conducted using tufts mounted on the wing tip and outboard portions of the forward and aft wings. Test results showed that the JWFD configuration had adequate stability and control characteristics for use as a flight demonstrator.

One potential aerodynamic issue of the JWFD that concerned the research team was the possible existence of significant jet effects on the rear wing and vertical tail. In a head-on view, the engine nacelles of the JWFD were about evenly placed above the forward wing and below the aft wing, mounted close in to the fuselage on stubs. The possibility therefore existed that the jet efflux could cause interference effects on the aft wing and tail, particularly at high angles of attack. The AFRL initiated a limited CFD investigation of powered effects using COBALT, a Navier-Stokes flow solver for unstructured grids, but none of the cases involved high angles of attack and the results were inconclusive relative to the suspected critical conditions.

Steve Krist's IPT team conducted CFD analyses with OVERFLOW, a Reynolds Averaged Navier-Stokes code for overset structured grids. CFD analyses of the joined-wing configuration were performed across the operating speed range, flow characteristics for the wing-body, wing vertical tail and fore/aft wing junctures were examined, and estimates of air loads were generated. Optimal engine inlet orientations were defined for good inflow to the fan face, and drag-rise characteristics were calculated to verify the aerodynamic efficiency of the joined wing design. Flow separation at the aft wing-vertical tail juncture of the JWFD was considerably improved over that discussed for the earlier EX configuration. This improvement resulted from the juncture of the aft-wing leading edge with the vertical tail being much further aft on the JWFD than on the EX. Flow visualizations at Mach numbers ranging from 0.4 to 0.81 indicated that extensive separation on the lower portion of the vertical tail appeared as early as Mach 0.6 on the EX, but not until Mach 0.78 on the JWFD.

Computational results indicated that the two areas on the JWFD providing the greatest potential for reduction in drag were the forward-wing/aft-wing juncture and the aft-wing/vertical tail juncture. Flow separation at the forward wing/aft wing juncture occurred primarily at high angle-of-attack subsonic conditions where the compressive effect of the aft-wing leading edge resulted in significant spanwise flow on the upper surface of the outboard forward wing and the wing tip. Procedures were developed for redesigning this juncture using OVERDISC, a computational design tool that couples the CDISC inverse design method developed by Richard L. Campbell of Langley with the OVERFLOW flow solver. Initial attempts at CFD designs at Mach 0.45 experienced great difficulty in controlling the local surface shape while meeting geometry constraints in this unconventional juncture.

Problems at the aft wing/vertical tail juncture arose at transonic conditions, resulting in a sharp drag rise at Mach 0.81. On the upper surface, a shock developed at the juncture and strengthened with increasing Mach number. On the lower surface, the separation at the trailing edge migrated forward with increasing Mach number until, at Mach 0.81, the flow on the lower portion of vertical tail separated just behind the aft wing leading edge. Procedures for using OVERDISC to design a fillet for this juncture—to mitigate the upper surface shock—are well developed, having been validated with Steve Krist's studies on the EX configuration. However, there was insufficient time in the JWFD project to fully explore this procedure.

The NASA-Boeing team made substantial progress in the structural definition and design of joined wings. A general structural arrangement for composite joined wings with integral radio frequency apertures was prepared, side-of-body connections for the engine nacelles were defined, and the required structural modifications to the S-3 airframe were identified. The team also generated detailed static and dynamic finite element models, performed loads and stress analysis of critical load cases to verify the wing-fuselage attachment concept, and performed initial structural element sizing. A conceptual design for an innovative fiber optic wing-shape sensing system was

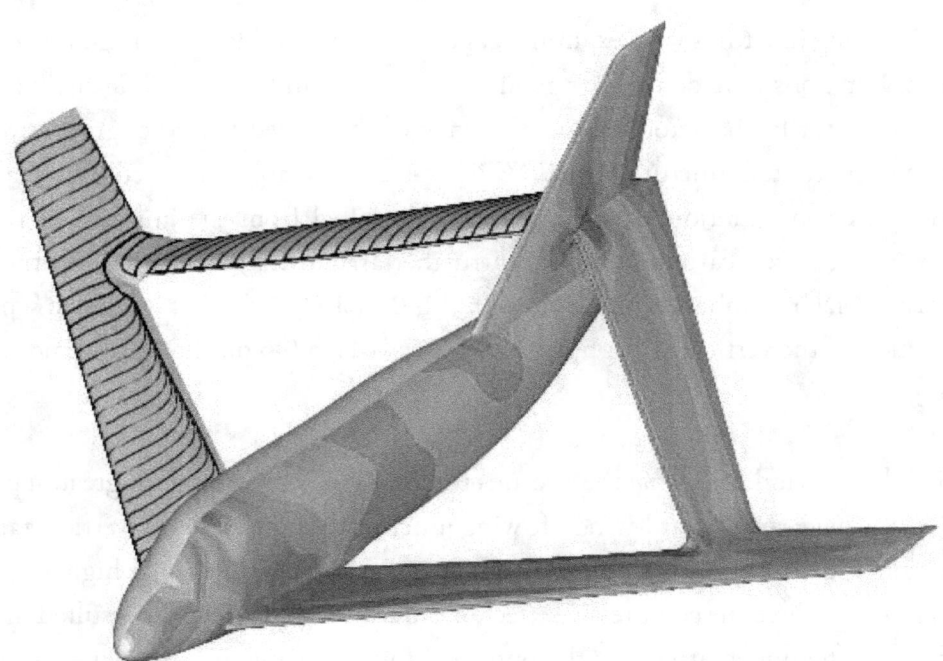

CFD predictions of pressures, streamlines, and flow over the Joined-Wing Flight Demonstrator for a Mach number of 0.7 and an angle of attack of 0 degrees.

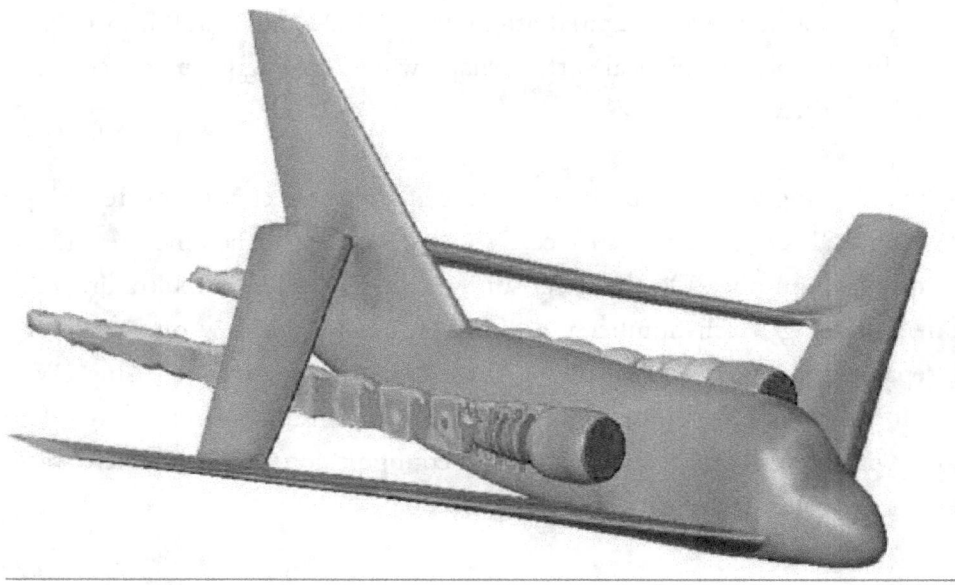

Computational simulation of power-induced flow on the RevCon Joined-Wing Flight Demonstrator.

also developed. The system would have computed in-flight wing deformations from fiber optic measured strains, thereby providing information needed for the phased-array application. The system would also have been a key building block for future in-flight health monitoring systems for other applications.

Boeing and NASA continued efforts to define the analytically redundant fly-by-wire control system of the JWFD, and simulations showed that the flight control system was robust. Boeing used the simulation to assess sensitivity to actuator sizing and rates.

In the area of electromagnetics, single array element models were built and tested, while analytical tools were developed and validated. A full-scale working model of the probe-fed element was built and chamber tested. Conceptual designs were completed for the flight control actuation, high lift, hydraulics, electrical, ECS, and fuel systems. The designs focused on using existing S-3 subsystems and components available from the existing inventory. In addition to component parts, the Navy identified a specific S-3 airplane for use by the JWFD project. Finally, the flight-test team developed a draft test plan that identified all required preflight qualification testing, indicated necessary flight-test instrumentation, and outlined an approach for obtaining airplane flight qualities, low-speed performance and flight-envelope expansion.

Brassboard testing of Joined-Wing Flight Demonstrator array element at Langley

The team submitted the final report on the phase I JWFD study results in April 2001. Unfortunately, funding priorities within the participating government agencies were directed elsewhere following the initiation of the RevCon Program. The Navy and the Air Force were unable to meet their shares of the required funding commitments, and NASA's portion of the funding was redirected to providing a return-to-flight capability for the NASA X-43A (Hyper X) Program following the X-43A accident on June 2, 2001. NASA terminated its RevCon Program on September 30, 2001.

Although the RevCon Program was terminated before phase II could be undertaken, the NASA-DoD-industry team significantly advanced the definition of a joined-wing aircraft system and developed a practical conceptual design for a manned flight demonstrator. Progress was made in a variety of risk reduction areas and the team developed a viable project plan, cost estimates, work breakdown structure, and definition of responsibilities between the partners. Because of this activity, the technical community now has a much better understanding of what it would take to design, build, and fly a joined-wing technology demonstrator of this type.

Status and Outlook

Joined wing aircraft application remains centered on surveillance, providing a means for integration of large apertures into compact aircraft for reduced cost and increased sensor performance. Current funding for concept development is being provided by the Air Force Research Lab (AFRL) as a part of its SensorCraft initiative. A series of contracts have been awarded to Boeing that focus on the viability of the joined wing concept as a sensor platform. The contracts include systems studies for concept refinement and for defining an advanced technology demonstration (ATD) of the concept; a contract for Aero Efficiency Improvements (AEI) that addresses the aerodynamic design of the joined wing sensor platform and the aero-elastic characterization of the joined wing structure; a contract that is part of the Very Affordable Advance Technology Engine (VAATE) program that addresses energy management, including secondary power, electrical power generation and thermal management; and a contract (a cooperative AFRL/Boeing program) that addresses the development of a structurally integrated X-band aperture and a full-scale wing conformal UHF aperture.

The Vortex Flap: Efficiency and Versatility

Concept and Benefits

Highly swept wings or other surfaces exhibit strong vortical flow over their upper surfaces during flight at moderate or high angle-of-attack conditions, such as those associated with takeoff, landing, or strenuous maneuvers. The vortical flow's beneficial influence on the integrated wing aerodynamic behavior results in greater lift for takeoff and maneuvers, better control of the aerodynamic center's location, and relatively similar flow fields over a wide range of angle of attack and Mach number. Many contemporary aircraft, including the Concorde supersonic transport and highly maneuverable fighters such as the F-16 and F/A-18, use vortex flows to enhance aerodynamic behavior through the mechanism of "vortex lift" across the range of operational conditions.

The F-16 (left) and the F/A-18 (right) use vortex lift for improved maneuverability.

Unfortunately, the generation of vortex lift by wing leading-edge flow separation also results in a very undesirable byproduct: a loss of aerodynamic leading-edge thrust (or leading-edge suction) that results in a dramatic increase in drag for a typical highly swept configuration. In contrast, wings of conventional aircraft having lower sweep exhibit leading-edge thrust produced by attached flow over the wing, thereby reducing aerodynamic drag. Rather than producing thrust, the leading-edge force for highly swept wings at high angles of attack is redirected to a position normal to the wing surface where it augments normal force, but no longer has a beneficial impact on drag.

The vortex-flap concept involves the use of specially designed wing leading-edge flaps that modify undesirable leading-edge flow separation behavior. This approach provides the aircraft designer with options to design highly swept wings with geometric features that recover a portion of the lost leading-edge thrust without compromising other aerodynamic characteristics, such as stability and control. Using this concept, the designer can reorient part of the vortex-force vector forward instead of directly normal to the chord plane.

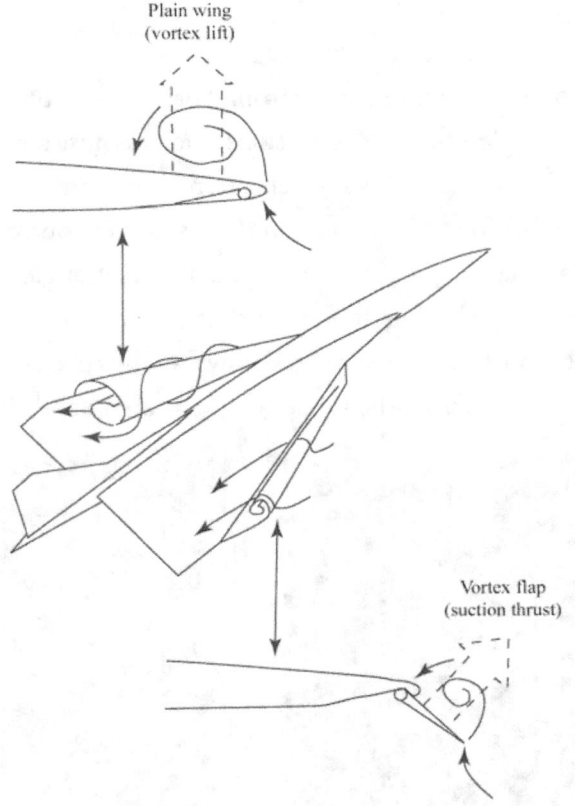

The vortex-flap concept.

The primary mechanism of the vortex flap is depicted in the sketch. Vortical leading-edge flows are depicted for a representative highly swept configuration having a conventional leading edge (left-wing panel) and a specially designed vortex flap that is deflected from underneath the wing leading edge about a pivot point on the lower surface (right-wing panel) at a high angle of attack. As indicated in the sketch, flow separates over the conventional left leading edge, inducing the previously discussed vortex-lift force component normal to the wing surface. On the right wing panel, the vortex flap reduces the vortex core's strength and size because of the leading-edge deflection (camber effect) and leads to a vortex path that is redirected along the leading edge. The result is a suction force that acts on the deflected flap in a forward, drag-reducing direction. Furthermore, the vortex also functions as a rotating fluid cylinder to turn the flow around the leading edge onto the wing upper surface, thereby promoting a smooth transition to attached flow on the wing.

Both civil and military aircraft can use the vortex-flap concept's potential benefits. For example, a supersonic transport or supersonic business jet that uses a wing with high leading-edge sweep for efficient supersonic cruise capability could use the improved L/D ratios provided by the flap for enhanced takeoff performance, thereby permitting the use of lower engine thrust settings and resulting in lower levels of community noise. Military aircraft could use the vortex flap's beneficial effects for significant improvements in maneuvering performance, particularly at transonic conditions where improvements in turning performance during high angle-of-attack maneuvers in close-in combat are extremely significant.

In addition to the vortex-flap concept's performance-enhancing potential, innovative applications of other vortex-flap configurations, such as upper-surface flaps, wing apex flaps, and differentially deflected leading-edge vortex flaps (for aircraft roll control), offer the potential for additional improvement of performance, stability, and control characteristics.

Challenges and Barriers

Before designers can apply this revolutionary concept for vortical flow control to production aircraft, numerous issues need to be addressed and resolved. Perhaps the most constraining barrier to the general application of the vortex flap is its inherent limitation for use on highly swept wings. Some of the other more important challenges and barriers involve aerodynamics, structural design and operational deployment issues, impacts on aircraft flying qualities, weight penalties, maintenance issues, and full-scale flight demonstrations of technology readiness.

Aerodynamic issues that have inhibited the application of vortex-flap technology begin with a fundamental understanding of the flow physics involved in the concept. Factors such as the sensitivity of vortical-flow physics to geometric wing design variables, including the effects of wing-sweep angle and leading-edge radius, must be defined and incorporated in robust design procedures. Relative stability of the vortical-flow pattern produced by the vortex flap must be predictable and consistent across the operational range of candidate aircraft. Thus, the aerodynamic maturity of the vortex flap concept must be ensured from the perspectives of fluid physics and operational applications at full-scale conditions involving large changes in the values of Mach and Reynolds number.

The vortex-flap concept's impact on aircraft stability, control, and handling qualities also demands in-depth research to ensure that undesirable behavior is not encountered in terms of changes in aircraft trim requirements, stability variations, control effectiveness, and aircraft maneuverability. For example, the use of differentially deflected leading-edge vortex flaps for roll control would not be acceptable if large amounts of adverse yawing moments (yawing moments that result in degraded roll response) are encountered. In addition, the potential for degradation of handling qualities because of vortex bursting or vortex instability due to aircraft dynamic motion effects must be evaluated.

Structural design barriers for the vortex flap include providing acceptable levels of complexity and weight for flap hinges, actuation devices, and structural loads. In particular, comparisons of results of performance or penalty trade studies between deflectable vortex flaps and other approaches, such as the use of fixed conical wing leading-edge geometries that do not use leading-edge devices (e.g., design approaches used by the F-106 and F-15), must be resolved in favor of the vortex-flap concept. Other associated challenges for military applications include the impact of leading-edge structural discontinuities and details on aircraft signature characteristics, such as radar cross section.

Langley Activities

Langley Research Center has a rich legacy of expertise in vortex-flow technology. Researchers at Langley had conducted brief studies of low-aspect-ratio delta wings in the 1930s; however, the prediction of extremely poor low-speed flying characteristics and the absence of propulsion systems for high-speed flight resulted in a loss of interest within the Center's research thrusts. During the latter stages of World War II, international research rapidly increased on the beneficial impact of wing sweep on aircraft performance at transonic speeds. By the war's end, renewed efforts of the NACA, industry, and military organizations were initiated and focused on the advantages and problems of swept-back and delta wings. As expected, major challenges ensued at takeoff and landing conditions because of the wing flow separation problems encountered as wing sweep was increased. Langley's research on the aerodynamics of swept and delta wings began to accelerate and intensify, leading in turn to pioneering research on vortical flows.

One interesting example of some early research being conducted at Langley on vortical-flow effects occurred during 1946 when the characteristics of the German Lippisch DM-1 glider were explored in the Langley 30- by 60-Foot (Full-Scale) Tunnel. This delta-wing research aircraft, which was captured by Allied forces and brought to the United States for analysis, had been designed to explore the low-speed handling characteristics of delta configurations. Langley's wind-tunnel testing indicated highly nonlinear lift variations with angle of attack, and studies of surface flows using wool tufts revealed peculiar swirling patterns that were ultimately attributed to the impingement of vortical flow fields on the wing's upper surface. Researchers found that the lift increase exhibited by the airplane at high angles of attack could be attributed to vortical flow actions, and that the lift augmentation could be intensified by modifying the relatively large leading-edge radius with a sharp-edged leading edge. This project was one of the first full-scale aerodynamic studies of delta wings at Langley.

The Lippisch DM-1 glider captured by the Allies (left) and undergoing tests in the Langley 30- by 60-Foot (Full-Scale) Tunnel (right).

Aerodynamic research on swept and delta wings at Langley reached a peak during the 1950s, with extensive efforts conducted in many wind tunnels at speeds from low subsonic conditions to supersonic speeds. These efforts were augmented by analytical studies, flight testing, and vastly increased intellectual knowledge of the flow physics associated with vortical flows. The Center attained international recognition for its expertise in this area, and when the Nation turned its attention to supersonic civil and military aircraft in the late 1950s, Langley was poised to make valuable contributions in the design and application of vortex flows.

Langley's participation in the U.S. SST Program of the 1960s and the NASA SCR Program in the 1970s provided additional opportunities to optimize highly swept configurations and advance the state of the art of vortex-flow technology.

In the late 1970s, the growing lethality of surface-to-air missile systems and the danger of deep-strike mission requirements led to intense interest in the U.S. Air Force for the development of supersonic cruise ("supercruise") fighter configurations. The Air Force awarded several industry contracts for studies of supercruise fighter designs. Stimulated by these contracts and the obvious application of highly swept configurations to the mission requirements, industry interacted with the Langley staff to share in the expertise and experiences gained by NASA with highly swept wing designs during the civil supersonic programs. Langley's staff had developed a research program known as the Supersonic Cruise Integrated Fighter (SCIF) Program under the leadership of Roy V. Harris, Jr., to extend its technology to this class of military aircraft. Langley researchers designed and tested several in-house supercruiser fighters across the speed ranges in Langley facilities. The objectives of SCIF were to focus in-house Langley aerodynamic and flight dynamic research toward feasible configurations for supercruiser applications and to provide coordinated activities with

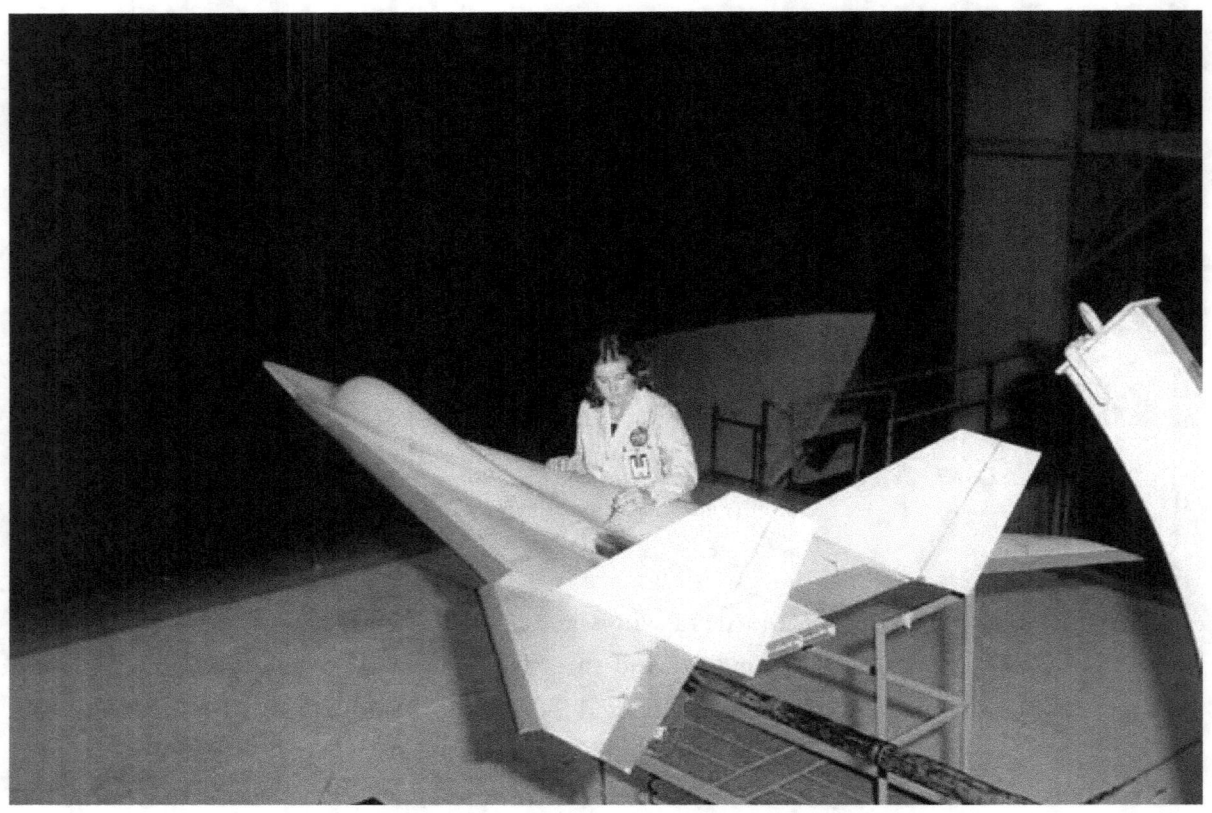

Model of a Langley-designed supersonic-cruise fighter concept (SCIF-IV) in the Langley Full-Scale Tunnel in 1977.

industry teams competing for leadership in supercruiser technology. Subsequent to the initiation of its SCIF program, Langley joined several industry partners in cooperative, nonproprietary studies of supercruiser configurations.

One of the earliest meetings to promote a cooperative supersonic wing design occurred in March 1977 when General Dynamics (now Lockheed Martin) met with Langley researchers to discuss a joint design effort involving several advanced supersonic wing candidates to be designed with NASA and tested in the supersonic Langley Unitary Plan Wind Tunnel and the transonic Langley High-Speed 7- by 10-Foot Tunnel. As part of the effort, General Dynamics assigned two engineers in residence at Langley for 4 months to interact in wing design methodology. Tests of new wing designs in 1978 indicated a supersonic performance improvement of about 30 percent compared with the basic F-16. At subsonic speeds, the modified configurations achieved the same performance as the F-16. Encouraged by these positive results, General Dynamics had committed to a Supersonic Cruise and Maneuver Prototype (SCAMP) concept that used a highly swept "cranked" (double-delta) wing planform for supersonic cruise efficiency. Refinement of this SCAMP concept later led to the development of the F-16XL prototypes by General Dynamics.

During the development of the final SCAMP configuration, several cooperative projects used the configuration as a focus. A wide range of topics was studied, including supersonic store carriage concepts, low-speed stability and control of highly swept configurations, and spin characteristics. One highlight of the 1978 research efforts was a study to provide transonic maneuvering lift at low drag. The research efforts focused on concepts to alter the drag produced by forming leading-edge vortices. Edward C. Polhamus led Langley's vortex research program, and his research group was within the Transonic Aerodynamics Division led by Percy J. (Bud) Bobbitt. Polhamus' group had gained industry's respect and close working relationships by making several significant contributions in cooperative programs as well as in specific aircraft development programs, such as the F-16 and the F/A-18. In a keynote activity, John E. Lamar, James F. Campbell, and their associates joined in a cooperative study with General Dynamics.

During the Langley tests, the NASA-General Dynamics team focused on wing design requirements for a 4-g transonic maneuver with a highly swept wing. Lamar conducted wind-tunnel and computational analyses to define the "optimum" camber and shape for such a wing, but his experiences with vortex flows suggested that a simpler, more versatile solution might be provided by vortex-control concepts. In exploratory testing, the team found that certain combinations of deflected full-span leading- and trailing-edge flaps on a planar (no camber) wing produced almost the same drag improvements at transonic speeds as a specially designed and transonically cambered wing. This early application of vortex-flap principles also produced nearly the same supersonic L/D as a supersonic designed wing (also better than the F-16), a subsonic cruise L/D nearly as good as the value for the F-16 (and better than the supersonic design), and transonic maneuver L/D was midway between that of the F-16 and the fixed supersonic wing. Results obtained with these simple flaps were very attractive from a practical design and fabrication standpoint and stimulated numerous other NASA studies. In-house and NASA-contracted projects included efforts that were focused on developing and validating the design methodology for the vortex-flap concept, as well as exploratory assessments of other innovative applications of vortex-control concepts using deflected flaps.

Neal T. Frink of Langley and his associates conducted extensive pioneering wind-tunnel tests to evaluate the effects of wing-sweep angle and other geometric characteristics on vortex-flap effectiveness. Frink's study provided a matrix of performance information for delta wings having sweep angles from 50 to 74 degrees with constant-chord vortex flaps and formed the key basis for an approach to the design process. Frink initiated and pursued complementary theoretical studies that led the way for predicting overall forces and moments as well as detailed pressures for vortex-flap configurations. His efforts culminated in development of a leading-edge vortex-flap design procedure in 1982.

Camber study model used to develop optimum camber.

Meanwhile, other NASA researchers and their industry peers pursued innovative applications of vortex-control technology based on lessons learned with the vortex-flap concept. NASA contractor Dhanvada M. Rao (initially of Old Dominion University and later ViGYAN Research Associates, Inc.) was particularly active in vortex-flap research. Rao demonstrated that reducing inboard length improved the flap's efficiency and that shaping the flap along the span improved flap efficiency and vortex formation. Rao and an independent team led by W. Elliott Schoonover, Jr., of Langley and W. E. Ohlson of Boeing showed that increasing the flap size delayed inboard movement of the vortex and reduced drag. Additional contributions by Rao included the use of flap segmentation to reduce flap area while achieving the same L/D as without segmentation. He also was the first to explore using vortex flap deflections on individual wing panels to produce roll control.

In 1981, Langley researchers Long P. Yip and Daniel G. Murri conducted studies of the effects of vortex flaps on the low-speed stability and control characteristics of generic arrow-wing configurations in a 12-ft low-speed tunnel at Langley. Although improved lateral stability and

L/D were obtained in the tests, an unacceptable nose-up pitching moment was caused by the flaps. The researchers investigated geometric modification impacts on the vortex-flap configuration, including the flap's spanwise length and the leading-edge geometry. A modified flap concept, which included a deflected "tab" on its leading edge, was found to alleviate the pitching-moment problem, and the flap configuration was then applied to SCAMP configuration models during the aircraft development program. Yip and Murri installed the tabbed vortex flap on a 0.18-scale free-flight model of the SCAMP (which had by then transformed into the F-16XL prototype) and conducted free-flight tests in the Langley 30- by 60-Foot (Full-Scale) Tunnel in 1982. Results indicated that the flap's performance benefits could be obtained with no degradation in flying characteristics or pitch problems.

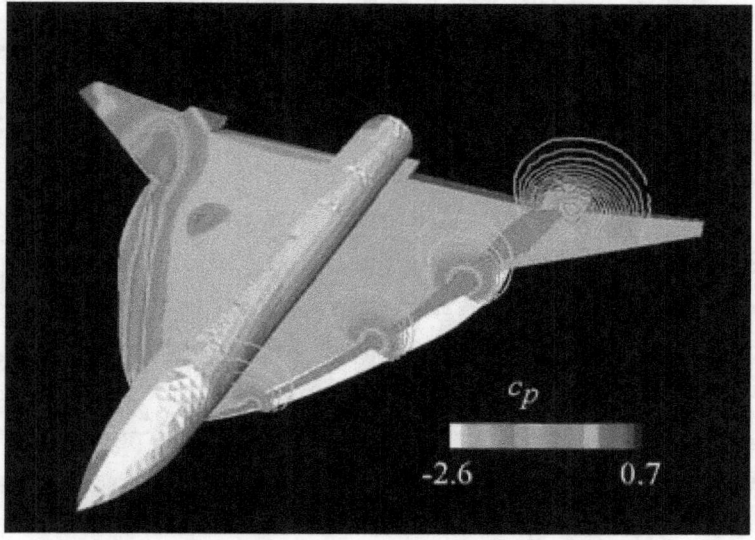

*Computational fluid dynamics study of vortex flap on a representative
high-speed civil transport.*

The vortex-flap concept's civil applications have centered on supersonic transports and supersonic business jets. As part of the NASA SCAR technology program, Paul L. Coe led several wind-tunnel studies of vortex flap effects on aerodynamic performance, stability, and control of representative supersonic transport designs. Coe also contributed vortex-flap studies during the NASA High-Speed Research Program, which focused on providing improved L/D for take-off operations of supersonic transports. During the program, improved low-speed aerodynamic performance was a major research focus, and the research team evaluated vortex-flap configurations in several Langley tunnels, including the 30- by 60-Foot Tunnel and the 14- by 22-Foot Tunnel. Kenneth M. Jones, Kevin Kjerstad, and Victor Lessard conducted computational studies of the aerodynamic characteristics of attached-flow leading-edge flaps and vortex-flap concepts at subsonic takeoff and landing conditions. Using the USM3D computer code developed at Langley, they obtained results that accurately predicted the primary vortex's reattachment line in good agreement with experimental flow visualization. Forces, moments, and surface pressures compared well with the experimental data.

Flight Research

F-106B

By 1983, research on the vortex-flap concept by Langley and its partners had progressed to the point that the next major step in technology maturation was required. Subscale models of generic aircraft configurations with vortex-flaps had been extensively evaluated in wind-tunnel and analytical studies; however, reliable extrapolation of model results to full-scale conditions and evaluations of potential effects of the concept on aircraft handling qualities were required. Following a review of vortex-flap technology progress, a joint NASA-AFWAL steering panel recommended a feasibility study for conducting a full-scale flight experiment using either an F-106, F-16XL, or the Advanced Flight Technology Integration (AFTI) F-111 research aircraft. James F. Campbell led a study team that examined the options and chose an F-106B airplane because of its wing geometry, flight characteristics, and accessibility to NASA researchers. NASA had used a two-place F-106B as a research aircraft for a variety of prior programs, including engine testing at NASA's Lewis (now Glenn) Research Center and severe storms and lightning assessments at Langley. At that time, the aircraft was based at Langley where engineering staff and fabrication shops could be used for aircraft modifications. With a wing leading-edge sweep of 60 degrees and transonic maneuver capability, as well as a second cockpit seat for observation of flow phenomena, the aircraft was ideally suited for an initial full-scale aerodynamic vortex flap flight assessment. The advocacy efforts of Joseph W. Stickle, Chief of the Low-Speed Aerodynamics Division at Langley, were also instrumental in the selection process.

In 1985, Langley held a national Vortex Flow Aerodynamics Conference to review the state of the art in vortex-flow technology under the joint sponsorship of NASA and AFWAL. At that meeting, several papers were presented on study results of vortex-flap applications to specific configurations, including the F-106.

The scope of studies required to implement and flight test the vortex flap on the F-106B included aerodynamic design (including wind-tunnel tests and analytical design), structural design and development of instrumentation, fabrication of flight hardware in Langley shops, installation of hardware and instrumentation by Langley aircraft technicians, development of simulation software, piloted simulator evaluations of aircraft handling qualities prior to flight, and flight tests of the modified aircraft at NASA's Wallops Flight Facility on Virginia's Eastern Shore. A particularly valuable aspect of the flight program was the use of unique on-surface and off-surface flow visualization techniques that the Langley staff developed and implemented.

Neal Frink led a team on the design of the vortex flap for the F-106B. Using his own design process, Frink arrived at the specific design to be flight tested on the airplane. An immediate project challenge was working with an existing old airframe design with specific load carrying capabilities. One critical result of the loads situation was that the vortex flap had to have a smaller chord than desired. If loads had permitted, a larger flap would have been used, resulting in improved performance. Researchers conducted numerous wind-tunnel tests to verify the flap design's effectiveness and obtain loads information prior to fabrication. A major problem for the austere project (Roy V. Harris, Jr., Director of Aeronautics, reprogrammed funding to accomplish this multiyear effort) was the unavailability of existing wind-tunnel models for the aged F-106 configuration. Following a nationwide search, Jim Campbell located a 1/20-scale high-speed test model of the F-106B that had been retired to the Smithsonian Air and Space Museum. Langley engineering support and brought the model out of mothballs, restoring it to testing condition.

James B. Hallissy, Jarrett K. Huffman, and Frink led the initial testing and analysis of the model in Langley's 7- by 10-Foot Tunnel. Unfortunately, for angles of attack of interest with the vortex flaps installed, the F-106B model was load limited in the wing leading-edge area and could only be tested up to Mach numbers of 0.5 in the atmospheric 7- by 10-Foot Tunnel. To obtain the necessary data, Langley researchers would have to conduct testing in a tunnel with reduced pressure and lower loads. The Langley 8-Foot Transonic Pressure Tunnel, with its capability to run at reduced pressures, would have been the obvious choice for this work, but was not available as it was heavily committed to Langley laminar-flow tests (discussed in a previous section). The Langley 16-Foot TDT was the only other transonic tunnel at Langley with the capability to test at stagnation pressures below atmosphere. Hallissy, Charles H. Fox, Michael H. Durham, and W. F. (Bill) Cazier took on a major challenge in this endeavor because the TDT was not set up for performance testing. The researchers confirmed that a significant performance increment could be achieved transonically, although the optimum flap deflections were different and the magnitude of the increment was somewhat reduced relative to the subsonic conditions. Hallissy further extended the data by conducting additional tests in the Ames Research Center 6- by 6-Foot Supersonic Tunnel.

While high-speed tunnel testing assessed transonic performance of the F-106B vortex-flap configuration, a team led by Long P. Yip conducted low-speed tests of a full-scale airframe in the Langley Full-Scale Tunnel. Because a full-scale F-106B could not be accommodated within the

Neal Frink inspects full-scale semispan model of F-106 equipped with his vortex-flap design.

tunnel's 30- by 60-ft test section dimensions, Yip and his team acquired a second, nonflightworthy F-106B aircraft and proceeded to physically slice the airplane down its centerline to create a semispan, full-scale F-106B test article. Referred to as the F-53 (half an F-106!), the semispan article was tested for flap loads and stability effects in 1985. Results of the tests indicated an apparent vortex-flow instability on the flap's inner portion near the fuselage intersection. In view of these results and additional guidance from CFD computations, the team increased the inner flap's local chord length.

The F-106B flap system's structural design was led by Joseph D. Pride, Garland O. Goodwin (Kentron Technologies, Inc.), and a team of in-house engineering personnel. The system consisted of a simplified ground-adjustable "bolt-on" flap that could be installed at different fixed deflection angles from 20 to 50 degrees. The flap was designed and constructed in spanwise segments to comply with structural loading and deflection issues. The actual fabrication included access straps that bridged leading-edge access areas between major segments located ahead of the wing spar. Langley's fabrication shops constructed the vortex-flap components.

Free-flight model of the F-106B modified with a vortex flap (left) and in flight in the Langley Full-Scale Tunnel (right).

Prior to flight test planning, no information on potential effects of a vortex flap on stability, control, and flying qualities of the F-106B was available. To assess this issue and to prepare pilots for the flight tests, Langley staff conducted a series of static and dynamic stability assessments and a piloted simulator study. Long P. Yip led dynamic stability testing in the Langley Full-Scale Tunnel, which included dynamic model force tests to obtain aerodynamic data for analysis of dynamic stability and for inputs to piloted simulators. Yip also led free-flight tests of the 15-percent scale model to assess the impact of vortex flaps on stability and control characteristics. One of the major concerns prior to these free-flight model tests was whether the leading-edge vortices on the vortex flaps would lift off the surface abruptly or discontinuously, causing undesirable aircraft responses. Assisted by Sue B. Grafton and Jay Brandon, Yip obtained free-flight model results demonstrating that vortex flaps did not significantly affect the damping characteristics of the configuration; and, with the exception of an acceptable reduction in longitudinal stability, the flaps did not degrade flying qualities.

Jay Brandon led a Langley team in gathering the necessary aerodynamic data for the development of a piloted simulator of the modified F-106B for pilot assessment and training using the Langley Differential Maneuvering Simulator (DMS). Langley research pilot Philip W. Brown was selected to be the primary evaluation pilot for the flight test program as he had accumulated significant flight time in the basic F-106B in previous Langley flight programs. Brown conducted several simulator assessments and concluded that the F-106B vortex-flap configuration would be expected to have satisfactory flying characteristics.

Project Manager for the F-106B flight-test program was Ronald H. Smith, who was assisted by James B. Hallissy. In addition to his managerial responsibilities, Hallissy was Principal Investigator for determining the flow-field characteristics and performance increments achieved with flaps on the airplane. Together with W. Elliott Schoonover, Hallissy contributed extensive efforts to prepare the airplane for performance measurements and postflight data analysis. The tasks were particularly challenging because the airplane lacked conventional instrumentation for performance tests (such as a calibrated engine) and pressures had to be obtained by upper-surface belts rather than pressure ports. Jay Brandon and Thomas D. Johnson (PRC-Kentron, Inc.) accompanied project pilot Phil Brown on flow-visualization flight tests.

A most informative aspect of the F-106B vortex-flap flight test program was the unique vapor-screen flow-visualization technique used to visualize details of the leading-edge vortex structure during actual flight tests. The visualization concept, which John E. Lamar conceived, involved a flight adaptation of an existing vapor-screen method for flow visualization commonly used in wind tunnels. During aircraft flight maneuvers, high relative humidity and low pressures in the flow around an aircraft will sometimes cause moisture to condense, providing a natural visualization of aerodynamic flow patterns. The Langley vapor-screen technique obviated the need for natural humidity by seeding the air stream with a heated propylene glycol vapor pumped from a missile-bay pallet and expelled through a probe placed under the left wing panel's leading edge. When exposed to cold temperatures, the clear glycol vapor became white, allowing for visualization of the flow field. In the first flight experiments, conducted in 1985 before the aircraft was modified, the vapor entrained by the vortices was illuminated by a thin light sheet that was projected across the wing in a fixed plane by a mercury-arc lamp behind a narrow slit in an apparatus mounted on the fuselage's side. Onboard video cameras were used to record flow patterns of the vortical flow within the fixed light sheet on the wing upper surface. The unmodified wing's flow-visualization flights began in February 1985 and were conducted on moonless nights to provide contrast and optimize the images produced. Joseph D. Pride and Tom Johnson were key members of Lamar's flow visualization team, which obtained detailed flow information from Mach 0.4 to Mach 0.9 during maneuvers for the basic F-106B.

Initial results from the visualization experiments showed the complicated flow field on the wing upper surface. Single leading-edge vortices were observed on each F-106B wing panel at angles of attack above about 20 degrees as expected, but at lower angles of attack, between 17 and 20 degrees, multiple leading-edge vortices appeared along the wingspan. This unexpected phenomenon warranted additional visualization studies (to be discussed later), which were conducted after the vortex-flap performance testing was completed.

In 1987, after nearly 3,000 hours of wind-tunnel model testing and computational studies, the Langley research team evaluated the vortex-flap concept in flight on the F-106B. Flight tests of the unmodified airplane were first conducted to establish a baseline for performance measurements, then the production wing leading edges were removed and the ground-adjustable vortex flaps installed. The right wing panel was instrumented to measure surface pressures, and the left wing was instrumented with accelerometers and strain gauges to monitor structural loads and deformation. The team made the first flight with the vortex flap on August 2, 1988, and continued testing for 93 research flights over the next 2-plus years. The flight program's primary objectives were to document detailed aerodynamic flow characteristics and compare them with wind-tunnel and computational predictions, and to assess the vortex flap's impact on aircraft performance and handling qualities, including takeoffs, landings, and transonic maneuvers. The research team designed the extensive pressure measurements and flow visualization tests to provide a database for design and analysis tool calibration for vortex-flap technology as well as generic experimental, computational, and flight test technology. Flights were conducted for vortex-flap angles of 30 and 40 degrees for Mach numbers from 0.3 to 0.9 and for altitudes up to 40,000 ft. Results were obtained in the form of incremental performance measurements from the basic F-106B, and parameter identification techniques were used to extract aircraft stability and control information. Tom Johnson and Jay Brandon flew as flight test engineers for all the performance flights and in the chase airplane for photos and coordination.

The NASA F-106B in flight with the vortex-flap modification.

The vortex flap's aerodynamic performance benefits as determined in the F-106B flight tests were extremely impressive. Improvements in the aircraft L/D resulted in very significant increases in sustained turn capability during maneuvers through the highest Mach number flown (0.9). For example, the airplane's achievable sustained g was increased by about 28 percent at a Mach number of 0.7. The quantitative performance results obtained in the flight program provided invaluable documentation and demonstration of the vortex-flap's potential benefits for highly swept military fighter aircraft. The calibration of analytical design methods and flow visualization data with flight data revealed complex flow fields that continued to challenge the capabilities of wind tunnels and computational fluid dynamics.

As previously discussed, initial F-106B in-flight flow-visualization results had indicated unexpected flow phenomena and multiple vortices. Therefore, the Langley researchers conducted a second series of flow visualization experiments beginning in 1990 to provide further information on the structure of multiple vortical flows for baseline and vortex-flap aircraft configurations. John Lamar and the engineering support staff conceived a refined flow-visualization system for these follow-up tests, providing much more research flexibility and integrated data over a larger viewing area by using a scanning light sheet source that was mounted in a streamline fairing atop the fuselage spine. Jay Brandon led the efforts for implementing the new visualization system on the aircraft. The flow visualization system included two video cameras (one located on the engine intake and one near the aircraft centerline aft of the canopy). Using this approach, researchers obtained flow information over a broad sector on and over the left wing during flight. In addition to the vapor screen information, they obtained on-surface results using oil flows and tufts on the wing upper surface. Brandon and Lamar served as co-investigators of the flight data.

The team obtained flow visualization results for vortex-flap settings of 30 and 40 degrees over a Mach number range of 0.3 to 0.9 in 1990. Once again, the researchers observed unexpected results for vortical flows. As had occurred previously for the basic wing, the vortex flap exhibited multiple vortices (on the flap surface), although this difference of the flow physics from that expected did not seem to result in degradation in predicted performance improvements.

The multiple vortices appeared to originate on the vortex flap and then migrate off the flap to run nearly streamwise over the wing as another vortex originated on the flap. This pattern was repeated many times down the wing depending on angle of attack. Oil-flow results confirmed existence of a highly complex flow pattern with multiple vortex systems as observed during the vapor-screen tests. Examinations of the research team's flight results showed that the multiple streamwise vortices observed above the wing originated at the flap leading edge where individual flap segments were joined. The team also conducted oil-flow studies with the joints sealed with fabric-backed tape to

The scanning vapor screen technique.

F-106B researchers prepare for a night mission.

prevent air leakage and observed essentially the same results. The oil-flow studies indicated that the small geometric perturbations along the leading edge were sufficient to generate a leading-edge flow that was very complex and significantly different in details than that observed previously in wind-tunnel model tests and in CFD calculations.

After observing the unpredicted vortex topologies in flight, Jim Hallissy, Elliot Schoonover, and Tom Johnson tested the F-106 wind-tunnel model in the Langley 7- by 10-Foot High-Speed Tunnel. All previous wind-tunnel tests had shown a single leading-edge vortex system along the flap, as predicted by the design methodology. During storage of the model, some minor damage in the form of small dents and nicks had occurred to the flaps. Because leading-edge discontinuities were believed to be at least partially responsible for the multiple vortices seen in flight, these dents were not repaired prior to the tests. During the test the dents, or the subsequent application of tape flow trips, provided sufficient perturbations for shedding of vortices and formation of the multiple vortex system seen in flight. The extremely small perturbation size indicated that with normal manufacturing tolerances, it might be impossible to avoid the multiple vortex patterns seen in flight on a full-scale airplane with leading-edge devices similar to vortex flaps.

In summary, the major purpose for developing the vortex flap was to improve L/D ratio at high

Surface-oil studies illustrate multiple vortex flows seen on the leading edge.

Langley researchers pose with the F-106B Vortex-Flap Research Aircraft in 1988.

maneuvering lift coefficients. Despite the strikingly different flow field details developed on the airplane compared with computational theory and wind-tunnel predictions, the flaps' overall effectiveness was very close to predictions, resulting in significant improvements in maneuver capability, such as sustained-turn characteristics.

Following the completion of vortex-flap flights, the NASA F-106B airplane was retired on May 17, 1991, in a formal ceremony at Langley. Later that year, the airplane was transferred to the Virginia Air and Space Center in Hampton, where it has been displayed to the public with the vortex-flap modification.

F-16XL Plans

Leadership of the High-Speed Research Program's integrated, NASA-wide high lift element was assigned to NASA Langley with Joseph R. Chambers, Chief of the Flight Applications Division, selected to lead the effort. A challenging problem facing a future supersonic transport is unacceptable takeoff noise caused by the high levels of thrust required to overcome inherently low lift and high drag of highly swept supersonic wings and poor subsonic cruise performance caused by the high induced drag of such wing shapes. Accordingly, wind-tunnel and computational efforts were undertaken to improve the subsonic L/D aerodynamic characteristics of candidate HSR configurations. The scope of research at Langley and the Ames Research Center included studies of various types of leading-edge designs, including fixed cambered configurations and deflectable cambered flaps and vortex flaps.

In 1993, Chambers advocated for flight testing of an appropriate airplane to obtain more detailed information on the impact of leading-edge devices, such as the vortex flap, for subsonic and high-lift conditions. NASA transferred one of its two F-16XL research aircraft from the NASA Dryden Flight Research Center to Langley for the proposed program. As previously discussed, Langley had conducted F-16XL low-speed and transonic vortex-flap tunnel tests during the early 1980s in concert with the aircraft's development. As the HSR Program interests in low-speed high-lift devices intensified, researcher David E. Hahne led wind-tunnel tests of an F-16XL model with several leading-edge flap configurations to begin the process of selecting candidate flaps to be flown on the airplane. In addition to aerodynamic studies, the research program was to include

F-16XL aircraft painted by Langley for flow visualization tests.

unconventional thrust management strategies to reduce power at certain takeoff conditions to further reduce noise. Noise level measurements would be made for the airplane with the flap modifications and throttle strategies. A piloted simulation of the F-16XL was implemented by Langley researchers in the Langley DMS in preparation for flight testing.

The F-16XL's upper surfaces were painted black to enhance the flow visualization studies planned for the flight tests. Unfortunately, changes in program priorities terminated the F-16XL flight effort within HSR before modifications for vortex-flap flight activities could begin. NASA did, however, support a series of basic aerodynamic vortex-flow studies on the F-16XL airplane led by John E. Lamar in a project known as the Cranked-Arrow Wing Aerodynamics Project (CAWAP). Lamar's team included Langley's Clifford J. Obara, Susan J. Rickard, and Bruce D. Fisher, as well as Dryden's David F. Fisher. The team focused on detailed measurements and analysis of the aircraft's exhibited vortical-flow characteristics, including wing pressures, boundary-layer measurements, and flow visualization on the upper wing surface using tufts. The results were correlated with computational results, providing a database for additional analyses and adding to Langley's contributions in vortical-flow technology.

Status and Outlook

To date, NASA, industry, and academia have accomplished much in the development of aerodynamic theories and exploratory aerodynamic applications of vortex-flap concepts. Enhanced aerodynamic performance has been measured for a wide range of slender-wing configurations, including full-scale flight tests. However, the technology maturation level for potential production applications has remained below the level required for low-risk implementation by industry.

Many barriers and challenges cited in the earlier discussion of this topic will need solving before applications can be expected. The ultimate demonstration of an "adaptive" vortex-flap design (deflections automatically controlled for maximum efficiency by flight computer) on a high-speed aircraft with production-type fabrication and tooling will be necessary before the concept can applied. Also, the systems-level impacts of the vortex flap (weight, maintenance, failure modes, etc.) must be assessed and compared with more conventional approaches currently used such as conical wing camber, or conventional leading-edge flaps.

Unfortunately, recent high-performance military configurations have used lower wing-sweep angles than those appropriate for the slender-wing vortex-flap applications. As a result, designers have chosen the use of conical camber, conventional leading-edge maneuver flaps and hybrid-wing (wing-body strake and relatively unswept outer wing) design options. Further, the dramatic reduction in new military aircraft programs has left few opportunities for injection of this technology. On the other hand, recent interest in uninhabited combat air vehicles that use delta and highly swept wing planforms might permit a renewed interest in the concept. Such concepts would be of even greater interest if the application of "smart" materials could permit the use of continuous outer mold lines, thereby resolving issues regarding the impact of vortex-flap physical discontinuities on stealth and radar observables.

From a civil aircraft perspective, the demise of the NASA HSR Program and a pessimistic international outlook for large supersonic transports in the future does not portend of opportunities for vortex-flap applications to that class of vehicle. However, growing interest in economically viable supersonic business jets could conceivably rekindle interest in vortex-flap technology, especially if the concept could help designers attack the known operational barriers of environmental noise issues.

Finally, it is appropriate to note that Langley's success in developing and demonstrating the benefits of vortex-flow control with the leading-edge vortex flap for performance inspired NASA and industry to focus on solutions to stability and control problems of contemporary fighter configurations caused by uncontrolled vortex flows at high angles of attack. Examples of follow-up research included the control of vortical flows shed by pointed slender forebodies, noncircular forebody cross sections, and nose strakes. In proof-of-concept experiments, most of these stability and control problems were demonstrated to be amenable to improvement by the use of innovative mechanical and pneumatic (blowing and suction) techniques for vortex management.

Innovative Control Effectors: Smart Muscles

Concept and Benefits

The challenge of providing satisfactory controllability and handling qualities for aircraft has been a crucial requirement throughout the history of aviation. Attempts to provide adequate levels of control have resulted in a wide variety of conventional control effectors, including empennage-mounted elevators and rudders; wing-mounted ailerons, elevons, rudders, and spoilers; fuselage-mounted canards; wing warping; mechanical engine thrust vectoring in pitch and yaw; and differential engine thrust for multiengine configurations. The overriding requirement that aircraft must exhibit satisfactory responses to control inputs for all phases of operational envelope, including off-design conditions, has driven the development of these various concepts.

Evolving requirements for flight mission capabilities and unconventional configurations have forced the technology "push" and the applications "pull" for advanced control effectors. For example, during the early days of heavier-than-air flight, designers attempted to meet the fundamental need to provide aircraft that could be successfully flown by a human pilot through relatively mundane maneuvers and very limited flight envelopes. As aircraft mission capabilities rapidly expanded to faster speeds and higher altitudes, new challenges—such as compressibility effects, structural flexibility, flutter, excessive hinge moments, pilot stick forces, pilot-induced oscillations, and control reversal—were encountered and researched. These efforts produced solutions that enabled the improved capabilities offered by unconventional configurations. Some unconventional configurations, such as flying wings, required innovative controls (e.g., wing tip-mounted split ailerons) that serve as ailerons and rudders.

Military aircraft, in particular, have been the recipients of extensive research on flight controls because of stringent maneuverability requirements and challenging off-design operations. During World War II, for example, NACA, the military services, and industry devoted continuous efforts to reducing stick forces and enhancing roll performance, thereby ensuring that the razor-thin combat advantage in close-in dogfights would belong to U.S. pilots. Following World War II, the advent of supersonic flight with its attendant compressibility effects resulted in the emergence of new control concepts: powered control systems, differentially deflectable stabilators for pitch and roll control at high speeds, and the use of spoilers for roll control.

In recent years, the ongoing changes in aircraft mission capabilities have continued to invigorate studies of new control effectors. For civil aircraft, commercial transport designers have directed their attentions to ensuring adequate controllability during high-subsonic cruise conditions where shock-induced separation may cause steady or unsteady aerodynamic phenomena that degrade control effectiveness. In addition, designers strive for efficient outer-wing aileron configurations (or middle-wing spoilers) that "free up" valuable inner-wing trailing-edge locations for high-lift flap devices. More efficient flaps permit the designer to reduce the wing's size, thereby reducing weight and improving overall mission capability. Finally, propulsive control for multiengine civil transport configurations received recent attention after the heroic flight crew efforts of the 1989 DC-10 crash at Sioux City, Iowa, following hydraulic power loss caused by an engine structural failure. NASA's Dryden Flight Research Center has conducted extensive research on the propulsive control technique.

For military aircraft, an aircraft's full use for strenuous maneuvers and other requirements demands control effector research on barrier problems. For example, providing satisfactory levels of control effectiveness and coordination at high-angle-of-attack flight conditions in the presence of extensive flow separation has promoted interest in thrust vectoring, vortex-flow control, and using forebody strakes for lateral-directional control. Automatic departure and spin prevention have been developed and applied using advanced control system architectures. Another controls challenge has arisen from the application of stealth technology for low-observable configurations. Stealth configurations require special consideration, or elimination, of control surface geometries such as gaps, hinges, and other details that can degrade radar or infrared signature characteristics.

Recent disciplinary advances in structures and materials have led to a new family of "smart" materials that respond to stimuli with shape changes that could be integrated into innovative control effectors. "Morphing" configurations, which adjust external shape as a function of flight conditions, vehicle health considerations, or other factors, would use such an approach to provide control.

Innovative control effectors can be used with advanced adaptive control system architectures that sense the changes in flight environment and automatically schedule the control gains, feedback, and mixing to promote more optimal response characteristics as well as improved aerodynamic efficiency. The sketch illustrates some of the innovative control effector research topics conducted by NASA.

Vortex-flow control effectors modify and control the powerful vortical flows generated by highly swept wings, wing-fuselage strakes (also called leading-edge extensions), and fuselage forebodies to provide control at high angle-of-attack conditions. At relatively high angles of attack, such devices provide significantly larger control effectiveness than conventional wing or tail-mounted controls, which are usually ineffective for highly separated flow conditions.

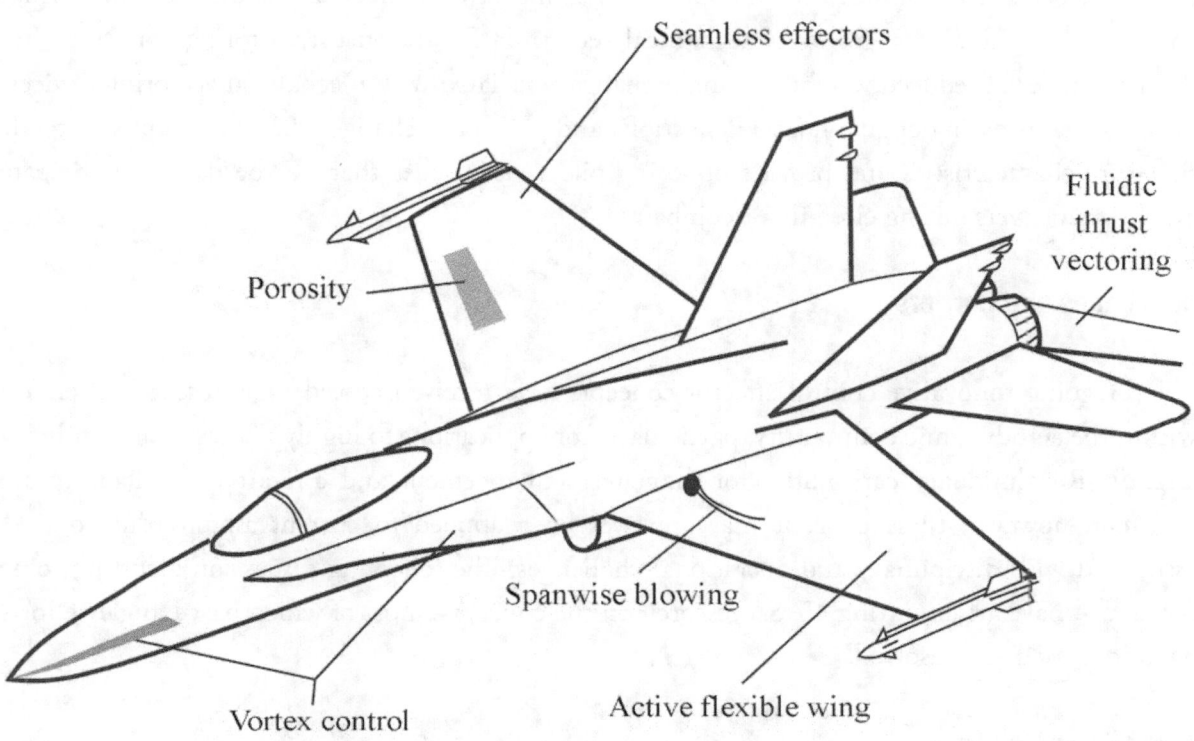

Innovative control effector concepts studied by NASA-Langley.

Langley has led international research on the application of passive porosity—the use of perforated regions on aircraft surfaces to control aerodynamic pressures and flow characteristics—to modify aerodynamic phenomena such as shock locations, separation, and lift for enhanced performance, stability, and control. This approach permits relatively large variations in aerodynamic behavior without constraints, such as hinge moments, normally encountered with conventional aerodynamic control surfaces.

Spanwise blowing concepts use compressed air derived from engine bleed or other sources to modify airflow over lifting surfaces. By modifying and creating vortical-type flows as a result of steady or pulsed blowing in a spanwise direction, airflow over the upper surface may be significantly influenced, to the extent that vortex lift and reattachment of separated flows can be effected. Thus, aerodynamic performance as well as stability and control can be enhanced, particularly for high angle-of-attack conditions.

Due to extensive research by NASA, industry, and DoD, various types of mechanical thrust vectoring concepts have now been implemented in current military aircraft for enhanced maneuverability and control. A concept previously applied to rockets and missiles, known as fluidic thrust vectoring, uses fluidic rather than mechanical means to redirect engine thrust for vectoring, and offers several advantages over mechanical vectoring. Fluidic thrust vectoring concepts offer the advantages of reduced weight and maintenance associated with mechanical vectoring devices, as well as eliminating engine nozzle deflections and geometric changes that compromise aircraft signature characteristics and provide opposing pilots visual cues that can be used to anticipate evasive maneuvers during close-in air combat.

Challenges and Barriers

The foregoing innovative control effector concepts have received considerable research attention within the aerodynamic community, particularly for applications to highly maneuverable military aircraft. Revolutionary capabilities for maneuver enhancement and aircraft controllability are apparent; however, these concepts have not yet been applied to current aircraft due to risks associated with disciplinary and operational challenges. The following discussion identifies some issues that have arisen during NASA research on the concepts, most of which past Langley efforts have addressed and resolved.

Disciplinary Challenges

Many control concepts face significant challenges in the areas of aerodynamics and structural weight, as well as aeroelasticity and flutter. For example, the aerodynamic effectiveness of the control concepts must be maintained across the flight envelope. For high-performance aircraft, this requirement is especially daunting because of compressibility effects, shock-induced separation, and unsteady flow phenomena. Fundamental effector characteristics must be established through extensive wind-tunnel and flight aerodynamic research efforts. The rapid maturation of advanced CFD methods is now contributing to the assessment and solution of problems involving the effects of Mach number and other flight variables on control effectors.

These concepts must be designed to eliminate undesirable aeroelastic characteristics, such as flutter or aeroelastically induced control reversals, along with maintaining adequate control effectiveness. Evaluations of these phenomena are normally conducted in unique wind tunnels, including the Langley 16-Foot TDT, or in carefully controlled flight experiments.

The magnitude and character of control moments produced by innovative control effectors must also provide satisfactory control response characteristics. In particular, the aerodynamic moments produced by control actuation must vary in a linear fashion with the pilot's control inputs for satisfactory aircraft response, and unsatisfactory by-products known as "cross-axis" moments (for example, yawing moments produced by a roll control) must be minimized. Additional constraints, such as excessive control hinge moments, must be avoided and the weight of control mechanisms and actuation devices must be acceptable.

Operational Challenges

Significant challenges to the applications of innovative control effectors also exist relative to operational and environmental issues. Paramount to all operational issues is the cost of implementation, maintenance, and replacement of control systems. These cost considerations primarily involve associated control system software requirements and other requirements, such as certification time and cost.

Operational challenges to advanced controls include a myriad of issues dominated by maintenance requirements, health monitoring, and failure modes. In-depth analysis of each factor is mandatory before the ultimate feasibility of advanced concepts can be established. In addition, environmental effects (e.g., icing and corrosion) must be assessed and resolved.

Special constraints are placed on innovative control effectors that use auxiliary air for flow control mechanisms. Weight and engine performance issues severely restrict the potential benefits if, for example, excessive engine bleed requirements are necessary.

Arguably, the most important operational challenge to new control effectors is the opinion of evaluation pilots relative to the crispness, predictability, and effectiveness of advanced control effectors on aircraft response characteristics. If significant nonlinearities, cross-axis interactions, and degraded effectiveness occur during critical phases of flight, the control system will be rejected for application.

Langley Activities

Langley Research Center has historically led the research community in advanced flight control systems development. In addition to extensive in-house aerodynamic studies coupled with control system architecture and failure analysis methodology, the Center has partnered with other NASA Centers, industry, and DoD in advancing the state of the art in flight controls. Coupled with legendary contributions in developing conventional aileron, elevator/stabilator, and spoiler control effectors, Langley researchers have pursued many innovative concepts yet to be applied. The following discussions briefly describe some of these pioneering efforts.

Spanwise Blowing

In the middle 1970s, an intense national interest arose in highly maneuverable military aircraft capable of flight at extreme angles of attack with controllable "care-free" characteristics. Associated with this activity was a mainstream of attention on vortical flows and the use of vortical flow control for enhanced lift and performance. As discussed in *Partners in Freedom*, Langley Research Center contributed directly to vortex technology in activities ranging from fundamental research to specific aircraft applications, including uses on the F-16 and F/A-18 aircraft. Edward C. Polhamus led a wide variety of vortex-control investigations conducted within the basic NASA research program, with industry and DoD partners, and with universities.

Within this environment of innovation and opportunity, Polhamus' group directed its efforts toward the potential use of several concepts to enhance the powerful aerodynamic effects of vortical flows for high-performance configurations. One concept was the use of a high-pressure jet blowing spanwise over a wing upper surface in a direction parallel to the leading edge to augment vortex lift and enhance favorable flow phenomena over the wing. Preliminary experiments indicated that spanwise blowing would aid in the formation and control of the leading-edge vortex shed by moderately swept wings. Polhamus assigned the lead role for spanwise blowing research in his group to James F. Campbell, who was assisted by researchers Gary E. Erickson, Jarrett K. Huffman, and Thomas D. Johnson, Jr.

Campbell and his associates accomplished exploratory spanwise blowing studies in 1974 during tests of simple wings with leading-edge sweep angles of 30 and 45 degrees in the Langley High Speed 7- by 10-Foot Tunnel. These early tests indicated that spanwise blowing significantly improved the aerodynamic characteristics of both wing models at high angles of attack. These tests also revealed that spanwise blowing generated large increases in lift at high angles of attack, improved the drag polars, and extended linear pitching moments to high lift conditions. The study also unveiled an important aspect of the spanwise blowing mechanism: full vortex suction lift was achieved at the inboard span station with a relatively small blowing rate, but higher blowing rates would be necessary to attain the full vortex-lift level at increased span distances.

Campbell and Erickson followed this exploratory study with additional wind-tunnel studies of increased sophistication and scope. They teamed for a study in 1977 involving tests of a 44-degree swept trapezoidal wing model for a range of angle of attack, jet momentum coefficients, and leading and trailing-edge flap deflection angles. They found blowing to be more effective at higher Mach numbers (0.5). The researchers found that spanwise blowing in conjunction with a deflected trailing-edge flap resulted in lift and drag benefits that exceeded the summation of the effects of

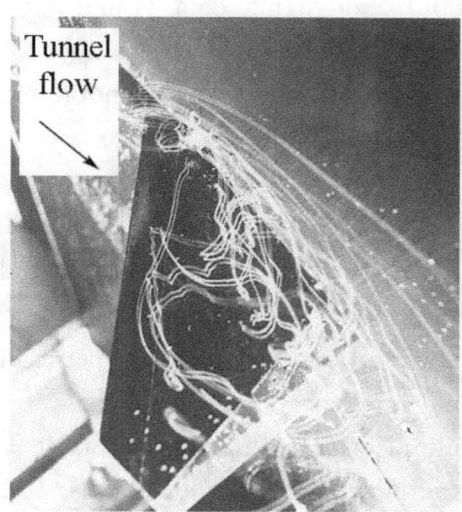 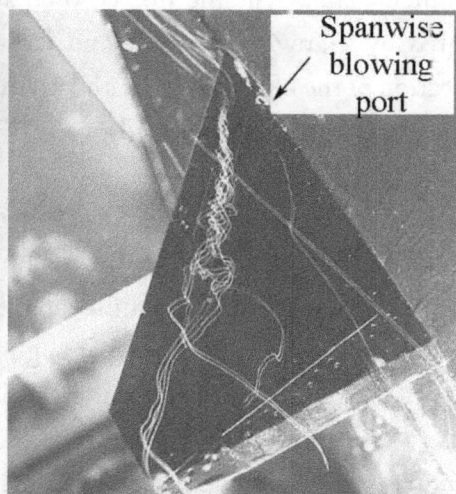

Blowing off; vortex breakdown.   *Blowing on; vortex breakdown delayed.*

Flow visualization of spanwise blowing on a swept trapezoidal wing for an angle of attack of 30 degrees.

each high-lift device acting alone. Of relevance to the current discussion, they found asymmetric blowing to be an effective lateral control device at the higher angles of attack.

While Campbell and Erickson were pursuing their fundamental aerodynamic studies of the impact of geometric and pneumatic variables on the effectiveness of spanwise blowing, a group under Joseph R. Chambers and Joseph L. Johnson, Jr., at the Langley 30- by 60-Foot (Full Scale) Tunnel began research on the effects of spanwise blowing on the dynamic flight behavior of dynamically scaled free-flight models. Using the remotely controlled free-flight test technique described in other sections of this document, this group assessed the impact of spanwise blowing on longitudinal and lateral-directional behavior for generic and specific aircraft configurations.

In 1978, Dale R. Satran and Ernie L. Anglin led free-flight tests of a general research fighter configuration (based on a modified F-5 configuration) to evaluate spanwise blowing effects of two different wing planforms. One configuration incorporated the wing of the baseline F-5 design (34-degree leading-edge sweep) and the second configuration used a 60-degree delta wing. Three blowing ports were located on each side of the fuselage, oriented parallel to each wing panel's leading edge. Emphasis was on determining dynamic lateral-directional characteristics, particularly in the stall and departure angle-of-attack range; however, effects of spanwise blowing on longitudinal aerodynamics were also determined. The tunnel tests included measurement of conventional static force and moment data, dynamic (forced-oscillation) aerodynamic data, visualization of airflow changes created by spanwise blowing, and free-flight model tests. The effects of blowing rate, chordwise location of the blowing ports, and asymmetric blowing on the conventional aerodynamic control characteristics were investigated.

In the angle-of-attack regions wherein spanwise blowing substantially improved the wing upper surface flow field (i.e., provided reattachment of the flow aft of the leading-edge vortex), improvements in both static and dynamic lateral-directional stability and control were observed. Rolling moment substantially increased at high angles of attack when asymmetric blowing was used for roll control. In fact, the magnitude of rolling moment was as large as that provided by the ailerons at low angles of attack. However, the results also showed that unacceptable large adverse yawing moments were associated with asymmetric blowing, to the extent that full deflection of the rudder would be required to trim out the undesirable yawing moments and coordinate the roll maneuver.

National interest in spanwise blowing continued to expand in the late 1980s. Industry and DoD efforts began to focus on flight testing of specific full-scale aircraft to extend the limited aerodynamic database available in wind tunnels to full-scale hardware. These efforts also provided detailed engineering information on blowing requirements, engine bleed and ducting characteristics, and other system-level features required to design and determine the concept's feasibility. In a 1984 study, McDonnell Douglas modified an F-4C Phantom II airplane under Air Force sponsorship to investigate spanwise blowing. The goal was to validate wind-tunnel data indicating that the F-4C's existing chordwise BLC system could be replaced with a more maintenance-free spanwise blowing system without degrading performance. The designers piped high-pressure bleed air from the F-4C's J-79 engine compressors forward along the inside of the fuselage and expelled the flow through a nozzle in the fuselage near the wing's leading edge and just above the surface. The flight-test results showed that the approach speed could be reduced by about 7 kts and maneuverability was noticeably improved. Because the configuration's leading-edge jet did not penetrate to the outer wing panel, it was suggested that further improvements would occur if some of the blowing were distributed over the outer wing panel of the F-4C.

Langley's Jim Campbell and Dryden's Theodore (Ted) Ayers advocated for a follow-up NASA flight test of the F-4C at Dryden, which was supported by the Air Force. At Langley, Jarrett K. Huffman, David E. Hahne, and Thomas D. Johnson, Jr., led tests in the Langley 7- by 10-Foot High-Speed Tunnel and the Langley 30- by 60-Foot (Full-Scale) Tunnel to determine the optimum location and orientation of the outer panel blowing ports and the effect of blowing on lateral-directional characteristics. Huffman used a 0.10-scale F-4C model for his studies in the 7-

Langley researcher David E. Hahne poses with F-4 free-flight model, shown in flight at high angles of attack in the Langley Full-Scale Tunnel.

by 10-foot tunnel, and Hahne used a 0.13-scale dynamically scaled free-flight model for flight and force tests in the 30- by 60-Foot Tunnel. Limitations in NASA resources prevented the planned flight tests at Dryden even though the static and free-flight test results were promising.

Langley and national interest in spanwise blowing waned following these 1980s studies and research on the concept's use in an asymmetric manner for roll control was terminated. Currently, it appears that the use of spanwise blowing for roll control still faces many fundamental issues, especially the level of engine bleed air required and the large adverse yawing moments produced by spanwise blowing for roll control.

Fluidic Thrust Vectoring

Maintaining air supremacy for the United States requires stealthy, supermaneuverable aircraft. Decades of national research on mechanical engine thrust vectoring techniques initiated in the 1970s were designed to meet the demand for fighter aircraft with increased agility. This research and development culminated in the application of thrust vectoring to the Air Force's F-22 design. In the 1990s, additional requirements for low-observable aircraft and for lower exhaust system weights were the catalysts for research on the use of fluidic concepts for thrust vectoring. Langley has been a leader in the evolving technology for fluidic vectoring due to extensive in-house and cooperative research with industry, DoD, and academia. Researcher Karen A. Deere has contributed an excellent summary of Langley contributions in this area, and the reader is referred to her publication for detailed information (see bibliography).

The concept of fluidic vectoring uses fluid control mechanisms to redirect the engine exhaust with no mechanical nozzle parts such as those used for mechanical nozzle vectoring concepts. Typically, the fluidic vectoring concepts use secondary air sources to create an off-axis deflection of the jet thrust. In the early 1990s, the staff of the Langley 16-Foot Transonic Tunnel, under the direction of Bobby Berrier, initiated a cooperative fluidic thrust vectoring program with the Air Force called Fluidic Injection Nozzle Technology (FLINT). David J. Wing led the NASA effort for the program. The results of the FLINT Program predicted that the potential benefits of fluidic thrust vectoring nozzles would be a 28- to 40-percent weight reduction by implementing fluidic throat area control, a 43- to 80-percent weight reduction by implementing fluidic throat area and exit area control, a 7- to 12-percent improvement in engine thrust-to-weight ratio, and a 37- to 53-percent reduction in nozzle procurement and life cycle costs. In addition to these considerations, fixed aperture nozzles would enhance low-observable characteristics by eliminating moving flaps, discontinuities, and gaps. Fluidic systems without moving external nozzle parts would also eliminate visual cues of vectoring control inputs that might be used by enemy pilots to anticipate an impending maneuver during close-in air combat.

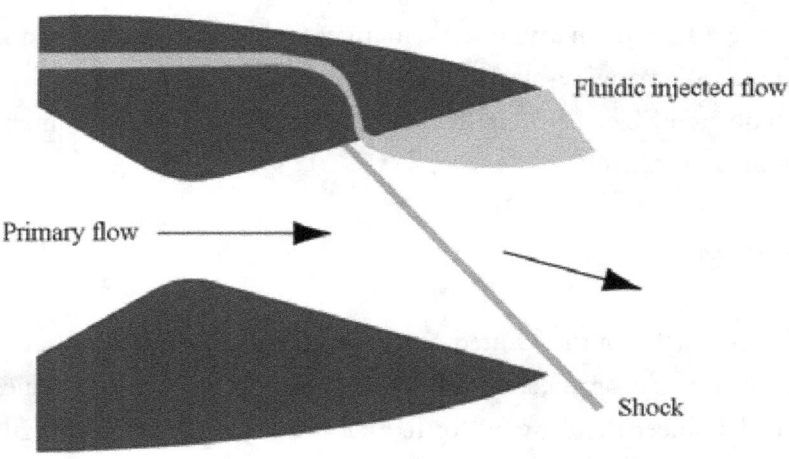

The shock-vector-control concept for fluidic thrust vectoring.

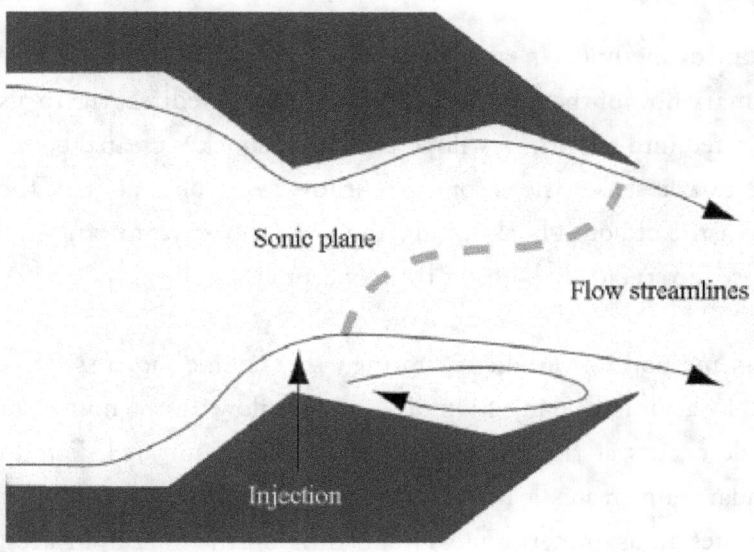

The throat-shifting concept for fluidic thrust vectoring.

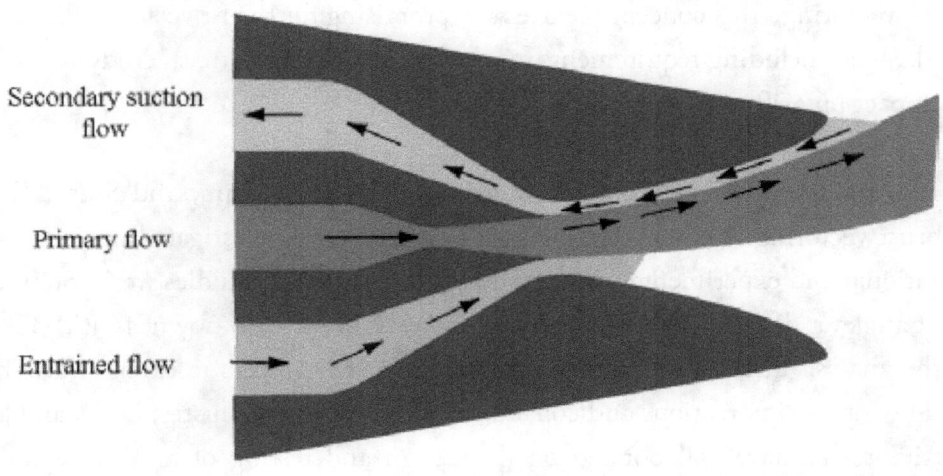

The counterflow concept for fluidic thrust vectoring.

Langley fluidic thrust vectoring concept studies are divided into three categories according to the method used for fluidic thrust vectoring: the shock-vector-control method, the throat-shifting method, and the counterflow method.

In the shock-vector-control method, an asymmetric injection of secondary air into the engine nozzle's supersonic primary flow of the divergent section is used to redirect the thrust angle. When the secondary air is injected into the primary flow, an oblique shock is created because the primary supersonic flow in the nozzle senses the secondary airflow as an obstruction. The primary flow is then directed through the oblique shock, producing large thrust vector angles. Unfortunately, thrust performance losses are typically high for this concept.

In the throat-shifting method of fluidic vectoring, the engine nozzle's effective throat is asymmetrically shifted by asymmetric injection of secondary flow. In the nonvectoring condition, the throat of the nozzle occurs at the nozzle's geometric minimum area. For thrust vectoring, the injection of secondary air creates a new skewed minimum area, which shifts the effective minimum area and creates an asymmetric pressure loading on the nozzle surfaces, resulting in a thrust deflection of the primary exhaust flow.

The counterflow method of fluidic thrust vectoring uses the approach of counterflowing the primary and secondary airstreams with the application of suction at a slot between the primary nozzle and collar. Mixing occurs in the shear layers between the aft-directed primary flow and the forward-directed suction flow, contributing to the establishment of asymmetric pressures that result in thrust vectoring. This concept is extremely promising for thrust vectoring but faces many technical challenges, including requirements for the suction supply source, aerodynamic hysteresis effects, and impact on airframe integration.

David J. Wing, Karen A. Deere, Bobby L. Berrier, Jeffery D. Flamm, and Stuart K. Johnson led fluidic thrust vectoring research conducted at Langley. They investigated promising concepts with computational and experimental tools, and supporting system studies were conducted when appropriate. Langley's development of a Navier-Stokes CFD code known as PAB3D played a key role in the analysis and design of fluidic vectoring methods. The research efforts have been characterized by intense interactions and collaborative studies with industry, DoD, and academia. The cooperative teams have collaborated on the design and testing of hardware, and Langley researchers have typically led experimental testing in the Langley Jet Exit Test Facility (JETF), a unique facility devoted to simulating propulsion systems at static (wind-off) conditions. The industry partners have generally led the nozzle's design, but Langley researchers originated and developed the most recent and promising dual throat nozzle designs.

The scope of fluidic vectoring concepts studied at Langley within the three primary types previously mentioned is extremely broad. Researchers conceptualized and evaluated variants of the types, adding to the basic knowledge and advances in the state of the art for thrust vectoring. Teaming has been extensive, including studies of the shock-vector-control concept with Rockwell, Rohr, Pratt & Whitney, General Electric, and Boeing. The throat-shifting concept has been explored with Pratt & Whitney and Lockheed Martin, and Langley joined Florida State University and the University of Minnesota to study the counterflow method.

Deere's summary publication provides results and details of the foregoing activities beyond the present publication's intended scope, and it is highly recommended for the interested reader. Briefly, results from Langley investigations of fluidic thrust vectoring concepts indicate that the most thrust efficient fluidic thrust vectoring concept is the throat-shifting method, but larger thrust-vector angles are obtained with the shock-vector-control method. However, the most recent throat-shifting nozzle designs developed by NASA and Lockheed researchers are now providing thrust vector angles equivalent to the shock-vector-control method with lower engine bleed requirements. The counterflow fluidic vectoring concept offers promise, but faces several significant technical issues. Langley's pioneering contributions and fluidic thrust vectoring technology are widely recognized and the Center is actively participating in and consulting on the continuous research on this topic.

Vortical Flow Control

As discussed in a previous section on the vortex-flap and spanwise-blowing concepts, as well as in *Partners in Freedom*, Langley has played a key role in fundamental research on vortical flow and its application to aircraft for enhanced performance, stability, and control. With the advent of long, pointed fuselage shapes and wing-body strakes, researchers identified vortical-flow mechanisms that generated large potential control moments, especially at high angles of attack. Beginning in the 1970s, Langley embarked on studies to control the powerful vortices shed by fuselage forebody shapes and wing-body strakes. Researchers discovered that they could produce large rolling and yawing moments for enhanced maneuverability by differentially deflecting these devices.

Dhanvada M. Rao and Langley's Daniel G. Murri were among the first to explore the feasibility of deflecting the wing-body strakes of configurations similar to the F-16 and F/A-18 in a differential manner to produce asymmetric vortex flow fields resulting in rolling moments. Their exploratory wind-tunnel results of this concept indicated very large rolling moments could be produced; however, research efforts on the concept were terminated because large adverse yawing moments, similar to those encountered for asymmetric spanwise blowing, were also produced.

Strong vortical flow emanating from the wing-body strake of the F/A-18 is clearly defined by natural condensation in air.

More productive Langley research on using vortical flow for vehicle control resulted from applications to high-performance aircraft for improved yaw control at high angles of attack. A primary control deficiency that limits the maneuverability of fighter aircraft is loss of rudder effectiveness when the vertical tails are submerged in the low-energy wake of the stalled wing at high angles of attack. Loss of yaw control at such conditions is especially critical for maneuverability because the primary source of rolling motions at extreme angles of attack is yaw control rather than conventional roll control. This phenomenon is a result of inertial distribution of the airplane's mass and the vehicle's relative responses to roll and yaw control inputs.

Researchers noted that naturally occurring, large asymmetric yawing moments developed on slender bodies at high angles of attack. They were therefore inspired to develop concepts that could precisely produce and control these potentially revolutionary levels of yaw control power. Initially, Langley staff demonstrated the use of jet blowing from a thin slot near the nose tip and proceeding along the typical fighter radar radome to be an effective controller for forebodies having geometric features known to promote strong vortex asymmetry effects. However, for forebody shapes not naturally prone to pronounced vortex asymmetry, a different vortex manipulation concept is required to generate an effective yaw control. In addition, the classic concerns over providing adequate levels of pneumatic blowing inhibited potential applications of the blowing concept.

Langley researchers developed a highly successful yaw control concept based on the use of deployable, differentially deflectable fuselage forebody strakes in the early 1980s. In cooperative research with the Air Force, Murri and Rao led the development of a pioneering wind-tunnel database that documented the fundamental flow physics associated with forebody strake controls for a variety of aircraft configurations. This early research indicated that differentially deflected forebody strakes could provide revolutionary levels of precision control for close-in air combat. Continuing evolution and refinement of the studies addressed potential effects of Reynolds number and providing a linear controller for the pilot.

In the middle 1980s, NASA launched its High-Angle-of-Attack Technology Program (HATP), which focused on the advancement of the state of the art for predicting and controlling aerodynamic phenomena for enhanced maneuverability at high angles of attack (see *Partners in Freedom*). Using the F/A-18 configuration as a baseline for wind-tunnel experiments, CFD predictions, and simulator and flight assessments, the HATP included an element to develop and evaluate the promising forebody-strake concept that the previous investigations had matured. Accordingly, a research project referred to as the Actuated Nose Strake for Enhanced Rolling (ANSER) flight experiment was planned. Dan Murri, Gautam H. Shah, and Daniel J. DiCarlo led the activities at Langley. The scope of activities at Langley required to define, assess, and optimize the strake configuration for the F/A-18 included conventional static wind tunnel force and moment tests across a range of Reynolds and Mach numbers, flow-visualization tests, free-flight model assessments of strake effectiveness, CFD studies, and piloted simulator studies of maneuverability and handling qualities on the Langley DMS. In addition to Murri, Shah, and DiCarlo, many other Langley researchers contributed to these efforts: Robert T. Biedron, Gary E. Erickson, Frank L. Jordan, Sue B. Grafton, and Keith D. Hoffler.

In conjunction with the ground tests of the HATP, NASA modified an F/A-18 fighter aircraft as its High Angle-of-Attack (Alpha) Research Vehicle (HARV) for a three-phased flight research program lasting from April 1987 until September 1996. The aircraft completed 385 research flights and demonstrated stabilized flight at angles of attack between 65 and 70 degrees using thrust vectoring vanes, a research flight control system, and the ANSER forebody strakes. The hardware's implementation on the HARV was a remarkable display of intercenter coordination and cooperation between Langley and the NASA Dryden Flight Research Center. Langley engineering and shop organizations designed and fabricated the ANSER forebody-strake hardware, and Dryden's staff completed the tasks of aircraft installation, verification, software control final design and development, and flight test evaluations. Flight assessment results were outstanding, demonstrating the effectiveness of this revolutionary control effector for advanced military aircraft.

Supporting tests for the Actuated Nose Strake for Enhanced Rolling experiment included free-flight model studies (top) and tests on a full-scale F/A-18 in the Ames 80-by 120-Foot Tunnel (above).

Computational results of the effect of forebody strake deflection on the F/A-18 forebody.

*Close-up view of the forebody strakes on the F/A-18 High
Angle-of-Attack Research Vehicle research airplane.*

In-flight pictures of the High Angle-of-Attack Research Vehicle showing the right strake deployed (left) and smoke-flow visualization of the vortex path shed by the strake (right)

With the extremely favorable results of the HARV flight tests in the HATP, Langley initiated a cooperative program known as Strake Technology Research Application to Transport Aircraft (STRATA) with Boeing in 1997, the objective being to evaluate forebody-strake technology applied to transport aircraft configurations for enhanced directional stability and control. Because the sizing requirement for vertical tail geometry of conventional transport aircraft is usually based on critical asymmetric flight conditions, such as engine-thrust loss during takeoff or high-crosswind landings, alternate concepts that can reduce the size requirements for vertical fin and rudder areas (and thereby reduced weight) are of interest.

Unlike fighter aircraft, the typical operational angle-of-attack range for transport aircraft is relatively low, with landing approach angles of attack typically around 8 degrees. Thus, substantially less shed vortex strength exists on the fuselage at those conditions in contrast to the extreme angles of attack used by fighters. The McDonnell Douglas DC-9 and its subsequent derivatives have used fuselage forebody strakes for years to enhance directional stability at angles of attack within the transport operational environment. However, the STRATA Program was formulated to provide more fundamental information on the detailed aerodynamic effects of fuselage strakes for transport aircraft, including differential deflection for yaw control.

Langley's Gautam H. Shah led STRATA tests of a generic commercial transport model using a low-mounted swept wing and a conventional tail arrangement. Shah's tests, which were conducted in the Langley 12-Foot Low-Speed Tunnel, covered a range of geometric strake variables, including span and chord, strake incidence angle, and the effectiveness of deploying a single strake as a directional control device. The angle-of-attack range covered in the investigation was up to 25 degrees. Unfortunately, study results indicated that the magnitude of yawing moments produced by a single strake was extremely low relative to those that can be generated by conventional rudder.

Although some applications, such as stability augmentation in yaw, might use low levels of control effectiveness, a larger issue surfaced when it was discovered that the yaw control provided by the single strake was extremely nonlinear, making any application as a control device more difficult and complex than a conventional control effector, such as a rudder.

Even though the results of the STRATA tests were generally negative regarding using fuselage strakes for yaw control with representative transport aircraft in normal flight conditions, additional research might provide valuable information if such devices could improve emergency out-of-control recovery capability for extreme attitude conditions at high angles of attack.

Model configuration tested in Strake Technology Research Application to Transport Aircraft project.

Passive Porosity

The passive porosity concept consists of a porous outer surface, a plenum, and a solid inner surface as shown in the illustration. Pressure differences between high-and low-pressure regions on the outer surface communicate through the plenum, thereby modifying the pressure loading on the outer surface. In addition, a small amount of mass transfer into and out of the plenum occurs that changes the effective aerodynamic shape of the outer surface. Using passive porosity began in the early 1980s as a means of shock-boundary layer interaction control. In the late 1980s and early 1990s, however, Langley researchers began a series of exploratory investigations to apply regions of porosity for aircraft stability and control enhancement. Richard M. Wood and Steven X. S. Bauer of Langley pioneered the initial control effector research that has since developed into a well-proven aerodynamic technology with a wide range of potential applications.

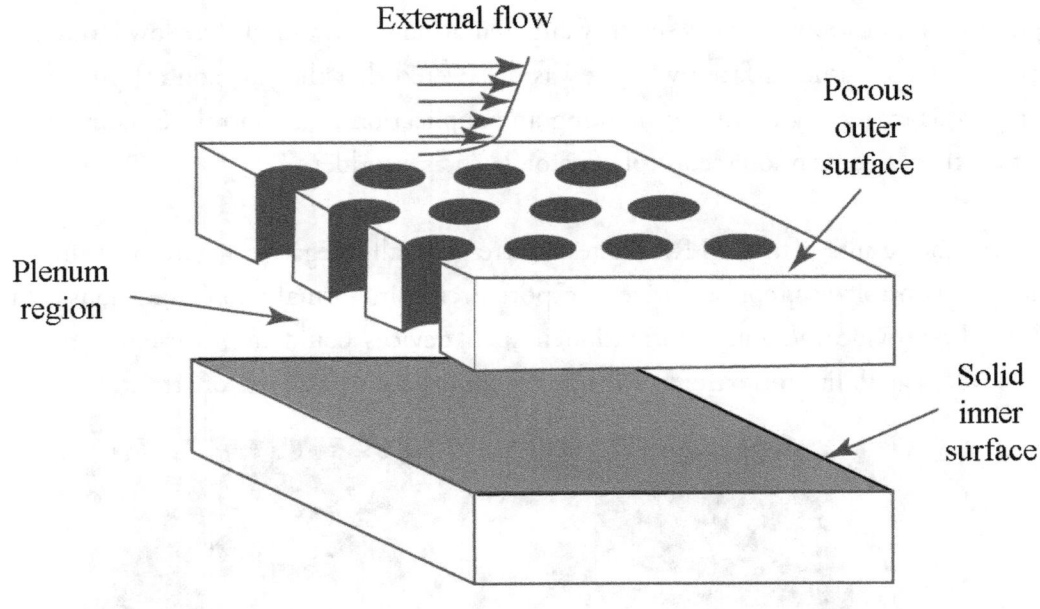

Passive porosity concept.

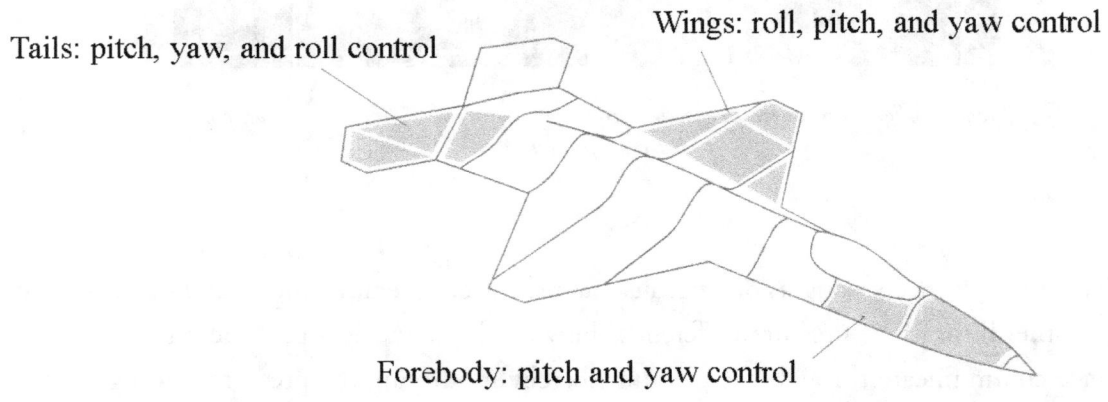

Potential passive porosity control effector arrangements.

Wood and Bauer's early research on the use of porosity to control aerodynamic moments included an effort with Michael J. Hemsch and Daniel W. Banks to evaluate the potential of porosity to alleviate large, uncommanded yawing moments generated by asymmetric vortex shedding on long pointed forebodies at high angles of attack. In the early 1990s, Bauer and Hemsch conducted an experimental wind-tunnel test in the Langley High-Speed 7- by 10-Foot Tunnel using porous and solid forebody models that demonstrated the ability of porosity to virtually eliminate such asymmetries. Wood and Banks immediately followed this test with a study in the 14- by 22-Foot Tunnel to couple forebody strakes with passive porosity to enhance the control authority of the both technologies.

In the early 1990s Wood led several teams involving Industry and DoD investigating advanced aerodynamic control effectors for military aircraft. Maturation of passive porosity technology was a major focus in both efforts. These programs resulted in the development of two fully porous wing models that underwent extensive testing in the 14- by 22-Foot Subsonic Tunnel and have served as the basis for most industry investigations.

Perhaps the best-known application of Langley passive porosity technology was to the U.S. Navy F/A-18E/F aircraft to help solve an unacceptable lateral "wing drop" characteristic that had been unexpectedly encountered during the early developmental flight testing of the preproduction aircraft. As discussed in *Partners in Freedom*, the availability of Langley's database and experience with passive porosity proved to be a critical contribution to the resolution of the problem and was incorporated in subsequent production aircraft. In this application, porosity was used by Navy and NASA engineers to stabilize flow separation phenomenon encountered during transonic maneuvers, ensuring symmetric stall behavior.

Langley researchers have also pursued the application of passive porosity for aircraft control effector systems. Applied to different areas of an aircraft, the use of porosity can permit the generation of a variety of control forces and moments. In applications, the porous cavities and interconnected plenums would be controlled and actuated by valves or other pneumatic control devices. Passive porosity has no external moving parts, preserves the vehicle outer mold lines, and provides a control force that varies linearly with vehicle lift in a predictable manner.

Langley's staff has also developed CFD methods to augment experimental studies by assisting in the analysis and design of passive porosity concepts. The CFD breakthrough was by Daryl L. Bonhaus in 1999, when he successfully reformulated the passive porosity boundary conditions. His efforts greatly improved the accuracy of passive porosity analysis and allowed for the design of passive porosity control effectors. The aerodynamic integration of passive porosity control effectors into revolutionary new configurations involves a departure from current aircraft design methods. Currently, aircraft airfoils are designed to maximize cruise performance, and then trailing-edge flaps (elevons, ailerons, etc.) are sized to provide sufficient moments to provide adequate control of the aircraft. With passive porosity concepts, the airfoils will be designed to generate a specified pressure distribution that can be modified by the actuation of the porosity device. Thus, the design of control effectors benefits greatly from the use of CFD. Modification, development, and validation of the highly successful Langley TetrUSS by Neal T. Frink, Daryl Bonhaus, Steve Bauer, and Craig A. Hunter has provided a powerful design tool for applications of the passive porosity technology.

As might be expected, the numerous potential applications of passive porosity have resulted in extensive, ongoing cooperative research between Langley, industry, and DoD. In one such activity, Craig Hunter, Sally A. Viken, Richard Wood, and Steve Bauer led a design and analysis study of the application of passive porosity control effectors to an advanced multimission tailless fighter configuration developed under the Air Force Aero Configuration/Weapons Fighter Technology Program. Focusing on the low-speed, high angle-of-attack flight regime, the team used TetrUSS to develop a series of longitudinal and lateral-directional controllers. Study results indicated that passive porosity effectors could produce large nose-down control at high angle of attack, equaling or exceeding the control authority provided by conventional elevons. As discussed in the previous section on forebody control concepts, yawing moment is especially critical for low-speed maneuvers, and the study identified several yaw control concepts that generated large yawing moments, with low levels of adverse rolling moments.

Demonstrated success of passive porosity application to the F/A-18E/F prototype wing-drop problem, and the rapidly maturing aerodynamic analyses of advanced control effectors, resulted in a significant level of interest currently existing in future applications of the technology. Cooperative studies with industry and DoD are continuing, and all indications point toward extremely effective, versatile flow control devices based on passive porosity for aircraft control. Yet to be demonstrated, however, is the application and assessment of the concept to full-scale hardware and risk-reduction flight testing.

Active Flexible Wing

The rapid emergence of advanced composite technology for wing design and fabrication stimulated significant national interest in the late 1980s and early 1990s in the potential integration of active control and flexible wings for weight savings. By reducing wing stiffness requirements and instead employing advanced control technologies to avoid aeroelastic problems, the innovative use of aeroelastic characteristics and control systems has promised a potential breakthrough in wing design. Designers of conventional civil and military aircraft are now constrained by aeroelastic and structural phenomena such as flutter and aeroelastic-induced aileron control reversal. With revolutionary composite and control design procedures, researchers are exploring the benefits of using, rather than avoiding, wing flexibility effects. Langley researchers have been working in this research area since the middle-1980s, with participation by Center experts in aeroelasticity, flutter, active controls, and advanced instrumentation.

In the 1980s, two researchers at Rockwell International Corporation (now Boeing), Gerald Miller and Jan Tulinius, conceived an active flexible wing concept for advanced fighter aircraft. The Rockwell concept exploited wing flexibility and active leading- and trailing-edge control surfaces to provide high-performance roll rates without the use of all-movable horizontal tails. Discussed in a previous section on control of aeroelastic response, a cooperative program among Langley, the U.S. Air Force, and Rockwell was formalized to research and demonstrate this active flexible wing

Multiple exposure photograph of the active flexible wing model mounted in the Langley 16-Foot Transonic Tunnel.

(AFW) concept. Active control concepts considered during the research effort included active flutter suppression and rolling-load maneuver alleviation. The active flutter suppression system's goal was to use multiple surfaces and sensors to prevent two flutter modes occurring simultaneously. For the rolling-maneuver load alleviation design, the goal was to reduce wing loads at multiple points on the wing while executing roll maneuvers representative of fighter aircraft. As the research efforts intensified, Langley researchers successfully completed additional tests in 1989 and 1991 involving more than 20 researchers. Contributors to the program included Boyd Perry, Stan Cole, Carey S. Buttrill, William M. Adams, Jr., Jacob A. Houck, Anthony S. Pototzky, Jennifer Heeg, Martin R. Waszak, Vivek Mukhopadhyay, and Sherwood H. Tiffany. Key accomplishments of this second AFW Program included single- and multiple-mode flutter suppression, load alleviation and load control during rapid roll maneuvers, and multi-input/multi-output multiple function active controls tests above the open-loop flutter boundary. A highlight of the effort was a special issue of the highly respected AIAA Journal of Aircraft for January-February 1995 that summarized the research details, findings, and conclusions of the project.

After decades of NASA, DoD, and industry research on actively controlled flexible wings in wind tunnels, the next major challenge, piloted full-scale aircraft flight demonstrations, was ready to be addressed. To meet the challenge, NASA, the Air Force, Boeing, and Lockheed Martin initiated and are now participating in an Active Aeroelastic Wing (AAW) Flight Program at NASA's Dryden Flight Research Center using a modified Boeing F/A-18 aircraft. The program goal is to demonstrate improved aircraft roll control through aerodynamically induced wing twist on a full-scale high performance aircraft at transonic and supersonic speeds. Data will be obtained to develop design information for blending flexible wing structures with control law techniques to obtain the performance of current day aircraft with much lighter wing structures. The flight data will include aerodynamic, structural, and flight control characteristics that demonstrate and measure the AAW concept in a comparatively low cost, effective manner.

Begun in 1996, the AAW Program completed the wing modifications required for the research program. In preproduction versions of the F/A-18, the wing panels were relatively light and flexible. During preproduction flight tests (particularly at high-speed, low altitude conditions), the wings were too flexible for the ailerons to provide the required roll rates. This unacceptable result occurred because the high aerodynamic forces against a deflected aileron and resulting wing torsion would cause the wing to deflect in the opposite direction, causing severe degradation of roll control in the intended direction. The F/A-18 production aircraft were subsequently fitted with stiffer wings to minimize the undesirable loss of roll control.

NASA's active aeroelastic wing F/A-18A research aircraft maneuvers during a test mission.

The wing panels on a Dryden F/A-18 research aircraft were modified for the AAW research program. Several of the existing wing skin panels along the wing's rear section just ahead of the trailing-edge flaps and ailerons have been replaced with thinner, more flexible skin panels and structure, similar to the preproduction F/A-18 wings. In addition, the research airplane's leading-edge flap has been divided into separate inboard and outboard segments, and additional actuators have been added to operate the outboard leading-edge flaps separately from the inboard leading-edge surfaces. By using the outboard leading-edge flap and the aileron to twist the wing, the aerodynamic force on the twisted wing will provide the rolling moments desired. As a result, the flexible wing will have a positive control benefit rather than a negative one. In addition to the wing modifications, a new research flight control computer has been developed for the AAW test aircraft, and extensive research instrumentation, including more than 350 strain gauges, has been installed on each wing.

Langley's Jennifer Heeg leads a team of dedicated and skilled professionals currently conducting research within the AAW Program on the development and validation of scaling methodology for reliable wind-tunnel projections to flight. The team is applying a Langley method called wind tunnel to atmospheric mapping (WAM) for scaling and testing a static aeroelastic wind-tunnel model of the AAW aircraft. The WAM procedure employs scaling laws to define a wind-tunnel model and wind-tunnel test points such that the static aeroelastic flight-test data and wind-tunnel data will be correlated throughout the test envelopes. The specific scaling is enabled by capabilities of the Langley TDT and by relaxation of scaling requirements present in the dynamic problem that are not critical to the static aeroelastic problem.

AAW flight tests began in November 2002 with checkout and parameter identification flights. New flight control software was then developed based on data obtained during 50 research flights over a 5-month period in 2003. The Langley-Dryden team evaluated the controls' effectiveness in twisting the wing at various speeds and altitudes. A second series of research flights planned to last into 2005 is scheduled to evaluate the AAW concept in a real-world environment. Obvious issues regarding flutter suppression, failure modes, and cost-benefit trades remain to be addressed before AAW concepts can be applied to production aircraft.

The NASA Smart Vehicle Program

Langley's researchers had aggressively pursued advanced control effector studies with industry and DoD in the 1990s. In one of the most important cooperative projects, Richard M. Wood served as team lead for a NASA-industry-Air Force military team (1990 to 1993) that included members from Langley, NASA Ames, McDonnell Douglas, and the DoD. The focused activity resulted in the conception and development of four advanced control effector technologies that were subsequently adopted by industry. The four patented technologies were passive porosity, advanced planforms, micro drag bumps, and advanced forebodies. Almost a decade later, an opportunity arose to carry some of the concepts to flight tests.

As discussed in other sections of this document, NASA initiated a program known as RevCon in 2000 to accelerate the exploration of high-risk, revolutionary technologies. The nine projects initially selected included a study known as the Smart Vehicle (SV) Program. The Smart Vehicle was envisioned by NASA and its partners to be an unmanned advanced technology demonstrator that would demonstrate the application of a set of novel aerodynamic effectors and an advanced adaptive vehicle management system to enhance the operational effectiveness of revolutionary air vehicles. Specifically, the demonstrator would use novel aeroeffectors as primary control devices in the research envelope, demonstrate the effectiveness of an adaptive closed loop vehicle control system that accommodates anomalies in the research envelope, and define the benefits of integrating an adaptive control system and novel actuators. The flight envelope was to include a design Mach number of about 0.8 at an altitude of 25,000 ft. The project would be conducted by a team led by Langley, with team members including Lockheed Martin Tactical Aircraft Systems (LMTAS), Physical Sciences, Inc., Tel Aviv University, Naval Air Systems Command, and NASA Dryden Flight Research Center.

Initially, the team considered several advanced control effector concepts, including passive porosity, spanwise blowing, seamless control effectors, inflatable flaps, pulsed jet vortex generators, oscillatory blowing, wing decamber "bumps," drooped leading-edge flaps, and reaction control systems. The team judged each of the foregoing effector concepts (and others) on research merit in terms of technology readiness level and disciplinary research and development required. The initial control effector concepts chosen were passive porosity, seamless control effectors, spanwise blowing, and decamber bumps. Fluidic thrust vectoring was identified as a desirable yaw control effector but was eliminated to reduce program cost.

Artist's sketch of the RevCon Smart Vehicle Demonstrator.

Under the teaming agreement of the RevCon SV project, Langley would be responsible for project lead; risk reduction studies in low- and high-speed wind tunnels including the TDT and 16-Foot Transonic Tunnel facilities; design and development assistance of the advanced control laws to be used by the vehicle; development of design criteria for the passive porosity and spanwise blowing concepts; and vehicle fabrication, assembly, integration, and selected ground testing. Following the SV demonstrator's fabrication and initial checkout at Langley, it would be shipped to Dryden for final assembly, preflight completion and checkout, and research flight tests.

Langley's Manager for the SV Program was Jean-Francois M. Barthelemy, assisted by Scott G. Anders and Henry S. Wright. Major Langley contributions to the program were provided by Bobby Berrier (lead for aerodynamic data base development), Steven X. S. Bauer (porosity), Thomas M. Moul (aerodynamics), Richard F. Catalano (system studies), Stuart Johnson (system studies), John V. Foster (wind-tunnel testing), Richard J. Re (wind-tunnel testing), and Richard DeLoach (wind-tunnel techniques).

By spring of 2001, the NASA-industry team had conducted a phase I study and developed an attractive project within 8 months. At that time, NASA had developed an interest and vision in "morphing aircraft" that would employ many of the concepts involved in the SV demonstrator, and

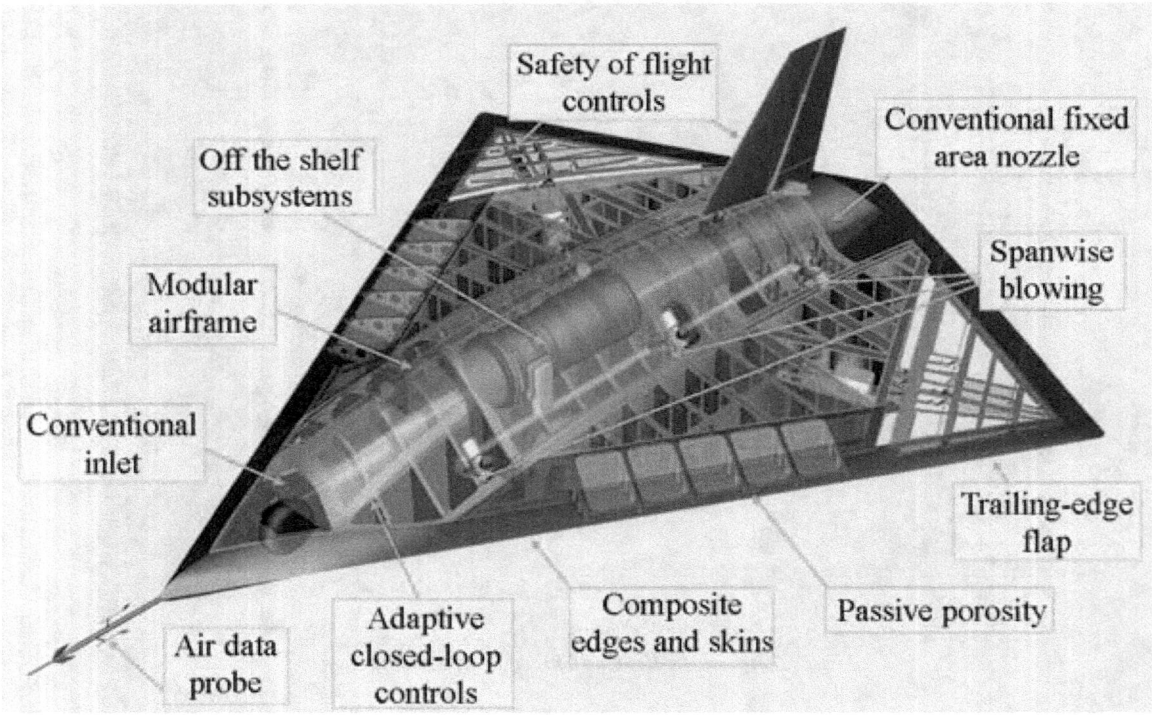

Elements of the Smart Vehicle.

the SV concept was viewed by many as a stepping-stone to future revolutionary aircraft activities. However, it was also recognized that many of the technologies required additional research and development, and that the anticipated costs of the demonstrator program would be large. During the phase I studies, the demonstrator vehicle design had accelerated: conceptual design, initial structural and flutter analyses, CFD calculations for performance, and 6-degree-of-freedom flight simulations had been accomplished; systems-level analyses of various proposed technologies had been completed; and two wind-tunnel entries had been accomplished (low-speed configuration screening and transonic performance and control effector evaluations).

Anticipated milestone events included a NASA RevCon go-ahead for phase II activities in the fall of 2001, shipping the vehicle to Dryden in the fall of 2004 and conducting the first flight tests during the summer of 2005. Unfortunately, NASA's funding for RevCon was redirected to providing a return-to-flight capability for the NASA X-43A (Hyper X) Program following the X-43A accident on June 2, 2001. NASA subsequently terminated its RevCon Program on September 30, 2001.

Following the cancellation of the RevCon Program, Langley provided advocacy and funding for a follow-up project within its Revolutionary Airframe Concepts Research (RACR) project. Known as the Aeronautical Flight Vehicle Technologies Demonstrator (AVTD) project, a similar SV vehicle configuration was retained based on a version of the LMTAS Innovative Control Effector Vehicle that had been used for several studies on advanced fighters, including an uninhabited air combat system. In addition, the new project's goal was changed to provide a robust, reusable, unmanned, modular high-performance flight demonstrator to serve as a test bed for maturation of advanced technologies, and the scope of applications was changed to emphasize potential civil as well as military applications.

By 2003, the research within AVTD had proceeded to include systems-level assessments of control effectors, vehicle conceptual design and cost estimates, wind-tunnel entries in the Langley 16-Foot Transonic Tunnel and the Langley 16-Foot TDT, supporting CFD studies for flow diagnostics and analysis, and simulation of flying qualities. Bobby L. Berrier, John Foster, Jerome H. Cawthorn, Richard J. Re, Craig A. Hunter, and Steve Bauer had conducted aerodynamic studies, and several options for vehicle configurations (wing planforms and configuration layouts) had been assessed.

A 7-week test program in the 16-Foot Transonic Tunnel at Mach numbers from 0.3 to 0.9 and angles of attack from –5 to 15 degrees significantly advanced the state of the art in advanced control effectors. The team conducted parametric studies and data were obtained on the effectiveness of passive porosity control effectors, seamless trailing-edge flaps, porosity and trailing-edge interactions, deployable bumps, and a rudder. The staff used CFD also to predict the effectiveness of deployable bumps and passive porosity, providing good qualitative trends.

Unfortunately, NASA cancelled the project in fiscal year 2004 because of resource constraints in the face of relatively large costs projected for SV flight tests.

Status and Outlook

The remarkable progress made by Langley researchers on advanced innovative control effectors continues to generate significant interest for applications, particularly within the military community. With the advent of highly lethal, signature-sensitive combat environments, designers are striving for unconventional approaches to maximize performance and handling qualities while maximizing stealth and reducing costs, maintenance, and vehicle weight. New technology fields are being pursued to maximize the effectiveness of advanced controls. For example, the introduction of controls allocation technology has reduced the concern about unwanted cross-axis moments from each effector. This technology was motivated by configurations that had several control effectors and could benefit from blending of the effector inputs to minimize control deflections, yet provide the desired control moments. Control law software has been developed that can provide the commanded control moment by combinations of control positions to optimize control strategy. One application is to provide proverse yawing moments with roll control while minimizing additional yaw control. At this time, however, the major barrier to implementation of many of the concepts is the lack of full-scale aircraft flight experience to resolve numerous application issues that cannot be resolved at model scale.

With the introduction of high-performance Uninhabited Air Combat Vehicles, many technology concepts conceived and developed at Langley are now appropriate for future applications. Langley's staff is continuing its quest to provide designers with valuable technology information for use in design and trade studies for future air vehicles. Undoubtedly, extensive demonstrations of the technologies discussed herein by manned or unmanned vehicles will occur in the near future.

Personal Air Transportation Concepts: On-Demand Revolution in Air Travel

Concept and Benefits

By the 1990s, demand for public air transportation in the United States had intensified to the point that widespread frustration over system shortcomings existed. Commercial flight delays due to the cascading effects of bad weather, inconvenient and indirect flight schedules, lack of physical comfort and overcrowding within airports and airplanes, and excessive "lost time" getting to and from remote airports had become more frequent. These frustrations stimulated the technical, regulatory, and political communities to consider and evaluate proposals for innovative modifications to the current air transportation system. Following the world changing events of September 11, 2001, the resulting adjustments to commercial aviation operations to ensure security, and the delays caused by security breaches, further aggravated the lost time and personal inconvenience of air travel.

Assessments of the current system's shortcomings have focused on the problems created since airline deregulation resulted in the centralized "hub and spoke" system now used by most major air carriers. Over 75 percent of aviation passenger traffic within the United States is conducted through only 30 major airports. Although the hub and spoke system will continue as a vital asset for long distance travel, it does not serve rural, regional, and intraurban travel very effectively. For travel distances of 100 to 500 miles, the public chooses to use automobiles 20 times more often than aircraft. Considering that the average home-to-destination auto speed for these trips is only 35 mph, and that projected highway congestion over the next 25 years will reduce this speed even further, a critical need exists for a revolutionary form of faster travel that can avoid the gridlock of either highways or hub and spoke airports. More frequently, analysis of the problem reveals that expanded use of over 5,000 public general aviation airports within the existing U.S. infrastructure might provide a solution to the anticipated future decline in public mobility.

Innovative concepts for personal air travel that have been periodically revisited over the past 80 years include personal-owner general aviation aircraft; personal air travel through distributed, on-demand air taxi operations; and even futuristic, self-operated personal air vehicles (PAV) capable of both roadable and airborne operations, as well as short-field operations from neighborhood roads and "at home" storage. Such concepts use the distributed air operational scenarios discussed previously and are compatible with an innovative and revolutionary vision of potential future air transportation. The unfulfilled perspectives within these visions would permit an unprecedented level of mobility for average citizens, resulting in significant improvement in productivity and quality of life.

Within its charter and mission to define and develop concepts to improve the quality of life for the U.S. public, and to conceive and mature technologies required for new systems and vehicles, NASA is conducting research designed to advance and accelerate the state of air mobility.

The Nation's underused public airports might provide significant public mobility.

To permit greater mobility and freedom in air travel, near-term NASA goals have been to develop and demonstrate technologies enabling the safe and cost-effective operations of today's small aircraft from the vast number of public airports. Additional efforts are underway to provide designers with methods of transforming today's personal-owner aircraft to eliminate extensive current public perceptions of unacceptable operational cost, lack of comfort, lack of safety (especially in adverse weather conditions), objectionable noise, and unacceptable training time and costs. Vehicle-oriented research goals are to develop technology for a small airplane that can fly out of small airports, to keep the cost less than $100,000 while being equipped for all weather operation, and to be unobjectionably quiet to the surrounding community. NASA is also conducting research on pilot-vehicle automation to make flying nearly as simple as driving a car.

Langley has also conducted research to enable the design of small (two passenger) personal-owner aircraft that have door-to-door travel capability, including the ability to travel in a limited roadable fashion on side streets, while taking off and landing at very small airfields. While the desire to have a true flying car is widespread and understandable, NASA researchers believe that the dream of the flying car will continue to be unfulfilled even 25 years from now. The problems that result when full highway roadability is coupled with flight capability will continue: vehicles that aren't very good cars, aren't very good aircraft, and are much more expensive than both.

Challenges and Barriers

An on-demand aviation system, with convenience, low costs, and proven safety, has been a dream of aviation innovators and futurists since the earliest days of flight. The proposed distributed air operational system's capacity to provide this capability has been firmly blocked in the past by a multitude of technical, regulatory, economic, and operational issues. The following discussion of challenges and barriers provides background on that which must be overcome to permit the successful implementation of a distributed, on-demand air system. The issues are addressed for two different vehicles: a near-term advanced general aviation-type aircraft designed for intercity and rural travel from nonradar-equipped small airports, and a futuristic roadable aircraft with ultra–short-field takeoff and landing capability designed for intraurban short trips. Both vehicles might be flown by either air-taxi pilots or by private owners. NASA and its partners are engaged in pioneering efforts to accelerate solutions to existing and anticipated challenges and barriers for both types of aircraft. Through its programs on general aviation technologies, small aircraft transportation systems, and personal air vehicles, the Agency is contributing significant stimuli toward this objective.

Disciplinary Challenges

The development of economically feasible air vehicles with satisfactory performance, flying qualities, and safety is a traditional NASA mission. Within the disciplines of aerodynamics, propulsion, stability and control, structures, and flight deck technology, NASA supplies advanced concepts and data for use by designers to ensure that the mission requirements of new vehicles can be met.

An advanced general aviation aircraft envisioned for intercity travel from rural airports creates technology requirements that are driven by cost, safety, security, environmental compatibility, and ease of use. Within the disciplines, these requirements translate into technical simplicity and innovative approaches to lower acquisition and operational costs. For example, reliable propulsion systems comparable with automotive systems will be mandatory. Structures and materials must be low cost, easily replaced, damage tolerant, and provide a high level of crashworthiness. For applications envisioned, aerodynamic characteristics of a vehicle are probably within the state of the art; however, the pilot-vehicle interface for flight planning, guidance, stability, and control will have to be exceptionally good to permit safe operations in marginal weather conditions by novice pilots.

The disciplinary requirements for a futuristic intracity aircraft are tremendously more demanding than an intercity vehicle. Envisioned as the ultimate personal-owner aircraft with ultra–short-field takeoff and landing capability and all-weather operations, the vehicle requires extensive advances in disciplinary technologies far beyond levels available today. Sophisticated powered-lift concepts, including morphing technologies like circulation control, will be required, as well as high power-to-weight engines, sophisticated control systems with extensive artificial stabilization, advanced navigation and guidance, and lightweight structures.

Operational Challenges

Arguably, the challenges and barriers to future personal air transportation are more dominant in the area of operations than those within the technical disciplines. For near-term aircraft, operating from small, nonradar equipped airports will pose stringent requirements on situational awareness and collision avoidance (both airborne and ground operations), guidance displays, and weather awareness. The far-term personal air vehicle faces even more issues. The complexity of flying an airplane in all-weather conditions (compared with driving a car), and the difficulty and costs involved in gaining a pilots license, create immediate barriers unpalatable to most of the public. The issues of regulatory requirements, certification, liability, and operational flexibility will require years of study, debate, and resolution before the dream can be realized.

Economic Challenges

No single factor affects the public's interest and willingness to use new technology more than cost. Acquisition cost of an excessively sophisticated vehicle (compared with automobiles) will immediately undermine advantages of new transportation capability and deter the application of advanced technology, no matter how impressive the benefits may be. Solutions regarding additional costs associated with pilot training and currency, maintenance, insurance, medical certificates, and other factors will require innovative approaches and perspectives.

In summary, the challenge of providing increased mobility and productivity to the public via advanced personal air transportation involves an extensive and complicated series of issues that ignite classical confrontations between technological, regulatory, and economic factors. Many argue that these same factors have faced every step of advancement in transportation, from sailing ships to locomotives to automobiles, yet when the barriers were ultimately addressed, new forms of transportation were adopted, and naysayers were proved wrong. In its role as an advanced research and development organization, NASA is addressing these issues.

Langley Activities

Although NASA has not conceived, developed, and demonstrated a vehicle appropriate for the futuristic personal air vehicle vision, early contributions include concepts, technology, and data in the areas of aerodynamics, flight dynamics, structures and materials, flight deck technology, propulsion, and controls, all key to the potential success and airworthiness of the vehicles. Past Langley research on relatively inexpensive concepts for individual airborne transportation are worthy of note.

During the 1950s and 1960s Langley conducted research on personal "flying platforms."

In the middle 1950s, considerable interest was expressed by the U.S. Army and associated industries over development of a general-purpose vertical takeoff and landing (VTOL) aircraft that could be operated by a single person and serve as a reconnaissance aerial vehicle. As envisioned, the vehicle would be able to hover or fly forward at speeds up to about 50 kts. Military versions would carry a payload of about 1,000 lb, and it was expected that the proposed vehicle would be simpler in construction and easier to operate and maintain than a small helicopter. Potential civil applications for the concept were quickly recognized, and studies of "flying platform" vehicles began to emerge.

With an extensive ongoing research program in rotorcraft and VTOL technology in the 1950s and 1960s, researchers at Langley conducted several investigations of the performance, stability, and control of such concepts. The necessity of minimizing the rotor diameter and slipstream velocity, and for providing protection for surrounding personnel and equipment, prompted the use of ducted fans—rather than rotors—which became a focal point of the Langley studies.

Under the direction of Marion O. McKinney, a team of Langley researchers conducted conventional static force and moment wind-tunnel tests as well as free-flight tests of several configurations incorporating either two- or four-duct arrangements. As early as 1954, McKinney's team started a series of free-flying model tests of ducted-fan flying platform configurations. Robert H. Kirby, Lysle P. Parlett and Charles C. Smith, Jr., led the research efforts, and early results revealed two serious problems inherent in any fixed-geometry ducted-fan configuration in forward flight. These problems are an undesirably large forward tilt angle of the platform required for trim at high speeds and nose-up pitching moment that increases rapidly with forward speed. Solutions to these two problems are imperative for practical operation of ducted-fan vehicles.

Parlett's test results indicated that a tandem two-fan arrangement exhibited less severe tilt angle and pitching moments than a side-by-side fan arrangement, but the tandem configuration required appreciably more power for forward flight. Analysis of these early results indicated that deficiencies might be alleviated by departing from the concept of ducted fans fixed with respect to the airframe and tilting the ducts for the forward flight condition. Subsequent tests by Smith with the tilting-duct arrangement showed that the problems have been minimized, providing pioneering research data that contributed to a rising interest in compact tilt-duct aircraft configurations, such as the Bell X-22 research aircraft in the 1960s. During the height of VTOL research at Langley in the 1960s, ducted-fan vehicles were studied in detail in the Langley 30- by 60-Foot (Full-Scale) Tunnel and the Langley 14- by 22-Foot (V/STOL) Tunnel. Current knowledge of the aerodynamic performance, stability, and control characteristics of this class of vehicle was contributed by research studies at Langley and at NASA Ames Research Center.

Langley has led the Agency's efforts in advancing the personal air transportation system capabilities for the past decade. The program's conception, planning and success to date can be attributed to the personal expertise, dedication, and leadership of Langley's Bruce J. Holmes, who is internationally recognized for his leadership and personal technical research for general aviation. Rising through the technical ranks, Holmes progressed from extensive technical contributions as a researcher for advanced general aviation aircraft configurations to a visionary program manager responsible for integrating and coordinating NASA, industry, FAA, and academic researchers on national-level programs to improve the national air transportation system. Most project activities discussed in this section (especially the near-term objectives and goals) were conceived or strongly influenced by his direction. Another NASA leader in visionary and futuristic perspectives on technology and personal air travel is Dennis M. Bushnell, Chief Scientist of Langley Research Center. His persistence in achieving the unthinkable, and his challenges to researchers to think beyond the envelope, has inspired numerous advances in innovation and revolutionary concepts at Langley. Bushnell's personal interest and managerial support for far-term PAVs of the future has provided the opportunities for Langley's researchers to pursue creativity and pioneering efforts in what is recognized as an exceedingly difficult research area.

The following discussion provides an overview of some of the most critical Langley research programs and contributions to the personal air transportation arena. Three research activities have been especially noteworthy in this topic. They are the NASA Advanced General Aviation Transport Experiments (AGATE) Program, which provided advanced technology to permit the domestic general aviation industry to remain a vibrant component of aviation; the NASA Small Aircraft Transportation System (SATS) Program, which is in the process of demonstrating the ability of advanced technology to permit routine operations of small aircraft from rural airports; and PAV studies, which explore the ability of technology in the near- and far-term years to provide revolutionary personal air transportation vehicles.

The NASA Advanced General Aviation Transport Experiments Program

Following the almost total collapse of the U.S. general aviation industry in the 1980s, the Nation searched for mechanisms to provide the resurgence required to reestablish this vital segment of the air transportation system. This decline included significant decreases in small aircraft deliveries, general aviation fleet size, flight hours, public use airports, pilot population, and new student pilots. At its peak in 1978, the general aviation industry delivered 14,398 aircraft. In 1994, the number of aircraft deliveries had fallen to an all-time low of 444. The average age of general aviation aircraft flying at that time was about 30 years. Flight deck technologies in use dated back as late as the 1950s, and piston propulsion technologies had remained unchanged for the past 40 years. Along with modifications of product liability issues, the potential impact of advanced technology to improve safety, reduce operating and training costs, and stimulate interest in general aviation was pursued.

Building on his long established relationship with the general aviation industry, Langley's Bruce Holmes took the lead in the formulation of cooperative planning with industry to create a new future for general aviation. Following Holmes' highly successful advocacy efforts within NASA and industry, in 1994 NASA created an AGATE Consortium under the general aviation element of the Advanced Subsonic Technology Program Office to revitalize national general aviation through the rapid development and fielding of new technologies, with a view toward providing an impetus for a new small aircraft transportation system.

Under the direction of Holmes and his Deputy Michael H. Durham, the AGATE team focused on goals that included the development of affordable new technologies, as well as new approaches to meeting industry standards and certification methods for airframe, cockpit, flight training systems, and airspace infrastructure for next generation single pilot, four to six seat, all-weather light airplanes. The AGATE alliance eventually grew to more than 50 members from industry, universities, the FAA, and other government agencies. Starting with NASA seed funding of $63 million in 1994, NASA, the FAA, the Small Business Innovation Research Program (SBIR), industry, and universities pooled nearly $200 million in combined resources among 39 cost sharing partners. About 30 other partners also joined the effort as noncost sharing, supporting members of the AGATE Consortium, totalling nearly 70 members.

The cumulative result of the AGATE alliance produced a revolution in the research and technology deployment capacity for all sectors of the general aviation industry. AGATE provided a voice for industry to provide national clarity and action on key technology development, certification, and standard-setting activities. During the AGATE Program, which ended in 2001, the general aviation industry research and technology capacity advanced from virtually nonexistent to world-class in avionics, engines, airframes, and flight training. Integrated with these advances was the rising advocacy for deployment of small aircraft at the Nation's distributed public airports for unprecedented advances in personal mobility and productivity for the public. Extensively cited as a classic example of NASA aeronautics at its best, the AGATE Program is viewed as the catalyst responsible for current interest in expanded use of small aircraft transportation systems.

The NASA Small Aircraft Transportation System Project

Following the highly successful AGATE Program, Bruce Holmes and his team of NASA-industry-academia-FAA partners turned attention to the next step in demonstrating potential benefits of a small aircraft transportation system for the U.S. public. Holmes began a difficult advocacy effort, which entailed a major step up in challenges from the technically focused AGATE Program. The new transportation system-focused program would entail numerous nontechnical factors not in the immediate control of NASA, such as local politics, community planning, and regulatory

The NASA SATS Program envisions the on-demand use of small aircraft from distributed public airports.

responsibilities. Despite outspoken critics and skeptics, Holmes and his team secured NASA funding in 2001 to begin a new program, called SATS, which would "put wings on America" and minimize the transportation woes and gridlock associated with clogged interstates and hub-and-spoke airports.

SATS highlighted the fact that, away from the congested hub-and-spoke airports, underused capacity at over 5,000 public use airports is abundant. Unfortunately, fewer than 10 percent of public airports have precision instrument guidance, communications, and radar coverage for safe and accessible near-all-weather operations. To move to the new paradigm of small aircraft operating as a key component of the proposed transportation system, flight deck and flight path technologies and operating procedures would have to be developed to provide the missing components.

Many enabling technologies from the AGATE Program and a related program, the General Aviation Propulsion (GAP) Program managed by the NASA Glenn Research Center were poised to contribute to this futuristic vision. These technology advances included:

- New turbine engines with revolutionary thrust-to-weight and cost metrics
- Commercial off the shelf (COTS)-based avionics with vast improvements in cost, reliability, and capabilities
- Highway-in-the-sky graphical pilot guidance systems
- New approaches to crashworthiness
- Streamlined composite airframe manufacturing techniques
- Ice protection technology
- Digital engine controls (for single-lever power control)
- Graphical weather information in the cockpit
- Advanced flight training and pilot certification processes

With such technology now available, the SATS vision is to provide the Nation with an alternative to existing road and airline choices for travel. Goals include hub-and-spoke-like airport accessibility to the smallest of neighborhood airports, without needing radar and control towers, and without needing more land for protection zones around small airports. Obviously, this travel alternative must be cost-competitive with existing choices and meet public expectations for safety and accessibility.

Early consumers of SATS would have access to air-taxi-like systems with hired pilot operations. The SATS project goal is to develop technologies and operating capabilities to enable affordable, on-demand, near all-weather access to even the smallest of markets. Scheduled services may also appear in more dense transportation markets as entrepreneurs discover effective ways to meet market demands.

The congressional budget appropriation for the SATS Program included a mandate to prove that the SATS concept works. This mandate includes demonstration of four operational capabilities enabled by the integration of emerging technologies from the AGATE and GAP Programs. These four capabilities are:

- Higher-volume operations at airports without control towers or terminal radar facilities
- Lower adverse weather landing minimums at minimally equipped landing facilities
- Integration of SATS aircraft into a higher en route capacity air traffic management system with complex flows and slower aircraft; and
- Improved single-pilot ability to function competently in complex airspace

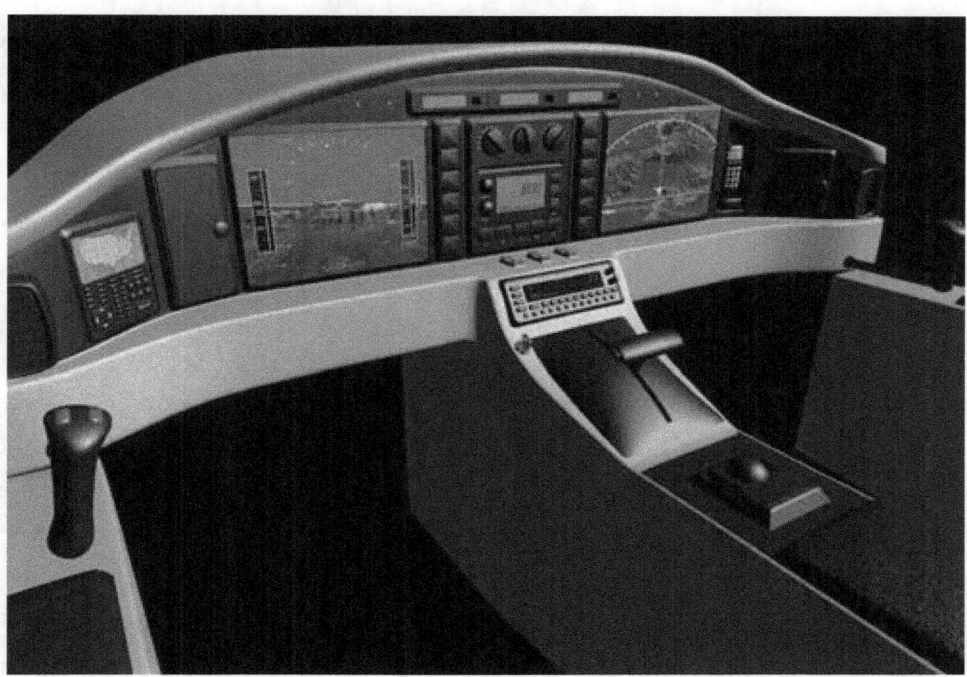

Representative cockpit display for Small Aircraft Transportation System applications.

Initial Langley planning for the SATS effort was led by Holmes and Durham within the General Aviation Program Office. Key NASA researchers included James R. Burley, David E. Hahne, Stuart Cooke, and Allen C. Royal. Later, a team of implementers was assigned to focus the SATS efforts, conduct the research, and ensure the success of the project objectives. Jerry N. Hefner was assigned as Project Manager for SATS, assisted by Langley researchers Guy Kemmerly, Sally C. Johnson, Mitchel E. Thomas, and Stuart A. Cooke, Jr., to conduct the SATS project in a public-private partnership with the FAA and the National Consortium for Aviation Mobility.

In view of the highly successful consortium-based approach used in AGATE, NASA facilitated the formation of a public-private alliance to encompass state-based partnerships for the execution of the SATS Program. These partnerships participate in continued technology development, system analysis and assessment, technology integration, and flight demonstrations of SATS operating capabilities.

In May 2002, NASA announced it had selected a partner for a joint venture to develop and demonstrate air mobility technologies for transportation using small aircraft and small airports. Known as the National Consortium for Aviation Mobility (NCAM), of Hampton, Virginia, NCAM leads a public-private consortium of more than 130 members. NCAM SATSLab members are: Maryland and Mid-Atlantic SATSLab (University of Maryland Research Foundation), North

Carolina and Upper Great Plains SATSLab (Research Triangle Institute), Southeast SATSLab (Embry Riddle Aeronautical University), Virginia SATSLab (Virginia Department of Aviation), Michigan SATSLab (Munro and Associates), and Indiana SATSLab.

The Langley Small Aircraft Transportation System Project Office, the FAA, and the NCAM SATSLabs became the driving forces behind SATS. The U.S. Congress approved $69 million for the 5-year proof-of-concept period.

Under Hefner's leadership, Langley, NCAM, and the FAA immediately worked toward a middle 2005 proof-of-concept demonstration of new operational capabilities geared toward technologically advanced small aircraft and small airports. The 2005 demonstration location was chosen to be Danville Regional Airport, Danville, Virginia. During the 3-day event, organizers planned to offer participants a look at the potential impacts that additional small aircraft traffic could have on the Nation's skies and the business prospects that could be available for air taxis and other services interested in capitalizing on a new air transportation system that would complement existing major airports.

Several technical concepts played a key role in the 2005 demonstration. The now well-known Global Positioning System (GPS) is an absolute necessity for SATS, providing critical data on aircraft position and track. Langley researchers worked to make GPS-based systems cheaper, smaller, and easier to install, particularly for retrofits to older aircraft. A system known as Automatic Dependent Surveillance-Broadcast was developed to emit a transmission every few seconds listing information such as location, speed, and destination of the aircraft. These data can be tracked by nearby pilots, air traffic controllers or others, providing airborne traffic awareness to others. A multifunction Cockpit Display of Traffic Information (CDTI) system compiles information transmitted by other aircraft emitters and give the pilot a visual representation of airborne activity and potential collision events. SATS also explored the use of enhanced vision concepts for improved visibility at airports without landing light systems. The highway in the sky display concept (discussed in another section of this document regarding synthetic vision) would use GPS and other sensors, such as Forward Looking InfraRed (FLIR), to create an animated flight path, displayed on a computer screen or even projected onto the inside of the windshield, for maneuver guidance and flight path information. Finally, NASA explored the use of single-pilot performance-enhancing systems that increase safety while reducing the need for two-person aircrews. In such systems, onboard computers monitor aircraft systems, warn of a malfunction, and even diagnose the problem and possibly offer a fix. Another computer-based aid is a concept for a virtual copilot that could handle tasks such as calling out altitudes or watching the flight path during the eventful final approach phase of a flight. If implemented on a laptop computer or even a personal data assistant, some of these concepts could conceivably be plugged into an older aircraft for low-cost retrofit. These technologies were developed and matured by researchers at Langley, industry, and at SATSLabs across the country.

The SATS 2005: A Transformation of Air Travel event was an impressive success at Danville on June 5-7, 2005. The three-day event attracted more than 3,000 aviation enthusiasts and was considered a great success in showcasing new aviation technologies. FAA Administrator Marion Blakey and NASA Administrator Michael Griffin presented keynote addresses stressing the value of the SATS vision. SATS personnel explained technologies and operating capabilities to a standing-room-only crowd with the help of live video feeds and pre-taped segment shows on a giant screen. During the live technical demonstration, six airplanes equipped with advanced cockpit displays were able to land safely and efficiently in a small airport that normally has no radar or air traffic control support.

Now that the 5-year proof-of-concept SATS project is complete, it is hoped that the SATS concept will continue with the development of federal regulations, airspace procedures, and industry products to accommodate SATS traffic.

Personal Air Vehicle Research

In addition to the relatively near-term objectives of the SATS Program and its focus on productive, on-demand use of existing small airports, NASA has taken a fresh look at innovative and revolutionary vehicle concepts that address the futuristic vision of PAVs. In accordance with NASA's mission to conduct long-term, high-payoff revolutionary research, Langley researchers assessed the potential of current-day and emerging technologies to enable the design of technically and economically feasible consumer-piloted vehicles. Langley researchers were extremely informed in and sensitive to the shortcomings of the many failed previous attempts to exploit the owner-operated light aircraft market and planned a relevant, phased program to develop technologies required for the concept. The scope of current Langley studies began with advanced vehicles that incorporated technologies developed within the AGATE and SATS Programs and extended the vision into the future with leapfrog vehicle capabilities, including limited roadability, super short field capabilities, and semiautonomous control and navigation.

Long a dream of frustrated motorists caught up in traffic jams and gridlock, the "flying car" is an extremely controversial topic that has been the target of innovators since the early 1920s. Over 70 individual designs for flying cars have been proposed during past years, with only two achieving FAA certification and none meeting DoT automotive regulations. In addition to the basic challenges of adequate consumer demand (necessary to lower production costs), the cost of pilot training and capability, massive liability issues, the issues of air traffic control, and an extensive number of skeptics, the PAV faces significant technical challenges. Integrating classical automobile and airplane configurations has so far resulted in unacceptable deficiencies and operational capabilities from both perspectives. The resulting vehicle is typically very heavy, slow, oversized, and much more expensive than automobiles and aircraft. However, in their efforts to alleviate the current and projected limitations of the Nation's transportation systems, researchers at NASA and industry explored new concepts that might be more appropriate for the envisioned missions.

At Langley, support for PAV studies initially came from seed money provided by NASA Headquarters and locally by the previously discussed creativity initiative stimulated by Dennis Bushnell. In 2003, NASA revamped its aeronautics program and created the NASA Vehicle Systems Program. Langley's Aerospace Vehicle Systems Technology Office (AVSTO) was an essential part of the Vehicle Systems Program. After extensive workshops with the aerospace community and NASA stakeholders were conducted to establish opportunities, goals, and approaches, NASA adopted an approach focused on vehicle sectors, including six different vehicle thrusts for the future. One such thrust was a Personal Air Vehicle Sector, led by Langley's Mark D. Moore. Moore's team included Andrew S. Hahn (PAV systems analyst), Russell H. Thomas (low-noise concepts), and Kenneth H. Goodrich (guidance and control concepts).

The overriding perspective of PAV research was that the market for small, single-engine general aviation airplanes has reached a plateau for many years and that "disruptive," revolutionary technologies are required to move into an era of aggressive new growth. The introduction of disruptive technologies and regulations into the existing market will change the customers, their requirements, and thus the components and vehicles. NASA projected the potential for a substantial market for a futuristic PAV that addresses customer preferences with regard to value of time, comfort, flexibility, and travel freedom.

Examples of the benefits of disruptive technology for PAV applications are indicated by comparisons with today's single-engine piston (SEP) airplane. Langley projected 15-year goals that were ambitious: the ease of piloting a small aircraft would change from the relative difficulty of today's SEP in IFR conditions to more relaxed semiautonomous operations; community and interior cabin noise would dramatically decrease to that of the typical automobile; acquisition cost (in 2004 dollars) would reduce from $300,000 to $100,000; fuel efficiency would increase from 13 mpg to 24 mpg; accidents would be reduced from today's 6.5 per 100,000 hr to 0.5 per 100,000 hr; and field length requirements would drop from 2,500 ft to about 250 to 500 ft.

As part of its PAV sector studies, Langley assessed the benefits of advanced technology for near-term missions involving rural and regional travel. Typical missions for this class of vehicle include a design range of 500 miles with a cruising speed of about 200 mph, a gross weight of about 3,400 lbs, and with IFR flight capability. With these missions in mind, goals of the research studies were to identify concepts to reduce training time and cost, community noise, and purchase price.

One of the most ambitious, and potentially high-impact, program goals was to identify approaches that reduce training time and cost by 90 percent from today's typical 45-day $10,000 experience to only 5 days at a cost of about $1,000. The technical breakthroughs to obtain this goal are rooted in the development, integration, and robustness of flight control systems and architectures that are both failsafe and reliable. Technical approaches pursued within the NASA program included development of a "naturalistic" flight control deck with control, guidance, sensing, avoidance, and an airborne internet.

Within Mark Moore's team, Ken Goodrich addressed the challenges of providing feasible approaches for automation to make flying small planes easier. The ultimate research goal was to develop vehicle concepts that are inherently "smart" and reduce demands for expertise and capability of the human pilot. Goodrich's efforts were part of a larger Langley project, known as the Autonomous Robust Avionics (Aura) project. Led by James R. Burley of the AVSTO, Aura invested in the areas of UAVs, PAV, and rotorcraft to enable smart vehicles that reduce the demands on human pilots.

Goodrich pursued a novel, futuristic approach to autonomous operations known as "H-mode" control. While H-mode is short for the technical term haptic, it can also be thought of as "horse mode." The concept involved is based on the fact that a horse, unlike a car, is more likely to be cognizant of obstacles, try to avoid collisions or other threats, and may even know how to find its way home without inputs from its rider. Likewise, within H-mode, the human pilot and the automation system physically "feel" one another's near-term intent, with intuitive monitoring and redirection. The concept's development and validation obviously entail detailed studies of sensors, system architecture and design, conflict detection and resolution, maneuver implementation, and failure modes. However, the benefits promised by such an approach would be remarkable, yielding a radical reduction in special piloting skills and training, loss of situational awareness, and pilot error.

Another ambitious PAV goal was to identify approaches that reduce community noise generated by small aircraft from today's levels of about 84 dBA at takeoff and landing conditions to only 60 dBA. Technical challenges to obtaining this goal include reducing the community (and cabin) noise generated by propulsion systems (propellers, exhausts, etc.) while meeting requirements for vehicle performance, reliability, and cost. Langley studies included the development of integrated and shielded ducted propeller systems with active wake control and acoustic suppression. Research in this area was led by Russell H. Thomas.

Progress toward a definition of the near-term PAV progressed to the point where a notional vehicle, known as the TailFan concept, served as a focus for assessing the benefits of advanced technology. The TailFan used an advanced ducted fan for low noise and safety, as well as an automotive engine and a dramatically simplified skin-stiffened structure to reduce manufacturing costs. The TailFan resembled current general aviation configurations in shape; however, it was specifically designed to address the minimum qualifications of ease, safety, noise, and comfort for PAV applications. In addition, it would be economically viable and environmentally friendly, enabling it to compete with alternative mobility choices of autos or airlines.

Notional "TailFan" personal aircraft.

The TailFan concept centered about an automotive V-8 engine (nominally the Corvette LS-1 engine) directly driving a reduced tip-speed ducted fan. The shorter fan blades generate higher frequency noise with the duct shielding absorbing the propulsor noise through embedded acoustical liners. Using an automobile engine with extensive muffling involves additional weight compared with aircraft engines, as does using a ducted fan compared with using a propeller. However, combining the two methods permits a total propulsion system cost reduction of over 60 percent while maintaining a reasonable time between overhaul and an extremely quiet integration.

The structure was radically simplified and designed for automated manufacturing, yielding a twelvefold reduction in labor. Use of a highly formed, skin stiffened structure reduces total part count (labor and inventory costs), while an unusually high degree of symmetry reduces unique part count (tooling costs). The all-aluminum structure uses automotive manufacturing methods with an untapered, skin-stiffened wing. The same parts are used for both sides of the wing simply by flipping the three spars and using the same four ribs and three skin panels. Rivets or laser welds are used in recessed troughs to attach the wing components under a strong polyester film wing covering for smoothness and weather protection. Identical vertical and horizontal tails use the same pressing molds for the same skin-stiffened construction. An axisymmetric tailcone is made with complex curvature, integral frames, and integral stringers pressed into each quarter panel. As the external skin is assembled, the internal structure is also assembled. The fan duct is made similarly of four identical sections. The combination of reduced tooling, assembly labor, and propulsion system costs are responsible for the much lower overall cost.

This $100,000 concept solution was based on a 2,000 unit per year production rate to permit affordability in the transition market between the current low production general aviation market and the high volume production of a future PAV market. Once a substantial market existed, and large production volumes are present, many performance compromises could be eliminated through investment in a higher tooling-based design and an optimum engine designed specifically for aircraft use.

Efforts included the demonstration of an LS-1 engine on the 150-hour FAA endurance test. Success has shown that it is possible for an automotive engine to perform the aircraft duty cycle. Also, NASA worked with the FAA to adapt rules for certifying quality assurance (QA) based products, instead of the current FAA certification standard of quality control (QC). The intent of QA-based certification was not to bypass the FAA's important role to ensure safety, but to permit certified processes (instead of parts) that enable safer small aircraft products. As long as small aircraft have to use specialty, small production volume, QC-based parts, there is little chance of small aircraft being affordable to the majority of mobility consumers.

Future Gridlock Commuter

While roadable aircraft have been attempted for over 50 years, a more practical dual-mode approach might be to require only side-street travel for limited distances in the equivalent of a safe taxi mode. This capability does not require full compliance with DoT regulations and safety standards. Instead, these dual-mode vehicles may meet a minimum set of standards that permit the vehicle to achieve a compact taxi mode with very few penalties. By meeting section 500 vehicle standards, these aircraft could travel at 25 mph on side streets, as long as the footprint can be limited to a 8.5-ft width and meet some additional relatively simple ground travel requirements. This mode of travel would require the addition of a wheel-drive concept, and although limited roadability does not overly penalize the air vehicle, it does involve some additional weight and complexity.

In Mark Moore's program, a notional Langley PAV concept known as the Spiral-Duct was conceived to combine highly integrated propulsion and aerodynamic lift in a lifting duct arrangement. The inner duct provided lift and thrust, while the outer panels provided control, even at very low takeoff and landing speeds. This vehicle would be capable of takeoff and landing in less than 250 ft. With folded wings, it could travel on the ground at speeds of 25 mph. Able to carry up to two passengers, this very compact and quiet vehicle would use an electric propulsion system as efficient as current compact cars.

Notional Spiral Duct Personal Air Vehicle.

The highly integrated propulsion-aerodynamic coupling would enable a 250-ft extreme short takeoff capability with no external high-lift system moving parts, such as wing flaps on conventional aircraft, and roll control would be achieved using moving outer wing panels. For the ducted propeller arrangement, yaw and pitch control would be enhanced through embedding the control surfaces into the propeller flow, and computerized active controls would be used to achieve outstanding stability and ease of control.

In 2004 and 2005 NASA redirected funding within its Vehicle Systems Program and the PAV activities at Langley were therefore terminated in 2005.

National Planning for Next-Generation Air Transportation System

Inspired to address the shortcomings of the present air system, and the challenges and opportunities of the future, the 108th Congress mandated the development of a national plan for the Next-Generation Air Transportation System. Legislation directed that this planning effort include experts in commercial aviation, general aviation, aviation labor groups, aviation research and development entities, aircraft and air traffic control suppliers, and the space industry. The parent organization for the study was known as the Joint Planning and Development Office (JPDO). Within the JPDO effort, a Futures Working Group (FWG) of over 150 stakeholders, U.S. Government employees, and contractors was formed under the Chairmanship of Langley's Bruce J. Holmes.

In May 2004, the FWG presented a set of 11 strategies derived from interviews and scenario-based planning. Due to the wide range of changes in the world situation, economy, and operating environment for air transportation envisioned between today and the study target year of 2025, the combination of strategies was aimed at transforming air transportation while addressing the Nation's needs in plausible futures that include a tripling (or shrinking) of the demand for air travel, fossil fuels becoming less available and more costly, a public that is increasingly concerned with the environment, an accelerating pace of production and distribution of goods, radically growing importance of international travel and commerce as the world becomes more interdependent, space travel becoming a reality, and conventional aircraft sharing the skies with uninhabited air vehicles that support safety, security, and national defense. Among the 11 strategies submitted to the JPDO, recommendations were made for a national transportation system that streamlines doorstep-to-destination travel to provide users with a wide range of options for managing efficiencies, costs, and uncertainties. In addition, priority was given to design, build, and deploy a network-centric, distributed air traffic management system to increase safety, scalability, capacity, efficiency, and opportunities for free-flight operations.

Status and Outlook

The highly successful demonstration of SATS technology in June 2005 was a critical milestone in NASA's vision of the future for small aircraft in air transportation. If, as hoped, the potential of advanced technology to open up the Nation's underused public airports is appreciated by the appropriate industrial, regulatory, and technical communities, there is no doubt that entrepreneurial interests will lead to a new generation of air-taxi capabilities.

The termination of NASA research on advanced PAV concepts in 2005 virtually eliminated Langley interest in this class of vehicle. Skeptics of the vision remain steadfast, and further maturation of the technical innovations that would enable such a revolutionary change in public transportation will require extensive, dedicated research efforts capped by convincing demonstrations of the technology's benefits.

The remarkable changes in culture and resources that have occurred at Langley as it approaches its ninetieth year in 2007 have shaped, encouraged, and influenced the Center's ability to identify and assess revolutionary concepts. In its earlier history as an NACA laboratory, the staff enjoyed a technical atmosphere characterized by immaturity in aeronautical science and technology, limited expertise and availability of facilities in industry, and took a major role in the shaping of aeronautics, the aviation industry, and national defense. Freedom to conduct research on new concepts was widespread, a rich environment of technical challenges stimulated the researcher, and the technical state of the art in aeronautics accelerated at a breath-taking rate. The legendary contributions of NACA and Langley stand as evidence of the innovation and dedication that pervaded the Center in that era.

With the coming of the Space Age and the evolution of NASA Centers, the role of Langley greatly expanded, and its focus broadened to include support activities for NASA's space program and new areas of concern to the Nation, such as atmospheric science. Budgetary issues rose to new levels as the Apollo Program and ensuing space exploration activities began to have an impact on the ability of researchers to conduct studies on revolutionary concepts that strayed markedly from the evolutionary. Aeronautics programs within NASA also became more focused on near-term goals, in part to pacify NASA's stakeholders and Congress, who wanted near-term payoff and highly focused activities. In more recent years, the aviation industry has put its own unique wind tunnels, laboratories, and computational centers into operation, with capabilities as good as, or exceeding, NASA's aging facilities. Foreign technology, facilities, and advanced aircraft are now keeping pace with, or surpassing, the aeronautical leadership of the United States. Finally, aeronautics itself has become a self-professed/self-fulfilling prophecy. That is, the world of aeronautics has become— according to Dennis Bushnell—"An asymptotic, barely evolutionary, mature science with only capacity, safety, and environmental issues." The reality of this perspective has led many to refer to aeronautics as a "sunset" technical area without excitement or fresh ideas.

Management at the NASA aeronautical centers (Langley, Ames, Glenn, and Dryden) recognized the constraints being placed on innovation and proceeded to implement new funding sources, known as the Center Director's Discretionary Fund (CDDF), as incubator mechanisms for fresh ideas. By providing resource and management support for selected efforts, the Centers protected and encouraged the potential for revolutionary studies. Specific advanced studies were judged and funded on a competitive basis with the participation of top management.

In 2001, Langley management reacted to a scenario wherein the Center's programs had become increasingly tightly controlled and out-of-the-box thinking and opportunities were becoming alarming constrained. Center Director Jeremiah F. Creedon and Associate Director for Research and Technology Competencies Douglas L. Dwoyer inaugurated a new program, known as the Creativity and Innovation (C&I) Initiative, to augment the existing CDDF resources and provide a competed opportunity for researchers to acquire a maximum amount of $300,000 per year for advanced ideas. The funding provides for research equipment, salaries, and travel, and an opportunity to impact the future of aeronautics and space technology. The program evolved from the mutual interests and advocacy of several senior managers, including Dennis M. Bushnell and Joseph Heyman. Heyman was the first manager of the C&I activity, later followed by Bushnell.

The C&I Initiative covers all technical elements of Langley's mission: aeronautics, atmospheric science, access to space, planetary and space exploration, and systems studies. The program stimulates and nurtures advanced ideas with minimal management and oversight. Proposals from Langley staffers are evaluated by a group of technical peers on the basis of technical content, inventive/creative content, and researcher capability.

Results of the C&I activity have been remarkably positive. Within the area of aeronautics, some topics receiving support have been: neural network flight controller, runway topography characterization, unconventional aircraft configurations, breakthrough noise suppression concepts, distributed propulsion, and circulation control/channel wing concepts.

Current Langley Director Roy D. Bridges, Jr., has embraced the spirit of the C&I Initiative, and the program has continued to thrive as a visible sign of the value placed on innovative ideas by management. The research community has taken notice and responded in excitement and interest, sparking continued growth of the legacy of Langley's contributions in advanced research.

In September 2004 Bridges announced a new Langley organizational structure which included a new element known as the Incubator Institute. Led by Richard R. Antcliff, the institute's mission is to stimulate new business and leading-edge research efforts for the Center. Antcliff's staff includes Dennis Bushnell (Chief Scientist), Mark J. Shuart (Associate Director for Transformation Projects). The name of the organization was subsequently changed to Innovation Institute to reflect its mission as a catalyst for fresh concepts and ideas. Antcliff and his staff face a daunting challenge in promoting and nurturing innovation during a chaotic atmosphere of change within the Agency's aeronautics program. Sweeping cultural and operational transitions are now occurring at Langley resulting in closure of many wind tunnels, severe reductions in funding for aeronautical research, and reductions in workforce. In addition, the fundamental method of securing resources for research is changing to a business mode of operation featuring competitive proposals and peer-reviewed awards for studies.

As Langley strives to align itself with the major thrusts and missions of the Agency, the benefactors of its leading-edge expertise and unique capabilities look forward to a continuation of this critical national asset and to the future U.S. leadership in aviation and aerospace technology.

BIBLIOGRAPHY

Innovation

Bushnell, Dennis M.: Civil Aeronautics: Problems, Solutions, and Revolutions. *Aerospace America*, April 2003. pp 28-32.

Bushnell, Dennis M.: Application Frontiers of "Designer Fluid Mechanics"--Visions Versus Reality or An Attempt to Answer the Perennial Question "Why Isn't It Used?" AIAA Paper 97-2110, June 1997.

Yaros, Steven F.; Sexstone, Matthew G.; Huebner, Lawrence D.; Lamar, John E.; McKinley, Robert E., Jr.; Torres, Abel O.; Burley, Casey L.; Scott, Robert C.; and Small, William J.: *Synergistic Airframe-Propulsion Interactions and Integrations. A White Paper Prepared by the 1996-1997 Langley Aeronautics Technical Committee.* NASA/TM-1998-207644, 1998.

Supersonic Civil Aircraft

Alford, W. J., Jr., and Driver, C.: *Recent Supersonic Transport Research.* NASA TM X-54982, 1965.

Alford, W. J., Jr., Hammond, A. D., and Henderson, W. P.: *Low-Speed Stability Characteristics of a Supersonic Transport Model with a Blended Wing-Body, Variable-Sweep Auxiliary Wing Panels, Outboard Tail Surfaces, and Simplified High-Lift Devices.* NASA TM X-802, 1963.

Anon.: National Aeronautical R&D Goals- Agenda for Achievement. Executive Office of the President, Office of Science and Technology Policy, 1987.

Anon.: *Proceedings of the Sonic Boom Symposium.* Acoustical Society of America. St. Louis, Missouri. November 3, 1965.

Anon.: *High-Speed Research: Sonic Boom Volume I.* NASA CP-10132, 1993.

Anon.: National Aeronautical R&D Goals- Technology for America's Future. Executive Office of the President, Office of Science and Technology Policy, 1985.

Anon.: Sonic Boom Experiments at Edwards Air Force Base. National Sonic Boom Evaluation Office Interim Report NSBEO-1-67, 1967.

Anon.: *Sonic Boom Research.* NASA SP-147, 1967.

Anon.: *Sonic Boom Research.* NASA SP-180, 1968.

Anon.: *Sonic Boom Research.* NASA SP-255, 1970.

Baals, D. D., and Corliss, W. R.: *Wind Tunnels of NASA.* NASA SP-440, 1981.

Baals, D. D.: *Summary of Initial NASA SCAT Airframe and Propulsion Concepts.* NASA TM X-905, 1963.

Baals, Donald D., Robins, A. Warner; and Harris, Roy V. Jr.: Aerodynamic Design Integration of Supersonic Aircraft. AIAA paper No. 68-1018, 1968.

Baize, Daniel G.: *1995 NASA High-Speed Research Program Sonic Boom Workshop.* NASA CP-3335, 1996.

Baize, Daniel G.: *1997 NASA High-Speed Research Program Aerodynamic Performance Workshop.* NASA CP-1999-209691, 1999.

Bales, T. T.: *SPF/DB Technology.* NASA CP-2160, 1980.

Bales, T. T.; Hoffman, E. L.; Payne, L.; and Reardon, L. F.: *Fabrication Development and Evaluation of Advanced Titanium and Composite Structural Panels.* NASA TP-1616, 1980.

Bales, Thomas T.: *Airframe Materials for HSR.* Presented at the First Annual High-Speed Research Workshop. Apr 1, 1992.

Behler, Robert F.: *Blackbird Diplomacy.* Aviation Week and Space Technology, January 24, 2005, pp. 53-54.

Bond, David: The Time is Right. *Aviation Week and Space Technology,* October 20, 2003. pp. 57-58.

Brown, C. E., McLean, F. E., and Klunker, E. B.: Theoretical and Experimental Studies of Cambered and Twisted Wings Optimized for Flight at Supersonic Speeds. *Proceedings of Second International Congress of the Aeronautical Sciences.* Volume 3, September 1960, pp. 415-431.

Carlson, H. W., Barger, R. W., and Mack, R. J.: *Application of Sonic Boom Minimization Concepts in Supersonic Transport Design.* NASA TN D-7218, 1973.

Carlson, H. W.: *The Lower Bound of Attainable Sonic-Boom Overpressure and Design Methods of Approaching This Limit.* NASA TN D-1494, Oct. 1962.

Carlson, H. W.; McLean, F. E.: The Influence of Airplane Configuration on the Shape and Magnitude of Sonic-Boom Pressure Signatures. AIAA Paper 65-803, 1965.

Carlson, Harry W.: *Supersonic Aerodynamic Technology. Presented at the NASA Conference on Vehicle Technology for Civil Aviation-The Seventies and Beyond.* NASA SP-292, 1971.

Coe, P. L., Jr.; Thomas, J. L.; Huffman, J. K.; Weston, R. P.; Schoonover, W. E., Jr.; and Gentry, G. L., Jr.: *Overview of the Langley Subsonic Research Effort on SCR Configurations. Presented at Supersonic Cruise Research '79.* NASA CP-2108, Part 1, 1980, pp. 13-33.

Condit, P. M., Eldridge, W. M., Schwanz, R. C., and Taylor, C. R.: *Simulation of Three Supersonic Transport Configurations with the Boeing 367-80 In-Flight Dynamic Simulation Airplane.* NASA CR-66125, 1966.

Conway, Erik M.: *High-Speed Dreams: NASA and the Technopolitics of Supersonic Transportation, 1945–1999.* Johns Hopkins University Press, 2005.

Darden, C. M., Clemans, A., Hayes, W. D., George, A. R., and Pierce, A. D.: *Status of Sonic Boom Methodology and Understanding.* NASA CR-3027, 1988.

Darden, C. M.: *Sonic Boom Minimization With Nose Bluntness Relaxation.* NASA TP 1348, 1979.

Darden, Christine M.: *High-Speed Research: Sonic Boom.* NASA CP-3172, 1992.

Darden, Christine M.; Powell, Clemans A.; Hayes, Wallace D.; George, Albert R.; and Pierce, Allan D.: *Status of Sonic Boom Methodology and Understanding.* NASA CP-3027, 1988.

Doggett, R. V., Jr.; and Townsend, J. L.: *Flutter Suppression by Active Control and Its Benefits.* Proceedings of the SCAR Conference, NASA CP-001, Part 1, 1976, pp. 303-333.

Driver, C., and Maglieri, D. J.: Some Unique Characteristics of Supersonic Cruise Vehicles and Their Effect on Airport Community Noise. AIAA Paper 80-0859, May 1980.

Driver, C.: Progress in Supersonic Cruise Technology. AIAA Paper 81-1687, August 1981.

Driver, Cornelius; Spearman, Leroy M.; and Corlett, William A.: *Aerodynamic Characteristics at Mach Numbers from 1.61 to 2.86 of a Supersonic Transport Model with a Blended Wing-Body, Variable-Sweep Auxiliary Wing Panels, Outboard Tail Surfaces, and a Design Mach Number of 2.2.* NASA TM X-1167, 1963.

Freeman, D. C., Jr.: *Low-Subsonic Longitudinal Aerodynamic Characteristics in the Deep-Stall Angle-Of-Attack Range of a Supersonic Transport Configuration With a Highly Swept Arrow Wing.* NASA TM X-2316, 1971.

Freeman, Delma C., Jr.: *Low Subsonic Flight and Force Investigation of a Supersonic Transport Model with a Highly Swept Arrow Wing.* NASA TN D-3887, 1967.

Freeman, Delma C., Jr.: *Low Subsonic Flight and Force Investigation of a Supersonic Transport Model with Double-Delta Wing.* NASA TN D-4179, 1968.

Freeman, Delma C., Jr.: *Low Subsonic Flight and Force Investigation of a Supersonic Transport Model with a Variable-Sweep Wing.* NASA TN D-4726, 1968.

Grantham W. D., and Smith P. M: *Development of SCR Aircraft Takeoff and Landing Procedures for Community Noise Abatement and Their Impact on Flight Safety.* Presented at Supersonic Cruise Research '79, NASA CP-2108, part 1, 1980, pp. 299-333.

Grantham, William D., and Deal, Perry L.: *A Piloted Fixed-Based Simulator Study of Low-Speed Flight Characteristics of an Arrow-Wing Supersonic Transport Design.* NASA TN D-4277, 1968.

Grantham, William D., Nguyen, Luat T., Deal, Perry L., Neubauer, M. J. Jr., Smith, Paul M., and Gregory, Frederick D.: *Ground-Based and In-Flight Simulator Studies of Low-Speed Handling Characteristics of Two Supersonic Cruise Transport Concepts.* NASA Technical Paper 1240, 1978.

Haering, Edward A., Jr.; Ehernberger, L. J.; and Whitmore, Stephen A.: *Preliminary Airborne Measurements for the SR-71 Sonic Boom Propagation Experiment.* NASA TM 104307, 1995.

Haering, Edward A., Jr.; Whitmore, Stephen A.; and Ehernberger, L. J.: *High-Speed Research: 1994 Sonic Boom Workshop. Configuration, Design, Analysis and Testing,* NASA/CP-1999-209699, 1999

Haering, Edward A., Jr.; Murray, James E.; Purifoy, Dana D.; Graham, David H.; Meredith, Keith B.; Ashburn, Christopher E.; and Stucky, Mark: *Airborne Shaped Sonic Boom Demonstration Pressure Measurements with Computational Fluid Dynamics Comparisons.* 43rd AIAA Aerospace Sciences Meeting and Exhibit, Reno, NV, 10-13 January 2005.

Hahne, David E.: *1999 NASA High-Speed Research Program Aerodynamic Performance Workshop.* NASA CP-1999-209704, 1999.

Hallion, Richard P.: *On the Frontier–Flight Research at Dryden, 1946-1981.* NASA SP-4303, 1984.

Hansen, James R.: *Engineer in Charge.* NASA SP-4305, 1987.

Harris, R. V., Jr.: *A Numerical Technique for Analysis of Wave Drag at Lifting Conditions.* NASA-TN-D-3586, 1966.

Harris, R. V., Jr.: *An Analysis and Correlation of Aircraft Wave Drag.* NASA TM X-947, 1964.

Heimerl, George J., and Hardrath, Herbert F.: *An Assessment of a Titanium Alloy for Supersonic Transport Operations.* Presented at the Conference on Aircraft Operating Problems, NASA Langley Research Center, May 10-12, 1965. (Paper 23 of NASA SP-83).

Henderson, W. P.: *Selected Results from a Low-Speed Investigation of the Aerodynamic Characteristics of a Supersonic Transport Configuration Having an Outboard Pivot Variable Sweep Wing.* NASA TM X-839, 1963.

Hergenrother, Paul M.: *Development of Composites, Adhesives and Sealants for High-Speed Commercial Airplanes.* SAMPE Journal 36:1 (January/February 2000), pp. 30-41

Hilton, David A., Huckel, Vera, Steiner, Roy, and Maglieri, Domenic J.: *Sonic-Boom Exposures During FAA Community-Response Studies Over a 6-Month Period in the Oklahoma City Area.* NASA TN D-2539, 1964.

Hilton, David A.; Huckel, Vera; and Maglieri, Domenic J.: *Sonic-Boom Measurements During Bomber Training in the Chicago Area.* NASA TN D-3655, 1966.

Hoffman, S.: *Bibliography of Supersonic Cruise Research (SCR) Program from 1977 to Mid-1980.* NASA-RP-1063, 1980.

Hoffman, S.: *Bibliography of Supersonic Cruise Aircraft Research (SCAR) Program from 1972 to Mid-1977.* NASA RP-1003, 1977.

Hortwich, M.: *Clipped Wings—The American SST Conflict.* The MIT Press, Cambridge, Massachusetts, 1982.

Houbolt, J. C., "Why Twin Fuselage Aircraft?" *Astronautics and Aeronautics*, April 1982, pp. 26-35.

Hubbard, Harvey H., Maglieri, Domenic J., Huckel, Vera, and Hilton, David A.: *Ground Measurements of Sonic-Boom Pressures for the Altitude Range of 10,000 to 75,000 Feet.* NASA TR R-198, 1964.

Jackson, E. Bruce, Raney, David L., Glaab, Louis J., and Derry, Stephen L.: *Piloted Simulation Assessment of a High-Speed Civil Transport Configuration.* NASA TP-2002-211441, 2002.

Johnson, Joseph L., Jr.: *Wind-Tunnel Investigation of Low-Subsonic Flight Characteristics of a Model of a Canard Airplane Designed for Supersonic Cruise Flight.* NASA TM X-229, 1960.

Kelly, T. C., Patterson, J. C., Jr., and Whitcomb, R. T.: *An Investigation of the Subsonic, Transonic, and Supersonic Aerodynamic Characteristics of a Proposed Arrow-Wing Transport Airplane Configuration.* NASA TM X-800, 1963.

Leatherwood, J. D., and Sullivan, Brenda M.: *Laboratory Studies of Effects of Boom Shaping on Subjective Loudness and Acceptability.* NASA TP-3269, 1992.

Mack, R. J., and Darden, C. M.: *A Wind-Tunnel Investigation of the Validity of a Sonic Boom Minimization Concept.* NASA TP-1421, 1979.

Mack, Robert J.: *A Supersonic Business-Jet Concept Designed for Low Sonic Boom.* NASA TM-2003-212435, 2003.

Mack, Robert J.: *An Analysis of Measured Sonic-Boom Pressure Signatures From a Langley Wind-Tunnel Model of a Supersonic-Cruise Business Jet Concept.* NASA TM-2003-212447, 2003.

Mack, Robert J.: *Wind-Tunnel Overpressure Signatures From a Low-Boom HSCT Concept With Aft-Fuselage-Mounted Engines.* High-Speed Research: 1994 Sonic Boom Workshop. Configuration, Design, Analysis and Testing. NASA/CP-1999-209699, , 1999.

Maglieri, D. J.; and Morris, G. J.: *Measurement of Response of Light Aircraft to Sonic Booms.* NASA TN D-1941, 1963.

Maglieri, D. J.; Carlson, H. W.; and Hubbard, H. H.: *Status of Knowledge on Sonic Booms.* NASA TM-80113, 1979.

Maglieri, Domenic J., and Hilton, David A.: *Significance of the Atmosphere and Aircraft Operations on Sonic-Boom Exposures.* Presented at the Conference on Aircraft Operating Problems, NASA Langley Research Center, May 10-12, 1965. (Paper 26 of NASA SP-83).

Maglieri, Domenic J., Hubbard, Harvey H., and Lansing, Donald L.: *Ground Measurements of the Shock-Wave Noise from Airplanes in Level Flight at Mach Numbers to 1.4 and at Altitudes to 45,000 Feet.* NASA TN D-48, 1959.

Maglieri, Domenic J.: *Outlook for Overland Supersonic Flight.* Presented at the SAE World Aviation Congress and Display, "Enabling Technologies-New Paradigm for Supersonic Aircraft." Phoenix, AZ, November 6, 2002.

Maglieri, Domenic J.; Sothcott, Victor E.; and Keefer, Thomas N.: *Feasibility Study on Conducting Overflight Measurements of Shaped Sonic Boom Signatures Using the Firebee BQM-34E RPV.* NASA Contractor Report 189715, 1993.

McCurdy, David A.: *High-Speed Research: 1994 Sonic Boom Workshop.* NASA CP-3279, 1994.

McLean, F. E.: *Supersonic Cruise Technology.* NASA SP-472, 1985.

McLean, F. Edward, Carlson, Harry W., and Hunton, Lynn W.: *Sonic-Boom Characteristics of Proposed Supersonic and Hypersonic Airplanes.* NASA TN D-3587, 1966.

McLemore, H. Clyde, and Parlett, Lysle P.: *Low-Speed Wind-Tunnel Tests of 1/10-Scale Model of a Blended-Arrow Advanced Supersonic Transport.* NASA TM X-72,671, 1975.

McMillin, S. Naomi: *1998 NASA High-Speed Research Program Aerodynamic Performance Workshop.* NASA CP-1999-209692, 1999.

Nichols, Mark R., Keith, Arvid L. Jr., and Foss, Willard E.: *The Second-Generation Supersonic Transport. Presented at the NASA Conference on Vehicle Technology for Civil Aviation-The Seventies and Beyond.* NASA SP-292, 1971.

Nichols, Mark R.: *Aerodynamics of Airframe -Engine Integration of Supersonic Aircraft.* NASA TN D-3390, 1966.

Nixon, Charles W.; and Hubbard, Harvey H.: *Results of USAF-NASA-FAA Flight Program to Study Community Responses to Sonic Booms in the Greater St. Louis Area.* NASA TN D-2705, 1965.

Owens, L. R., Wahls, R. A., Elzey, M. B., and Hamner, M. P.: Reynolds Number Effects on the Stability & Control Characteristics of a Supersonic Transport. AIAA Paper 2002-0417, 2002.

Owens, L.R., and Wahls, R.A.: Reynolds Number Effects on a Supersonic Transport at Subsonic, High-Lift Conditions. AIAA Paper 2001-0911, January 2001.

Pawlowski, Joseph W., Graham, David H., Boccadoro, Charles H., Coen, Peter G., and Maglieri, Domenic J.: Origins and Overview of the Shaped Sonic Boom Demonstration Program. AIAA Paper 2005-0005, 2005.

Peterson, J. B., Jr.; Mann, M. J.; Sorrells, R. B., III; Sawyer, W. C.; Fuller, D. E.: *Wind-Tunnel/Flight Correlation Study of Aerodynamic Characteristics of a Large Flexible Supersonic Cruise Airplane (XB-70-1) 2: Extrapolation of Wind-Tunnel Data to Full-Scale Conditions.* NASA-TP-1515, 1980.

Proceedings of the SCAR Conference. NASA CP-001, 1976.

Proceedings of the SCAR Conference. NASA-CP-001, Jan., 1976.

Ray, E. J., and Henderson, W. P.: *Low-Speed Aerodynamic Characteristics of a Highly Swept Supersonic Transport Model with Auxiliary Canard and High-Lift Devices.* NASA TM X-1694, 1968.

Ray, E. J.: *NASA Supersonic Commercial Air Transport (SCAT) Configurations: A Summary and Index of Experimental Characteristics.* NASA TM X-1329, 1967.

Ribner, H. S.; and Hubbard, H. H., ed.: Proceedings of the Second Sonic Boom Symposium. Acoustical Society of America. Houston, Texas. November 3, 1965.

Rivers, Robert A., Jackson, E. Bruce, Fullerton, C. Gordon, Cox, Timothy H., and Princen, Norman H.: *A Qualitative Piloted Evaluation of the Tupolev Tu-144 Supersonic Transport.* NASA TM-2000-209850, 2000.

Robins, A. W., and Carlson, H. W.: High-Performance Wings with Significant Leading-Edge Thrust at Supersonic Speeds. AIAA Paper 79-1871, 1979.

Robins, A. W.; Carlson, H. W.; and Mack, R. J.: *Supersonic Wings with Significant Leading-Edge Thrust at Cruise.* Presented at Supersonic Cruise Research' 79, NASA CP-2108, Part 1, 1980, pp. 229-246.

Robins, A. W.; Harris, R. V., Jr.; Carlson, H. W.; McLean, H. W.; and Middleton, W. D.: *Supersonic Aircraft.* U.S. Patent 3,310,262, March 21, 1067.

Robins, A. Warner, Dollyhigh, Samuel M., Beissner, Fred L., Jr., Geiselhart, Karl, Martin, Glenn L., Shields, E. W., Swanson, E. E., Coen, Peter G., and Morris, Shelby J., Jr.: *Concept Development af a Mach 3.0 High-Speed Civil Transport.* NASA TM-4058, 1988.

Runyan, L.J.; Middleton, W. D.; and Paulson, J. A.: *Wind Tunnel Test Results of a New Leading Edge Flap Design for Highly Swept Wings—A Vortex Flap.* Presented at Supersonic Cruise Research' 79, NASA CP-2108, Part 1, 1980, pp. 131-147.

Scott, Robert C., Silva, Walter A., Florance, James R., and Keller, Donald F.: Measurement of Unsteady Pressure Data on a Large HSCT Semispan Wing and Comparison With Analysis. AIAA Paper 2002-1648.

Serling, Robert J.: *Legend and Legacy-The Story of Boeing andIts People.* St. Martin's Press, NY, 1992.

Shepherd, K. P., and Sullivan, B. M.: *A Loudness Calculation Procedure Applied to Shaped Sonic Booms*. NASA TP-3134, 1991.

Shepherd, K. P., Brown, S. A., Leatherwood, J. D., McCurdy, D. A., and Sullivan, B. M.: *Human Response to Sonic Booms--Recent NASA Research*. Presented at INTER-NOISE 95--The 1995 International Congress on Noise Control Engineering , July 10-12, 1995.

Silsby, Norman S., McLaughlin, Milton D.,and Fischer, Michael C.: *Effects of the Air Traffic Control System on the Supersonic Transport*. Presented at the Conference on Aircraft Operating Problems, NASA Langley Research Center, May 10-12, 1965. (Paper 19 of NASA SP-83).

Spearman, M. Leroy: *The Evolution of the High-Speed Civil Transport*. NASA TM 109089, 1994.

Staff of the Langley Research Center: *Determination of Flight Characteristics of Supersonic Transports During the Landing Approach with a Large Jet Transport In-Flight Simulator*. NASA TN D-3971, 1967.

Staff of the Langley Research Center: *The Supersonic Transport—A Technical Summary*. NASA TN D-423, 1960.

Sullivan, Brenda M.: *Human Response to Simulated Low-Intensity Sonic Booms*. NoiseCon 04 Summer Meeting and Exposition, Baltimore, Maryland, July 12-14, 2004.

Supersonic Cruise Research '79. NASA CP-2106, 1979.

Sweetman, Bill: Supersonic Contenders Enter the Ring. *Professional Pilot Magazine*, December, 2004.

Wahls, R.A., Owens, L.R., and Rivers, S.M.B.: Reynolds Number Effects on a Supersonic Transport at Transonic Conditions. AIAA Paper 2001-0912, January 2001.

Whitcomb, R. T.: *The Transonic and Supersonic Area Rule*. NASA TM X-57755, 1958.

Whitehead, Allen H., Jr.: Overview of Airframe Technology in the NASA High-Speed Research Program. AIAA Paper 91-3100, 1991.

Whitehead, Allen H., Jr.: *First Annual High-Speed Research Workshop*. NASA-CP-10087, 1992.

Wilson, J. W., Goldhagan, P., Maiden, D. L., and Tai, H: High Altitude Radiations Relevant to the High Speed Civil Transport (HSCT). Presented at the International Workshop on Cosmic Radiation, Electromagnetic Fields, and Health Among Aircrews, Charleston, South Carolina , February 5-7, 1998

Wilson, J. W., Goldhagen, P., Raffnson, V., Clem, J., De Angelis, G., and Friedberg, W.: Overview of Atmospheric Ionizing Radiation (AIR) Research: SST-Present. World Space Congress 2002, Houston, Texas, October 10-19, 2002,

Wlezien, R., and Veitch, L.: Quiet Supersonic Platform Program. AIAA 2002-0143, Jan. 2002.

Wood, Richard M., Bauer, Steven X. S., and Flamm, Jefferey D.: Aerodynamic Design Opportunities for Future Supersonic Aircraft. ICAS Paper 2002-8.7.1 Presented at the 23rd International Congress of Aeronautical Sciences,Toronto, Canada, September 8-13, 2002.

Wood, Richard M.; Bauer, Stephen X. S.; and Flamm, Jeffrey D.: Aerodynamic Design Opportunities for Future Supersonic Aircraft. Presented at the 23rd International Congress of Aeronautical Sciences. Toronto, Canada, September 8-13, 2002.

Blended Wing Body

Campbell, Richard L.; Carter, Melissa B.; Pendergraft, Odis C., Jr.; Friedman, Douglas M.; and Serrano, Leonel: Design and Testing of a Blended Wing Body with Boundary Layer Ingestion Nacelles at High Reynolds Number. AIAA Paper 2005-0459, 2005.

Clark, Lorenzo R.; and Gerhold, Carl H.: Inlet Noise Reduction By Shielding for the Blended-Wing-Body Airplane. AIAA Paper 99-1937, 1999.

Giles, Gary L.: Equivalent Plate Modeling for Conceptual Design of Aircraft Wing Structures. AIAA Paper No. 95-3945, 1995.

Hilburger, M. W.; Rose, C. A.; and Starnes, J. H. Jr.: Nonlinear Analysis and Scaling Laws for Noncircular Composite Structures Subjected to Combined Loads. AIAA 2001-1335, 2001.

Liebeck, R.: Design of the Blended-Wing-Body Subsonic Transport, AIAA Paper No. 2002-0002, 2002.

Liebeck, Robert, H.; Page, Mark, A.; Rawdon, Blaine K.; Scott, Paul W.; and Wright Robert A.: Concepts for Advanced Subsonic Transports. NASA CR-4624, 1994.

Mukhopadhyay, V.; Sobieszczanski-Sobieski, J.; Kosaka, I; Quinn, G.; and Charpentier, C.: Analysis Design and Optimization of Non-cylindrical Fuselage for Blended-Wing-Body (BWB) Vehicle. AIAA Paper 2002-5664, 2002.

Mukhopadhyay, Vivek: Interactive Flutter Analysis and Parametric Study for Conceptual Wing Design. AIAA Paper 95-3943, 1995.

Mukhopadhyay, Vivek: Structural Concepts Study of Non-Circular Fuselage Configurations. Paper No. WAC-67. Presented at the SAE/AIAA World Aviation Congress, October 22-24, 1996.

Potsdam, Mark A.; Page, Mark A.; and Liebeck, Robert H.: Blended Wing Body Analysis and Design. AIAA Paper 97-2317, 1997.

Roman, D.; Allen, J. B.; and Liebeck, R. H. : Aerodynamic Design Challenges of the Blended-Wing-Body Subsonic Transport. AIAA Paper No. 2000-4335, 2000.

Spellman, Regina L.: Analysis and Test Correlation of Proof of Concept Box for Blended Wing Body-Low Speed Vehicle. NAFEMS World Congress 2003, Orlando, FL.

The Boeing Company, Long Beach, CA.: *Blended-Wing-Body Technology Study.* NAS1-20275 Final Report, prepared for NASA-Langley Research Center, October 30, 1997.

Tigner, Benjamin; Meyer, Mark J.; Holdent, Michael E.; Rawdon, Blaine K.; Page, Mark A.; Watson, William; and Kroo, Ilan: Test Techniques For Small-Scale Research Aircraft, AIAA Paper No. 98-2726, 1998.

Vicroy, Dan D.: NASA's Research on the Blended-Wing-Body Configuration. Presented at World Aviation Congress, October 10, 2000, San Diego, CA.

Wahls, R. A.: The National Transonic Facility: A Research Retrospective. AIAA 2001-0754, 2001.
Watkins, A. Neal, et.al.: Flow Visualization at Cryogenic Conditions Using a Modified Pressure Sensitive Paint Approach. AIAA Paper 2005-0456, 2005.

Synthetic Vision

Adams, J. J.: *Simulator Study of Pilot-Aircraft-Display System Response Obtained With a Three-Dimensional-Box Pictorial Display.* NASA NASA-TP-2122 , 1983.

Avionic Pictorial Tunnel-/Pathway-/Highway-In-The-Sky Workshops. Russell V. Parrish, Compiler. NASA/CP-2003-212164, March, 2003.

Bailey, Randall E.; Parrish, Russell V.; Kramer, Lynda J.; Harrah, Steve D.; and Arthur, J. J., III : Technical Challenges in the Development of a NASA Synthetic Vision System Concept. Presented at NATO RTO Enhanced and Synthetic Vision Systems, Ottawa, Canada, September 9-12, 2002.

Glaab, Louis J.; and Hughes, Monica F.;: Terrain Portrayal for Head-Down Displays Flight Test. Presented at the 22nd Digital Avionics Systems Conference, Indianapolis, Indiana, October 12-16, 2003.

Glaab, Louis J.; and Takallu, Mohammad A.: Preliminary Effect of Synthetic Vision Systems Displays to Reduce Low-Visibility Loss of Control and Controlled Flight Into Terrain Accidents. Presented at the 2002 SAE General Aviation Technology Conference & Exhibition, Wichita, Kansas, April 16-18, 2002.

Harrison, Stella V.; Kramer, Lynda J.; Bailey, Randall E.; Jones, Denise R.; Young, Steven D.; Harrah, Steven D.; Arthur, Jarvis J.; and Parrish, Russell V.: *Initial SVS Integrated Technology Evaluation Flight Test Requirements and Hardware Architecture* . NASA/TM-2003-212644, 2003.

Jones, Denise R.: Runway Incursion Prevention System Simulation Evaluation. Proceedings of the 21st Digital Avionics Systems Conference, Irvine, CA, Oct. 27 – 31, 2002.

Jones, Denise R.; Quach, Cuong C.; and Young, Steven D.: Runway Incursion Prevention System – Demonstration and Testing at the Dallas/Fort Worth International Airport. Proceedings of the 20th Digital Avionics Systems Conference, Daytona Beach, FL, Oct. 14 – 18, 2001.

Kramer, Lynda J.; Prinzel, Lawrence J., III; Arthur, Jarvis J.; and Bailey, Randall E.: Pathway Design Effects on Synthetic Vision Head-Up Displays. Presented at SPIE Defense & Security Symposium 2004, Orlando, Florida, April 12-16, 2004.

Kramer, Lynda J.; Prinzel, Lawrence J., III; Bailey, Randall E.; Arthur, Jarvis J.: Synthetic Vision Enhances Situation Awareness and RNP Capabilities for Terrain-Challenged Approaches. AIAA Paper 2003-6814, 2003.

Kramer, Lynda J.; Prinzel, Lawrence J., III; Bailey, Randall E.; Arthur, Jarvis J.; and Parrish, Russell V.: *Flight Test Evaluation of Synthetic Vision Concepts at a Terrain Challenged Airport* . NASA/TP-2004-212997, 2004.

Parrish, Robert V.; Busquets, Anthony M.; Williams, Steven P.; and Nold, Dean E.: *Evaluation of Alternative Concepts for Synthetic Vision Flight Displays With Weather-Penetrating Sensor Image Inserts During Simulated Landing Approaches.* NASA/TP-2003-212643, 2003.

Prinzel, Lawrence J., III; Hughes, Monica F.; Arthur, Jarvis J.; Kramer, Lynda J.; Glaab, Louis J.; Bailey, Randall E.; Parrish, Russell V.; and Uenking, Michael D.: Synthetic Vision CFIT Experiments for GA and Commercial Aircraft: A Picture is Worth a Thousand Lives. Presented at the Human Factors and Ergonomics Society 47th Annual Meeting, Denver, Colorado, October 13-17, 2003.

Scott, William B.: Clearing the Fog. *Aviation Week and Space Technology*, August 9, 2004. pp 48-51.

Synthetic Vision Workshop 2. Lynda J. Kramer, Compiler. NASA/CP-1999-209112, March 1999.

Uijt De Haag, M., Young, S., and Gray, R.: A Performance Evaluation of Elevation Database Integrity Monitors for Synthetic Vision Systems. 8th International Conference on Integrated Navigation Systems, Saint Petersburg, Russia, May 28-30, 2001

Young, S., and Jones, D.: Flight Testing of an Airport Surface Movement Guidance, Navigation, and Control System. Proceedings of the Institute of Navigation's National Technical Meeting, January 21-23, 1998.

Young, S., and Jones, D.: Runway Incursion Prevention: A Technology Solution. Proceedings of the Joint Meeting of the Flight Safety Foundation's 54th Annual International Air Safety Seminar, the International Federation of Airworthiness' 31st International Conference, and the International Air Transport Association, Athens, Greece, November 5-8, 2001.

Young, S., Jones, D., Eckhardt, D., and Bryant, W.: Safely Improving Airport Surface Capacity. *Aerospace America*, May 1998.

Laminar Flow Control

Arcara, P. C., Jr.; Bartlett, D. W.; and McCullers, L. A.: *Analysis for the Application of Hybrid Laminar Flow Control to a Long-Range Subsonic Transport Aircraft.* SAE Paper 912113, 1991.

Bhutiani, P. K.; Keck, D. F.; Lahti, D. J.; Stringas, M. J.: Investigating the Merits of a Hybrid Laminar Flow Nacelle. *The Leading Edge, General Electric Co.*, Spring, 1993, pp. 32–35.

Bobbitt, Percy J.; Ferris, James C.; Harvey, William D.; and Goradia, Suresh H.: *Hybrid Laminar Flow Control Experiment Conducted in NASA Langley 8-Foot Transonic Pressure Tunnel.* NASA TP-3549, 1996.

Braslow, A.L.; Maddalon, D. V.; Bartlett, D. W.; Wagner, R. D.; and Collier, F. S., Jr.: *Applied Aspects of Laminar-Flow Technology.* Viscous Drag Reduction in Boundary Layers, Dennis M. Bushnell and Jerry Hefner, eds., AIAA, pp. 47–78, 1990.

Braslow, Albert L.: *A History of Suction-Type Laminar-Flow Control with Emphasis on Flight Research.* NASA Monographs in Aerospace History Number 13, 1999.

Braslow, Albert L.; Burrows, Dale L.; Tetervin, Neal; and Visconti, Fioravante: *Experimental and Theoretical Studies of Area Suction for the Control of the Laminar Boundary Layer on an NACA 64A010 Airfoil.* NACA Rep. 1025, 1951.

Bushnell, D. M.; and Malik, M. R.: *Supersonic Laminar Flow Control.* Research in Natural Laminar Flow and Laminar-Flow Control, Jerry N. Hefner and Frances E. Sabo, compilers, NASA CP-2487, Part 3, pp. 923–946. 1987.

Bushnell, Dennis M.; and Tuttle, Mary H.: *Survey and Bibliography on Attainment of Laminar Flow Control in Air Using Pressure Gradient and Suction.* NASA RP-1035, 1979.

Collier, F. S., Jr.: An Overview of Recent Subsonic Laminar Flow Control Flight Experiments. AIAA-93-2987, 1993.

Harris, Roy V., Jr.; and Hefner, Jerry N.: *NASA Laminar-Flow Program—Past, Present, Future.* Research in Natural Laminar Flow and Laminar-Flow Control, Jerry N. Hefner and Frances E. Sabo, compilers, NASA CP-2487, Part 1, pp. 1–23. 1987.

Hastings, Earl C.: *Status Report on a Natural Laminar-Flow Nacelle Flight Experiment.* Research in Natural Laminar Flow and Laminar-Flow Control, Jerry N. Hefner and Frances E. Sabo, compilers, NASA CP-2487, Part 3, pp. 887–890. 1987.

Hefner, Jerry N.: *Laminar Flow Control: Introduction and Overview.* Natural Laminar Flow and Laminar Flow Control, R. W. Barnwell and M. Y. Hussaini, eds., Springer-Verlag, pp. 1–22, 1992.

Joslin, Ronald D.: *Overview of Laminar Flow Control.* NASA TP-1998-208705, 1998.

Lynch, Frank K.; and Klinge, Mark D.: *Some Practical Aspects of Viscous Drag Reduction Concepts.* SAE Paper 912129, 1991.

Maddalon, D. V.; Collier, F. S., Jr.; Montoya, L. C.; and Land, C. K.: Transition Flight Experiments on a Swept Wing With Suction. AIAA Paper No. 89-1893, 1989.

Runyan, L. J.; Bielak, G. W.; Behbehani, R.; Chen, A. W.; and Rozendaal, R. A.: *757 NLF Glove Flight Test Results.* Research in Natural Laminar Flow and Laminar-Flow Control, Jerry N. Hefner and Frances E. Sabo, compilers, NASA CP-2487, Part 3, pp. 795–818. 1987.

Streett, C. L.: Designing a Hybrid Laminar-Flow Control Experiment—The CFD-Experiment Connection. AIAA Paper 2003-0977, 2003.

Tuttle, Mary H.; and Maddalon, Dal V.: *Laminar Flow Control (1976-1991)–A Comprehensive, Annotated Bibliography.* NASA TM 107749, 1993.

Wagner, R. D.; Bartlett, D. W.; and Collier, F. S., Jr.: Laminar Flow-The Past, Present, and Prospects. AIAA-89-0989, 1989.

Wagner, R. D.; Fischer, M. C.; Collier, F. S., Jr.; and Pfenninger, W.: Supersonic Laminar Flow Control on Commercial Transports. 17th Congress of the International Council of the Aeronautical Sciences, ICAS Proceedings, 1990.

Wagner, R. D.; Maddalon, D. V.; and Fisher, D. F.: Laminar Flow Control Leading-Edge Systems in Simulated Airline Service. *Journal of Aircraft*, Vol. 27, No. 3, March 1990, pp 239-244.

Wagner, R. D.; Maddalon, D. V.; Bartlett, D. W.; and Collier, F. S., Jr.: *Fifty Years of Laminar Flow Flight Testing.* SAE Paper 881393, 1988.

Wagner, R. D.; Maddalon, Dal V.; Collier, F.S.; and Braslow, A. L.: *Laminar Flow Flight Experiments. Presented at Transonic Symposium: Theory, Application, and Experiment held at Langley Research Center.* NASA CP 3020, 1988.

Wie, Y. S.; Collier, F. S., Jr.; Wagner, R. D.; Viken, J. K.; and Pfeninnger, W.: Design of a Hybrid Laminar Flow Control Engine Nacelle. AIAA-92-0400, 1992.

Wilkinson, Stephan: "Go With the Flow". *Air & Space/Smithsonian Magazine*, June/July, 1995.
Zalovcik, John A.; Wetmore, J. W.; and Von Doenhoff, Albert E.: *Flight Investigation of Boundary-Layer Control by Suction Slots on an NACA 35-215 Low-Drag Airfoil at High Reynolds Numbers.* NACA WR L-521. 1944 (Formerly NACA ACR 4B29.)

Upper-Surface Blowing

Borchers, Paul F.; Franklin, James A.; and Fletcher, Jay W.: *Flight Research at Ames: Fifty-Seven Years of Development and Validation of Aeronautical Technology.* NASA SP-1998-3300, 1998.

Cochrane, J.A. and Riddle, D.W.: Quiet Short-Haul Research Aircraft-The First Three Years of Flight Research. AIAA Paper 81-2625, 1981.

Collection of Papers from the AIAA 2002 Biennial International Powered Lift Conference and Exhibit. Williamsburg, Virginia, November 5-7, 2002.

Deckert, W. H.; and Franklin, J. A.: *Powered-Lift Aircraft Technology.* NASA SP-501, 1989.

Hassell, James L., Jr.: *Results of Static Test of a ¼-Scale Model of the Boeing YC-14 Powered-Lift System.* NASA Conference on Powered-Lift Aerodynamics and Acoustics. May 24-26, 1976.

Johnson, Joseph L., Jr.; and Phelps, Arthur E., III: *Low-Speed Aerodynamics of the Upper-Surface Blown Jet Flap.* Society of Automotive Engineers Air Transportation Meeting; Dallas, TX. April 30-May 2, 1974.

Maglieri, D. J.; and Hubbard H. H.: *Preliminary Measurements of the Noise Characteristics of Some Jet-Augmented-Flap Configurations.* NASA Memo 12-4-58L, 1959.

Phelps, A. E., III; Johnson, J. L., Jr.; and Margason, R. J.: *Summary of Low-Speed Aerodynamic Characteristics of Upper-Surface-Blown Jet-Flap Configurations.* NASA Conference on Powered-Lift Aerodynamics and Acoustics. May 24-26, 1976.

Phelps, Arthur E., III,: *Wind-Tunnel Investigation of a Twin-Engine Straight-Wing Upper-Surface Blown Jet-Flap Configuration.* NASA TN D-7778, 1975.

Phelps, Arthur E.; Letko, William; and Henderson, Robert L.: *Low-Speed Wind-Tunnel Investigation of a Semispan STOL Jet Transport Wing-Body with an Upper-Surface Blown Jet Flap.* NASA TN D-7183, 1973.

Powered-Lift Aerodynamics and Acoustics. Conference held at Langley Research Center, Hampton, VA, NASA CP-406, May 24-26, 1976.

Riddle, Dennis W.; Stevens, Victor C.; and Eppel, Joseph C.: *Quiet Short-Haul Research Aircraft: A Summary of Flight Research Since 1981.* SAE Technical Paper 872315, Dec. 1987.

Shivers, James P.; and Smith, Charles C., Jr.: *Static Test of a Simulated Upper Surface Blown Jet-Flap Configuration Utilizing a Full-Size Turbofan Engine.* NASA TN D-7816, 1975.

Sleeman, William C., Jr.: *Low-Speed Wind-Tunnel Investigation of a Four-Engine Upper Surface Blown Model Having a Swept Wing and Rectangular and D-Shaped Exhaust Nozzles.* NASA TN D-8061, 1975.

Smith, Charles C., Jr.; Phelps, Arthur E., III; and Copeland, W. Latham: *Wind-Tunnel Investigation of a Large-Scale Semispan Model with an Unswept Wing and Upper-Surface Blown Jet Flap.* NASA TM D-7526, 1974.

Staff of the Langley Research Center: *Wind-Tunnel investigation of the Aerodynamic Performance, Steady and Vibratory Loads, Surface Temperatures and Acoustic Characteristics of a Large-Scale Twin-Engine Upper-Surface Blown Jet-Flap Configuration.* NASA TM X-72794, 1975.

Thomas, J. L.; Hassell, J. L., Jr; and Nguyen, L. T.: *Aerodynamic Characteristics in Ground Proximity.* NASA Conference on Powered-Lift Aerodynamics and Acoustics. May 24-26, 1976.

Turner, T. R.; Davenport, E. E.; and Riebe, J. M.: *Low-Speed Investigation of Blowing From Nacelles Mounted Inboard and on the Upper Surface of an Aspect Ratio 7.0 35° Swept Wing With Fuselage and Various Tail Arrangements.* NASA Memo 5-1-59L, 1959.

Wimpress, John K.; and Newberry, Conrad F.: The YC-14 STOL Prototype: Its Design, Development, and Flight Test. AIAA Case Study Series, 1998.

Zuk, John; Callaway, Robert K.; and Wardwell, Douglas A.: Adaptive Air Transportation System-a Catalyst for Change. AIAA Paper 2002-5956, 2002.

Control of Aeroelastic Response

Abel, I.: *Research and Applications in Aeroelasticity and Structural Dynamics at the NASA Langley Research Center.* NASA TM-112852, 1997.

Abel, I.; and Sandford, Maynard C.: *Status of Two Studies on Active Control of Aeroelastic Response.* NASA TM X-2909, 1973.

Abel, I.; Doggett, R. V.; Newsom, J. R.; and Sandford, M.: *Dynamic Wind-Tunnel Testing of Active Controls by the NASA Langley Research Center.* AGARD Ground and Flight Testing for Aircraft Guidance and Control; December 1984.

Abel, I.; Perry, B., III; and Newsom, J. R.: *Comparison of Analytical and Wind-Tunnel Results for Flutter and Gust Response of a Transport Wing with Active Controls.* NASA TP-2010, 1982.

Bennett. R. M.; Eckstrom, C. V.; Rivera, J. A., Jr.; *Dansberry, B. E.; Farmer, M. G.; and Durham, M. H.: The Benchmark Aeroelastic Models Program-Description and Highlights of Initial Results.* NASA TM-104180, 1991.

Cazier, F. W., Jr.; and Kehoe, M. W.: Flight Test of a Decoupler Pylon for Wing/Store Flutter Suppression. AIAA Paper 86-9730, 1986.

Chambers, Joseph R.: *Concept to Reality: Contributions of the NASA Langley Research Center to U.S. Civil Aircraft of the 1990's.* NASA SP-2003-4529, 2003.

Chambers, Joseph R.: *Partners in Freedom: Contributions of the NASA Langley Research Center to U.S. Military Aircraft of the 1990's.* NASA SP-2000-4519, 2000.

Doggett, R.V., Jr.; Abel, I; and Ruhlin, C. L.: *Some Experiences Using Wind-Tunnel Models in Active Control Studies.* Presented at the Symposium on Advanced Control Technology and Its Potential for Future Transport Aircraft. NASA TM X-3409, 1976, pp. 831-892.

Ethell, Jeffrey L.: *Fuel Economy in Aviation.* NASA SP-462, 1983

Foughner, J. T. Jr.; and Besinger, C. T.: *F-16 Flutter Model Studies of External Wing Stores.* NASA TM-74078, 1977.

Galea, S. C.; Ryall T. G.; Henderson, D. A.; Moses, R. W.; White, E. V.; and Zimcik, D. G.: Next Generation Active Buffet Suppression System. AIAA Paper 2003-2905, 2003.

Goodman, Charles; Hood, Mark; Reichenbach, Eric; and Yurkovich, Rudy: An Analysis of the F/A-18C/D Limit Cycle Oscillation Solution. AIAA Paper 2003-1424, 2003.

Heeg, J.: *Analytical and Experimental Investigation of Flutter Suppression by Piezoelectric Actuation.* NASA TP-3241, 1993.

Heeg, J.: *Flutter Suppression via Piezoelectric Actuation.* NASA TM 104120, 1991.

Hood, R. V., Jr.: The Aircraft Energy Efficiency Active Controls Technology Program. AIAA Paper 77-1076, 1977.

McGowan, A-M. R.; Heeg, J.; and Lake, R. C.: Results from Wind-Tunnel Testing from the Piezoelectric Aeroelastic Response Tailoring Investigation. AIAA Paper 96-1511, April 1996.

Middleton; D. B.; Bartlett, D. W.; and Hood, R. V., Jr.: *Energy Efficient Transport Technology: Program Summary and Bibliography.* NASA RP-1135, 1985.

Moses, R. W.: *NASA Langley Research Center's Contributions To International Active Buffeting Alleviation Programs.* Presented at the NATO-RTO Workshop on Structural Aspects of Flexible Aircraft Control, Ottawa, Canada, October 18-21, 1999.

Moses, R. W.: *Vertical Tail Buffeting Alleviation Using Piezoelectric Actuators – Some Results of the Actively Controlled Response of Buffet-Affected Tails (ACROBAT) Program.* Presented at SPIE's 4th Annual Symposium on Smart Structures and Materials, Industrial and Commercial Applications of Smart Structures Technologies, Conference 3044, March 4-6, 1997, San Diego, CA.

Moses, R. W.; and Huttsell, L.: Fin Buffeting Features of an Early F-22 Model. AIAA Paper 2000-1695, 2000.

Moses, R. W.; and Pendleton, E.: *A Comparison of Pressure Measurements Between a Full-Scale and a 1/6-Scale F/A-18 Twin Tail During Buffet.* NASA TM-110282, 1996.

Mukhopadhyay, V.: *Transonic Flutter Suppression Control Law Design, Analysis and Wind-Tunnel Results.* Paper presented at the International Forum on Aeroelasticity and Structural Dynamics 1999, Williamsburg, VA, June 22-25, 1999.

Murrow, H. N.; and Eckstrom, C. V.: Drones for Aerodynamic and Structural Testing (DAST). A Status Report. AIAA Journal of Aircraft, Vol. 16, No. 8, August 1979, pp. 521-526.

Nissim, E: Design of Control Laws for Flutter Suppression Based on the Aerodynamic Energy Concept and Comparisons with Other Design Methods. AIAA Paper 89-1212, 1989.

Perry, B. P., III; Cole, S. R.; and Miller, G. D.: *A Summary of the Active Flexible Wing Program.* NASA TM-107655, 1992.

Perry, B. P., III; Mukhopadhyay, V.; Hoadley, S. T.; Cole, S. R.; Buttrill, C. S.; and Houck, J. A.: *Digital-Flutter-Suppression-Investigations for the Active Flexible Wing Wind-Tunnel Model.* NASA TM 102618, 1990.

Perry, Boyd, III; Noll, Thomas E.; and Scott, Robert C.: Contributions of the Transonic Dynamics Tunnel to the Testing of Active Control of Aeroelastic Response. AIAA Paper 2000-1769, 2000.

Phillips, W. H.; and Kraft, C. C., Jr.: *Theoretical Study of Some Methods for Increasing the Smoothness of Flight Through Rough Air.* NACA TN 416, 1951.

Pinkerton, J. L.; McGowan, A-M R.; Moses, R. W.; Scott, R. C.; and Heeg, J.: *Controlled Aeroelastic Response and Airfoil Shaping Using Adaptive Materials and Integrated Systems.* Paper Presented at the SPIE 1996 Symposium on Smart Structures and Integrated Systems. San Diego, California, February 26-29, 1996.

Redd, L. T.; Gilman, J., Jr.; and Cooley, D. E.: A Wind-Tunnel Investigation of a B-52 Model Flutter Suppression System. *Journal of Aircraft*, Vol. II, No. 11. Nov. 1074, pp. 659-663

Redd, L. T.; Hanson, P. W.; and Wynne, E. C.: *Evaluation of a Wind-Tunnel Gust Response Technique Including Correlations With Analytical and Flight Test Results.* NASA-TP-1501, 1979.

Reed, Wilmer H. III , and Abbott, F.T., Jr.: *A New "Free-Flight" Mount System for High-Speed Wind Tunnel Flutter Models.* Proceedings of Symposium on Aeroelastic and Dynamic Modeling Technology. RT-TDR-66-4197, Pt.1, U.S. Air Force. March 1964, pp.169-206.

Reed, Wilmer H., III; Foughner, Jerome P., Jr.; and Runyan, Harry L., Jr.: Decoupler Pylon: A Simple, Effective Wing/Store Flutter Suppressor. *Journal of Aircraft*, Vol. 17, No. 3, March 1980, pp. 206-211.

Sandford, M. C.; Abel, I.; and Gray, D. L.: *Development and Demonstration of a Flutter Suppression System Using Active Controls*. NASA-TR-R-450, 1975.

Scott, R. C.; Wieseman, C. D.; Hoadley, S.T.; and Durham, M.H.: Pressure and Loads Measurements on the Benchmark Active Controls Technology Model. AIAA Paper 97-0829, 1997.

Sheta, E. F.; Moses, R. W.; Huttsell, L. J.; and Harrand, V. J.: Active Control of F/A-18 Vertical Tail Buffeting Using Piezoelectric Actuators. AIAA Paper 2003-1887, 2003.

Stewart, E. C.; and Redd, L. T.: *A Comparison of the Results of Dynamic Wind-Tunnel Tests with Theoretical Predictions for an Aeromechanical Gust-Alleviation System for Light Airplanes*. NASA TN D-8521, 1977.

Trame, L. W.; Williams, L. E.; and Yurkovich, R. N.: Active Aeroelastic Oscillation Control on the F/A-18 Aircraft. AIAA Paper 85-1858, 1985.

Waszak, M. R.: Modeling the Benchmark Active Control Technology Wind-Tunnel Model for Application to Flutter Suppression. AIAA Paper No. 96-3437, 1996.

Woods-Vedeler, J. A.; Pototzky, A. S.; and Hoadley, S. T.: *Active Load Control During Rolling Maneuvers*. NASA TP 3455, 1994.

Joined Wing

Frink, N.T., Pirzadeh, S.Z, Parikh, P.C., Pandya, M.J., and Bhat, M.K.: The NASA Tetrahedral Unstructured Software System. *The Aeronautical Journal*, Vol. 104, No. 1040, October 2000, pp. 491-499.

Gallman, J.W.; and Kroo, I.M.: Structural Optimization for Joined-Wing Synthesis. *Journal of Aircraft*, Vol. 33, no. 1, 1996. pp 214-223.

Gallman, J.W.; Smith, S.C.; and Kroo, I.M.: Optimization of Joined-Wing Aircraft. *Journal of Aircraft*, Vol. 30 no. 6, 1993. pp 897-905.

Smith, S.C.; and Stonum, R.K.: *Experimental Aerodynamic Characteristics of a Joined-Wing Research Aircraft Configuration*. NASA TM 101083, 1989.

Smith, S.C.; Cliff, S.E.; and Kroo, I.M.: The Design of a Joined-Wing Flight Demonstrator Aircraft. AIAA Paper 87-2930, 1987.

Tirpak, John A.: Wings to Come. *Air Force Magazine*, September 2001, pp. 33-39.

Turner, C. D.; and Ricketts, R. H.: Aeroelastic Considerations for Patrol High Altitude Surveillance Platforms. AIAA Paper 83-0924, 1983.

Wolkovitch, J.: The Joined Wing: An Overview. AIAA Paper 85-0274, 1985.

Vortex Flap

Bare, E. A.; Reubush, D. E.; and Haddad, R. C.: *Flow Field Over the Wing of a Delta-Wing Fighter Model With Vortex Control Devices at Mach 0.6 to 1.2.* NASA TM-4296, 1992.

Bobbitt, P. J.: Modern Fluid Dynamics of Subsonic and Transonic Flight, AIAA Paper 80-0861, 1980.

Brandon, J. M.; Brown, P. W; and Wunschel, A. J.: *Impact of Vortex Flaps on Low-Speed Handling Qualities of a Delta Wing Aircraft.* NASA TP-2747, 1987.

Brandon, Jay M.; Hallissy, James B.; Brown, Philip W.; and Lamar, John E.: *In-Flight Flow Visualization Results of the F-106B with a Vortex Flap.* Paper Presented at the RTO AVT Symposium on Advanced Flow Management: Part A – Vortex Flows and High Angle of Attack for Military Vehicles, held in Loen, Norway, 7-11 May 2001, and published in RTO-MP-069(I).

Campbell, James F.; Osborne, Russell F.; and Foughner, Jerome T., Jr.,eds.: Vortex Flow Aerodynamics, 1986.

Erickson, Gary E.: Application of Free Vortex Sheet Theory to Slender Wings with Leading-Edge Vortex Flaps. AIAA-83-1813, July 1983.

Frink, N. T.; Huffman, J. K.; and Johnson, T. D., Jr.: *Vortex Flow Reattachment Line and Subsonic Aerodynamic Data for Vortex Flaps on 50° to 74° Delta Wings on Common Fuselage.* NASA TM-84618, 1983.

Frink, Neal T.: *Concept for Designing Vortex Flap Geometries.* NASA TP-2233, 1983.

Hallissy, James B.; Schoonover, W. Elliott, Jr.; Johnson, Thomas D., Jr.; and Brandon, Jay M.: *Wind-Tunnel Investigation of the Multiple Vortex System Observed in Flight Tests of the F-106B Vortex Flap Configuration.* NASA TP-3322, 1993.

Lamar, J. E.; Bruce, R. A.; Pride J. D., Jr., Smith, R. H.; Brown, P.W.; and Johnson, T. D., Jr.: In-Flight Flow Visualization of F-106B Leading-Edge Vortex Using the Vapor Screen Technique. AIAA- Paper No. 86-9785, April 1986.

Lamar, John E.: Subsonic Vortex-Flow Design Study for Slender Wings. *Journal of Aircraft*, Vol. 15, No. 9, September 1978, pp. 611-617.

Lamar, John E.; and Campbell, James F.: Vortex Flaps – Advanced Control Devices for Supercruise Fighters. *Aerospace America*, pp. 95-99, January 1984.

Lamar, John E.; Brandon, Jay; Stacy, Kathryn; Johnson, Thomas D., Jr.; Severance, Kurt; and Childers, Brooks A.: *Leading-Edge Vortex-System Details Obtained on F-106B Aircraft Using a Rotating Vapor Screen and Surface Techniques*. NASA TP-3374, 1993.

Lamar, John E.; Hallissy, James B.; Frink, Neal T.; Smith, Ronald H.; Johnson, Thomas D., Jr.; Pao, Jenn-Louh; and Ghaffari, Farhad: Review of Vortex Flow Projects on the F-106B. AIAA Paper No. 87-2346, August 1987.

Lamar, John E.; Obara, Clifford J.; Fisher, Bruce D.; and Fisher, David F.: *Flight, Wind-Tunnel, and Computational Fluid Dynamics Comparison for Cranked Arrow Wing (F-16XL-1) at Subsonic and Transonic Speeds*. NASA TP-2001-210629, 2001.

Lamar, John E.; Schemensky, R. T.; and Reddy, C. S.: Development of a Vortex-Lift Design Procedure and Application to a Slender Maneuver-Wing Configuration. *Journal of Aircraft*, vol. 18, No. 4, pp. 259-266, April 1981.

Murman, E. M.; Powell, K. G.; and Miller, D.S.: Comparison of Computations and Experimental Data for Leading-Edge Vortices – Effects of Yaw and Vortex Flaps. AIAA Paper No. 86-0439, January 1986.

Papers of the Vortex Flow Aerodynamics Conference, Jointly Sponsored by NASA Langley and USAF Wright Aeronautical Laboratories, Hampton, Virginia, October 1985, NASA CP-2417, 1986.

Rao, D. M.: *Leading Edge Vortex-Flap Experiments on a 74 Deg Delta Wing*. NASA CR-159161, 1979.

Rao, D. M.: Vortical Flow Management for Improved Configuration Aerodynamics – Recent Experiences. AGARD CP-342, Paper No. 30, 1983.

Schoonover, W. E., Jr.; and Ohlson, W. E.: Wind-Tunnel Investigation of Vortex Flaps on a Highly Swept Interceptor Configuration, ICAS Paper 82-6.7.3, 1982.

Schoonover, W. Elliott, Jr.; Frink, Neal T.; Hallissy, James B.; and Yip, Long P.: *Subsonic/Transonic Development of Vortex Flaps for Fighter Aircraft, Langley Symposium on Aerodynamics, Vol. 2.* Sharon H. Stack, Compiler, NASA CP-2 398, 1986, pp. 163-178.
> Vol.I, NASA CP-2416
> Vol. II, NASA CP-2417
> Vol. III, NASA CP-2418

Yip, Long P.: *Investigations of Vortex Flaps on the F-106B Configuration in the Langley 30 X 60-Foot Tunnel.* NASA CP-2417, 1986.

Yip, Long P.: *Wind-Tunnel Free-Flight Investigation of a 0.15-Scale Model of the F-106B Airplane with Vortex Flaps.* NASA TP-2700, 1987.

Yip, Long P.; and Murri, Daniel G.: *Effects of Vortex Flaps on the Low-Speed Aerodynamic Characteristics of an Arrow Wing.* NASA TP-1914, 1981.

Innovative Control Effectors

Anglin, E. L.; and Satran, D. R.: Effects of Spanwise Blowing on Two Fighter Airplane Configurations. AIAA Paper 79-1663, 1979.

Bauer, Steven X. S.; and Hempsch, Michael J.: Alleviation of Side Force on Tangent-Ogive Forebodies Using Passive Porosity. AIAA Paper 92-2711, 1992.

Buttrill, Carey S.; Bacon, Barton J.; Heeg, Jennifer; Houck, Jacob A., and Wood, David V.: *Aeroservoelastic Simulation of an Active Flexible Wing Wind Tunnel Model.* NASA Technical Paper 3510, 1996.

Campbell, J. F.: Augmentation of Vortex Lift by Spanwise Blowing. AIAA Paper 75-993, 1975.

Campbell, J. F.; and Erickson, G. E.: *Effects of Spanwise Blowing on The Surface Pressure* Distribution and Vortex-Lift Characteristics of a Trapezoidal Wing-Strake Configuration. NASA-TP-1290 , 1979.

Deere, Karen A.: Summary of Fluidic Thrust Vectoring Research Conducted at NASA Langley Research Center. AIAA Paper 2003-3800, 2003.

Erickson, G. E.; and Campbell, J. F.: *Improvement of Maneuver Aerodynamics by Spanwise Blowing.* NASA TP-1065, 1977.

Frink, Neal T.; Bauer, Steven X. S.; and Hunter, Craig A.: Simulation of Flows with Passive Porosity. ICAS Paper 2002-2.10.2., 2002.

Heeg, Jennifer; Spain, Charles V., and Rivera, J. A.: Wind Tunnel to Atmospheric Mapping for Static Aeroelastic Scaling. AIAA Paper 2004-2044, 2004, pp. 12.

Huffman, J. K.; Hahne, D. E.; and Johnson, T. D., Jr.: Experimental Investigation of the Aerodynamic Effects of Distributed Spanwise Blowing on a Fighter Configuration. AIAA Paper 84-2195, 1984.

Hunter, C.A.; Viken, S.A.; Wood, R. M.; and Bauer, S. X. S.: Advanced Aerodynamic Design of Passive Porosity Control Effectors. AIAA Paper 2001-0249, 2001.

Lamar, J. E.; and Campbell, J. F.: Recent Studies at NASA-Langley of Vortical Flows Interacting with Neighboring Surfaces. AGARD Conference Proceedings No. 342, July 1983. ISBN 92-835-0334-1, pages 10-1 to 10-32.

Mukhopadhyay, Vivek: Flutter Suppression Digital Control Law Design and Testing for the AFW Wind-Tunnel Model. AIAA Paper 92-2095, 1992.

Murri, Daniel G.; and Rao, Dhanvada M.: Exploratory Studies of Actuated Forebody Strakes for Yaw Control at High Angles of Attack. AIAA Paper 87-2557, 1987.

Murri, Daniel G.; Fisher, David F.; and Lanser, Wendy R.: *Flight Test Results of Actuated Forebody Strake Controls on the F-18 High-Alpha Research Vehicle*. NASA CP-1998-207676, Part 2, June, 1998.

Murri, Daniel G.; Shah, Gautam H.; DiCarlo, Daniel J.; and Trilling, Todd W.: Actuated Forebody Strake Controls for the F-18 High-Alpha Research Vehicle. *Journal of Aircraft*, Vol. 32, No. 3, May-June 1995, pp. 555-562.

Noll, Thomas E.; Perry, Boyd, III; Tiffany, Sherwood H.; Cole, Stanley R.; Buttrill, Carey S.; Adams, William M., Jr.; Houck, Jacob A.; Srinathkumar, S.; Mukhopadhyay, Vivek; Pototzky, Anthony S.; Heeg, Jennifer; McGraw, Sandra M.; Miller,Gerald; Ryan, Rosemary; Brosnan, Michael; Haverty, James; and Klepl, Martin: *Aeroservoelastic Wind-Tunnel Investigations Using the Active Flexible Wing Model—Status and Recent Accomplishments*. NASA TM-101570, 1989.

Paulson, J. W., Jr.; Quinto, P. F.; and Banks, D. W.: *Investigation of Trailing-Edge-Flap, Spanwise-Blowing Concepts on an Advanced Fighter Configuration*. NASA-TP-2250 , 1984.

Perry, Boyd, III; Cole, Stanley R.; and Miller, Gerald D.: A Summary of the Active Flexible Wing Program. AIAA Paper 92-2080, 1992.

Perry, Boyd, III; Mukhopadhyay, Vivek; Hoadley, Sherwood Tiffany; Cole, Stanley R.; and Buttrill, Carey S.: Digital-Flutter-Suppression-System Investigations for the Active Flexible Wing Wind-Tunnel Model. AIAA Paper 90-1074, 1990.

Rao, D. M.: Vortical Flow Management for Improved Configuration Aerodynamics - Recent Experiences. AGARD Conference Proceedings No. 342, July 1983. pp 30-1 to 30-14.

Satran, D. R.; Gilbert, W. P.; and Anglin, E. L.: *Low-Speed Stability and Control Wind-Tunnel Investigations of Effects of Spanwise Blowing on Fighter Flight Characteristics at High Angles of Attack*. NASA-TP-2431, 1985.

Shah, Gautam H.; and Granda, J. Nijel: Application of Forebody Strakes for Directional Stability and Control of Transport Aircraft. AIAA Paper 98-444, 1998.

Waszak, M. R.; and Srinathkumar, S.: Flutter Suppression for the Active Flexible Wing: Control System Design and Experimental Validation. AIAA Paper 92-2097, 1992.

Wood, R. M. and Bauer, S. X. S.: *Advanced Aerodynamic Control Effectors*. SAE Paper 1999-01-5619, 1999.

Wood, R. M.; Banks, D. W.; and Bauer, S.X.S.; Assessment of Passive Porosity with Free and Fixed Separation on a Tangent Ogive Forebody. AIAA 92-4494, 1992.

Personal Air Transportation Concepts

Anders, Scott G.; Asbury, Scott C.; Brentner, Kenneth S.; Bushnell, Dennis M.; Glass, Christopher E; Hodges, William T.; Morris, Shelby J., Jr.; and Scott, Michael A.: *The Personal Aircraft: Status and Issues*. NASA TM-109174, 1994.

Flemisch, Frank O.; Adams, Catherine A.; Conway, Sheila R.; Goodrich, Ken H.; Palmer, Michael T.; and Schutte, Paul C.: *The H-Metaphor as a Guideline for Vehicle Automation and Interaction*. NASA TM-2003-212672, 2003.

McKinney, M. O.: *Stability and Control of the Aerial Jeep*. Preprint No. 10S, SAE Annual Meeting, 1959.

Moore, Mark D.: Personal Air Vehicles: A Rural/ Regional and Intra-Urban On-Demand Transportation System. AIAA Paper 2003-2646, 2003.

Moore, Mark D.; and Hahn, Andrew S.: Highly Affordable Personal Air Vehicles. *Contact Magazine.* July/August 2002.

Parlett, Lysle P.: *Wind-Tunnel Investigation of a Small-Scale Model of an Aerial Vehicle Supported by Ducted Fans.* NASA TN D-377, 1960.

INDEX

NASA History Series

This is a list of the various titles published in the NASA History Series. Many of these works, unfortunately, are no longer in print. Copies, however, are available in many Government Documents Sections at major university and public libraries through out the United States. The out of print books are noted.

Publications that are in print are available for sale from the NASA Information Center, Code CI-4, NASA Headquarters, Washington, DC 20546, or by calling 202-358-0000. To view a complete list of publications available through the NASA Information Center, go to http://history.nasa.gov/series95.html. Some NASA History books are also available for purchase through the Government Printing Office and are noted as such. NASA History Monographs are available at no cost with a self-adddressed, stamped envelope.

NASA Dryden also distributes its titles directly. A listing of these titles and information on ordering them may be found at http://www.nasa.gov/centers/dryden/history/Publications/index.html

Links are included to those publications which we have put on-line. Also included in this list are titles in the New Series in NASA History that John Hopkins University Press publishes.

Many of these publications are available through the Government Printing Office (GPO) or the NASA Center for AeroSpace Information (CASI) as well. The ordering information for the GPO can be found at GPO's web site at www.gpo.gov. The ordering information for CASI can be found here or at the Science and Technical Information (STI) web site at www.sti.nasa.gov.

Please visit www.klabs.org for additional materials on engineering and digital logic technologies for space flight applications. Among other topics, this site focuses on Apollo, the Space Shuttle, and Skylab.

To view online versions, go to http://history.nasa.gov/series95.htmll.

Reference Works, NASA SP-4000:

*   Grimwood, James M. *Project Mercury: A Chronology*. NASA SP-4001,1963.
    Out of print. Online version available.
*   Grimwood, James M., and Barton C. Hacker, with Peter J. Vorzimmer. *Project Gemini Technology and Operations: A Chronology*. NASA SP-4002,1969. **Out of print**. Online version available.
*   Link, Mae Mills. *Space Medicine in Project Mercury*. NASA SP-4003, 1965. **Out of print**.
    Online version available.
*   *Astronautics and Aeronautics, 1963: Chronology of Science, Technology,and Policy*. NASA SP-4004, 1964.
    Out of print.
*   *Astronautics and Aeronautics, 1964: Chronology of Science, Technology,and Policy*. NASA SP-4005, 1965.
    Out of print.
*   *Astronautics and Aeronautics, 1965: Chronology of Science, Technology,and Policy*. NASA SP-4006, 1966.
    Out of print.
*   *Astronautics and Aeronautics, 1966: Chronology of Science, Technology,and Policy*. NASA SP-4007, 1967.
    Out of print.

*   *Astronautics and Aeronautics, 1967: Chronology of Science, Technology, and Policy.* NASA SP-4008, 1968. **Out of print**.

*   Ertel, Ivan D., and Mary Louise Morse. *The Apollo Spacecraft: A Chronology,Volume I, Through November 7, 1962.* NASA SP-4009, 1969. **Out of print**. Online version available.

*   Morse, Mary Louise, and Jean Kernahan Bays. *The Apollo Spacecraft: A Chronology, Volume II, November 8, 1962-September 30, 1964.* NASA SP-4009, 1973. **Out of print**. Online version available.

*   Brooks, Courtney G., and Ivan D. Ertel. *The Apollo Spacecraft: A Chronology, Volume III, October 1, 1964-January 20, 1966.* NASA SP-4009, 1973. **Out of print**. Online version available.

*   Ertel, Ivan D., and Roland W. Newkirk, with Courtney G. Brooks, *The Apollo Spacecraft: A Chronology, Volume IV, January 21, 1966-July 13, 1974.* NASA SP-4009, 1978. **Out of print**. Online version available.

*   *Astronautics and Aeronautics, 1968: Chronology of Science, Technology, and Policy.* NASA SP-4010, 1969. **Out of print**.

*   Newkirk, Roland W., and Ivan D. Ertel, with Courtney G. Brooks, *Skylab: A Chronology.* NASA SP-4011, 1977. To purchase a copy of this book contact the NASA Headquarters Information Center at (202) 358-0000. Cost: $42.95 NASA HQ Info. Center. Online version available.

*   Van Nimmen, Jane, and Leonard C. Bruno, with Robert L. Rosholt. *NASA Historical Data Book, Vol. I: NASA Resources, 1958-1968.* NASASP-4012, 1976, rep. ed. 1988. To purchase a hardcover copy of this book contact the NASA Headquarters Information Center. Cost: $19.00 NASA HQ Info. Center A set of the first three NASA Historical Data Books is available for $55.00

*   Ezell, Linda Neuman. *NASA Historical Data Book, Vol II: Programs and Projects, 1958-1968.* NASA SP-4012, 1988. To purchase a hardcover copy of this book contact the NASA Headquarters Information Center. Cost: $19.00 NASA HQ Info. Center A set of the first three NASA Historical Data Books is available for $55.00

*   Ezell, Linda Neuman. *NASA Historical Data Book, Vol. III: Programs and Projects, 1969-1978.* NASA SP-4012, 1988. To purchase a hardcover copy of this book contact the NASA Headquarters Information Center. Cost: $19.00 NASA HQ Info. Center A set of the first three NASA Historical Data Books is available for $55.00. Online version available.

*   Gawdiak, Ihor, with Helen Fedor. *NASA Historical Data Book, Vol. IV: NASA Resources, 1969-1978.* NASA SP-4012, 1994. To purchase a hardcover copy of this book contact the HASA Headquarters Information Center. Cost: $28.00 NASA HQ Info. Center. Online version available.

*   Rumerman, Judy A. *NASA Historical Data Book, Vol. V: NASA LaunchSystems, Space Transportation, Human Spaceflight, and Space Science, 1979-1988.* NASA SP-4012, 1999. To purchase a hardcover copy of this book contact the Government Printing Office. Order GPO stock number: 033-000-01211-0.Cost: $43.00. GPO Order Form. Online version available.

*   Rumerman, Judy A. *NASA Historical Data Book, Vol. VI: Space Applications, Aeronautics, and other topics, 1979-1988.* NASA SP-4012,1999. To purchase a hardcover copy of this book contact the Government Printing Office. Order GPO stock number: 033-000-01222-5. Cost: $46.00. GPO Order Form. Online version available.

*   *Astronautics and Aeronautics, 1969: Chronology of Science, Technology, and Policy.* NASA SP-4014, 1970. **Out of print**.

*   *Astronautics and Aeronautics, 1970: Chronology of Science, Technology, and Policy.* NASA SP-4015, 1972. **Out of print**.

* *Astronautics and Aeronautics, 1971: Chronology of Science, Technology, and Policy.* NASA SP-4016, 1972. **Out of print**.

* *Astronautics and Aeronautics, 1972: Chronology of Science, Technology, and Policy.* NASA SP-4017, 1974. To purchase a paperback copy of this book contact the NASA Headquarters Information Center. Cost: $9.50 NASA HQ Info. Center

* *Astronautics and Aeronautics, 1973: Chronology of Science, Technology, and Policy.* NASA SP-4018, 1975. To purchase a paperback copy of this book contact the NASA Headquarters Information Center. Cost: $9.50 NASA HQ Info. Center

* *Astronautics and Aeronautics, 1974: Chronology of Science, Technology, and Policy.* NASA SP-4019, 1977. To purchase a paperback copy of this book contact the NASA Headquarters Information Center. Cost: $9.50 NASA HQ Info. Center

* *Astronautics and Aeronautics, 1975: Chronology of Science, Technology, and Policy.* NASA SP-4020, 1979. To purchase a paperback copy of this book contact the NASA Headquarters Information Center. Cost: $9.50 NASA HQ Info. Center

* *Astronautics and Aeronautics, 1976: Chronology of Science, Technology, and Policy.* NASA SP-4021, 1984. To purchase a paperback copy of this book contact the NASA Headquarters Information Center. Cost: $10.50 NASA HQ Info. Center

* *Astronautics and Aeronautics, 1977: Chronology of Science, Technology, and Policy.* NASA SP-4022, 1986. To purchase a paperback copy of this book contact the NASA Headquarters Information Center. Cost: $10.50 NASA HQ Info. Center

* *Astronautics and Aeronautics, 1978: Chronology of Science, Technology, and Policy.* NASA SP-4023, 1986. To purchase a paperback copy of this book contact the NASA Headquarters Information Center. Cost: $11.50 NASA HQ Info. Center

* *Astronautics and Aeronautics, 1979-1984: Chronology of Science, Technology, and Policy.* NASA SP-4024, 1988. To purchase a paperback copy of this book contact the NASA Headquarters Information Center. Cost: $11.50 NASA HQ Info. Center

* *Astronautics and Aeronautics, 1985: Chronology of Science, Technology, and Policy.* NASA SP-4025, 1990. To purchase a paperback copy of this book contact the NASA Headquarters Information Center. Cost: $12.50 NASA HQ Info. Center

* Noordung, Hermann. *The Problem of Space Travel: The Rocket Motor.* Edited by 1Ernst Stuhlinger and J.D. Hunley, with Jennifer Garland. NASA SP-4026, 1995. This **Out of print** . Online version available.

* Gawdiak, Ihor Y., Ramon J. Miro, and Sam Stueland, comps. *Astronauticsand Aeronautics, 1986-1990: A Chronology.* NASA SP-4027, 1997. To purchase a paperback copy of this book contact the Government Print ing Office. Order GPO Stock Number #033-000-01180-6 $21.00 GPO Order Form

* Gawdiak, Ihor Y. and Shetland, Charles. *Astronautics and Aeronautics, 1990-1995: A Chronology.* NASA SP-2000-4028, 2000. To purchase a paperback copy of this book contact the Government Printing Office. Order GPO Stock Number #033-000-01230-6 $43.00 GPO Order Form

* Orloff, Richard W. *Apollo by the Numbers: A Statistical Reference.* NASA SP-2000-4029, 2000. To purchase a paperback copy of this book contact the Government Printing Office. Order GPO Stock Number #033-000-01236-5 $40.00 GPO Order Form. Online version available. The online version includes all the extensive text and useful tables of the hard copy edition. The author has also made a number of corrections to the data in the hard copy edition. The online version does not include the original photos.

Management Histories, NASA SP-4100:

* Rosholt, Robert L. *An Administrative History of NASA, 1958-1963*. NASASP-4101, 1966. **Out of print**.

* Levine, Arnold S. *Managing NASA in the Apollo Era*. NASA SP-4102,1982. To purchase a copy of this book contact the NASA Headquarters Information Center.Cost: $10.00 NASA HQ Info. Center. Online version available.

* Roland, Alex. *Model Research: The National Advisory Committee for Aeronautics,1915-1958*. NASA SP-4103, 1985. To purchase a hard cover copy of this book contact the NASA Headquarters Information Center. Cost:$32.00 NASA HQ Info. Center . Online version available.

* Fries, Sylvia D. *NASA Engineers and the Age of Apollo*. NASA SP-4104,1992. **Out of print**. Online version available.

* Glennan, T. Keith. *The Birth of NASA: The Diary of T. Keith Glennan*. Edited by J.D. Hunley. NASA SP-4105, 1993. To purchase a hardcover copy of this book contact the Government Printing Office. Order GPO Stock Number #033-000-01131-8 $32.00 GPO Order Form

* Seamans, Robert C. *Aiming at Targets: The Autobiography of Robert C. Seamans*. NASA SP-4106, 1996. This hardcover book is available by calling the Government Printing Office at 202-512-1800 and ordering Stock Number 033-000-01175-0 $25.00 GPO Order Form.

* Garber, Stephen J., editor. Looking Backward, Looking Forward: Forty Years of Human Spaceflight Symposium. NASA SP-2002-4107. This hardcover book is available by calling the Government Printing Office at 202-512-1800 and ordering Stock Number 033-000-01249-7 $25.00 GPO Order Form. Online version available.

* Mallick, Donald L. with Peter W. Merlin. *The Smell of Kerosene: A Test Pilot's Odyssey* . NASA SP-4108. This hardcover book is available by calling the Government Printing Office at 202-512-1800 and ordering Stock Number 033-000-01270-5 $22.00 GPO Order Form.

* Iliff, Kenneth W. and Curtis L. Peebles. *From Runway to Orbit: Reflections of a NASA Engineer* . NASA SP-2004-4109. This hardcover book is available by calling the Government Printing Office at 202-512-1800 and ordering Stock Number 033-000-01267-5 $55.00 GPO Order Form.

* Chertok, Boris. *Rockets and People, Volume 1.* (NASA SP-2005-4110). Please order by contacting the NASA Center for Aerospace Information at 7121 Standard Drive Hanover, Maryland 21076, (301) 621-0390 or Online version available. Please mention the title and NASA Report #NASASP20054110. The price code is A03 ($27.50 within the U.S. plus $2 shipping and handling.)

* Laufer, Alexander, Post, Todd, and Hoffman, Edward. *Shared Voyage: Learning and Unlearning from Remarkable Projects* (NASA SP-2005-4111). Please order by contacting the NASA Center for Aerospace Information at 7121 Standard Drive Hanover, Maryland 21076, (301) 621-0390 or . Online version available. Please mention the title, and NASA Report #NASASP20054111. The price code is A03 ($27.50 within the U.S. plus $2 shipping and handling.)

Project Histories, NASA SP-4200:

* Swenson, Loyd S., Jr., James M. Grimwood, and Charles C. Alexander. *This New Ocean: A History of Project Mercury*. NASA SP-4201, 1966, reprinted 1999. . Online version available.This book is also available by calling the Government Printing Office at 202-512-1800 and ordering Stock Number 033-000-01210-1. $46.00 GPO Order Form

* Green, Constance McLaughlin, and Milton Lomask. *Vanguard: A History*. NASA SP-4202, 1970; rep. ed. Smithsonian Institution Press, 1971. **Out of print**. Online version available.

* Hacker, Barton C., and James M. Grimwood. *On Shoulders of Titans: A History of Project Gemini.* NASA SP-4203, 1977, reprinted 2002. This book is available by calling the Government Printing Office at 202-512-1800 and ordering Stock Number 033-000-01242-0. $47.00 GPO Order Form . Online version available.

* Benson, Charles D. and William Barnaby Faherty, *Moonport: A History of Apollo Launch Facilities and Operations.* NASA SP-4204, 1978. The SP edition is **Out of print**, but the University Press of Florida has republished the book in two volumes, *Gateway to the Moon* and *Moon Launch!* . Online version available.

* Brooks, Courtney G., James M. Grimwood, and Loyd S. Swenson, Jr. *Chariots for Apollo: A History of Manned Lunar Spacecraft.* NASA SP- 4205, 1979. **Out of print**. . Online version available.

* Bilstein, Roger E. *Stages to Saturn: A Technological History of the Apollo/Saturn Launch Vehicles.* NASA SP-4206, 1980 and 1996. This SP version is **Out of print**, but it has been reprinted by the University Press of Florida, please see below. Online version available.

* Compton, W. David, and Charles D. Benson. *Living and Working in Space:A History of Skylab.* NASA SP-4208, 1983. To purchase a hardcover copy of this book contact the NASA Headquarters Information Center. Cost: $20.00 NASA HQ Info. Center. Online version available.

* Ezell, Edward Clinton, and Linda Neuman Ezell. *The Partnership: A History of the Apollo-Soyuz Test Project.* NASA SP-4209, 1978. **Out of print**. Online version available.

* Hall, R. Cargill. *Lunar Impact: A History of Project Ranger.* NASA SP-4210, 1977. **Out of print**. Online version available.

* Newell, Homer E. *Beyond the Atmosphere: Early Years of Space Science.* NASA SP-4211, 1980. Online version available.To purchase a paperback copy of this book contact the NASA Headquarters Information Center. Cost: $15.00 NASA HQ Info. Center

* Ezell, Edward Clinton, and Linda Neuman Ezell. *On Mars: Exploration of the Red Planet, 1958-1978.* NASA SP-4212, 1984. **Out of print**. Online version available.

* Pitts, John A. *The Human Factor: Biomedicine in the Manned Space Program to 1980.* NASA SP-4213, 1985. To purchase a paperback copy of this book contact the NASA Headquarters Information Center. Cost: $19.00 NASA HQ Info. Center. This book is also available online.

* Compton, W. David. *Where No Man Has Gone Before: A History of Apollo Lunar Exploration Missions.* NASA SP-4214, 1989. . Online version available.To purchase a paperback copy of this book contact the Government Printing Office. Order GPO Stock Number #033-000-01047-8$25.00 GPO Order Form

* Naugle, John E. *First Among Equals: The Selection of NASA Space ScienceExperiments* NASA SP-4215, 1991. Online version available. To purchase a copy of this book contact the NASAHeadquarters Information Center. Cost: $8.00 NASA HQ Info. Center

* Wallace, Lane E. *Airborne Trailblazer: Two Decades with NASA Langley's737 Flying Laboratory.* NASA SP-4216, 1994. . Online version available. To purchase a paperback copy of this book contact the Government Printing Office. Order GPO Stock Number #033-000-01140-7$27.00 GPO Order Form

* Butrica, Andrew J. *Beyond the Ionosphere: Fifty Years of Satellite Communications.* NASA SP-4217, 1997. To purchase a hardcover copy of this book contact the Government Printing Office. Order GPO Stock Number#033-000-01178-4. $31.00 GPO Order Form. Online version available.

* Butrica, Andrew J. *To See the Unseen: A History of Planetary Radar Astronomy.* NASA SP-4218, 1996. To purchase a hardcover copy of this book contact the Government Printing Office. Order GPO Stock Number #033-000-01163-6$26.00 GPO Order Form

* Mack, Pamela E., ed. *From Engineering Science to Big Science: The NACA and NASA Collier Trophy Research Project Winners*. NASA SP-4219,1998. To purchase a copy of this book contact the Government Printing Office. Order GPO Stock Number #033-000-01199-7 $35.00 GPO Order Form. Online version available.

* Reed, R. Dale. *Wingless Flight: The Lifting Body Story*. NASASP-4220, 1998. To purchase a hardcover copy of this book contact the Government Printing Office. Order GPO Stock Number #033-000-01191-1 $25.00 GPO Order Form *Wingless Flight* is now also available in paperback from The University Press of Kentucky. Click here to order.

* Heppenheimer, T. A. *The Space Shuttle Decision: NASA's Search for a Reusable Space Vehicle*. NASA SP-4221, 1999. To purchase a copy of this book contact the Government Printing Office. Order GPO Stock Number #033-000-01215-2 $23.00 GPO Order Form. Online version available.

* Hunley, J. D., ed. *Toward Mach 2: The Douglas D-558 Program*. NASA SP-4222, 1999. Online version available. To purchase a copy of this book contact the Government Printing Office. Order GPO Stock Number #033-000-01208-0 $18.00 GPO Order Form

* Swanson, Glen E., ed. *"Before This Decade is Out..." Personal Reflections on the Apollo Program*. NASA SP-4223, 1999. To purchase a copy of this book contact the Government Printing Office. Order GPO Stock Number #033-000-01216-1$38.00 GPO Order Form. Alternately, this book has also been printed by the University Press of Florida and can be ordered here as well. Online version available.

* Tomayko, James E. *Computers Take Flight: A History of NASA's Pioneering Digital Fly-By-Wire Project* NASA SP-4224, 2000. To purchase a copy of this book contact the Government Printing Office. Order GPO Stock Number#033-000-01220-9 $26.00 GPO Order Form. Online version available.

* Leary, William M. *We Freeze to Please: A History of NASA's Icing Research Tunnel and the Quest for Safety*. NASA SP-2002-4226, 2002. To purchase a copy of this book contact the Government Printing Office. Order GPO Stock Number #033-000-01244-6 $28.00. GPO Order Form.

* Mudgway, Douglas J. *Uplink-Downlink: A History of the Deep Space Network, 1957-1997* . NASA SP-2001-4227. To purchase a copy of this book contact the Government Printing Office. Order GPO Stock Number #033-000-01241-1 $26.00 GPO Order Form. Online version available.

* Dawson, Virginia P. and Mark D. Bowles. *Taming Liquid Hydrogen: The Centaur Upper Stage Rocket, 1958-2002* . NASA SP-2004-4230. To purchase a copy of this book contact the Government Printing Office. Order GPO Stock Number #033-000-01271-3 $28.00 GPO Order Form.

Center Histories, NASA SP-4300:

* Rosenthal, Alfred. *Venture into Space: Early Years of Goddard Space Flight Center*. NASA SP-4301, 1985. **Out of print**.

* Hartman, Edwin, P. *Adventures in Research: A History of Ames Research Center, 1940-1965*. NASA SP-4302, 1970. **Out of print**. Online version available.

* Hallion, Richard P. *On the Frontier: Flight Research at Dryden, 1946-1981*. NASA SP-4303, 1984. To pur chase a hardcover copy of this book contact the NASA Headquarters Information Center.
  Cost: $18.00 NASA HQ Info. Center

* Muenger, Elizabeth A. *Searching the Horizon: A History of Ames Research Center, 1940-1976*. NASA SP-4304, 1985. To purchase a paperback copy of this book contact the NASA Headquarters Information Center. Cost:$13.00 NASA HQ Info. Center. . Online version available.

* Hansen, James R. *Engineer in Charge: A History of the Langley Aeronautical Laboratory,1917-1958*. NASA

SP-4305, 1987. To purchase a paperback copy of this book contact the NASA Headquarters Information Center. Cost: $30.00 NASA HQ Info. Center. Online version available.

* Dawson, Virginia P. *Engines and Innovation: Lewis Laboratory and American Propulsion Technology.* NASA SP-4306, 1991. To purchase a paperback copy of this book contact the Government Printing Office. Order GPO Stock Number #033-000-01095-8 $16.00 GPO Order Form. Online version available.

* Dethloff, Henry C. *"Suddenly Tomorrow Came...": A History of the Johnson Space Center, 1957-1990.* NASASP-4307, 1993. To purchase a hardcover copy of this book contact the Government Printing Office. Order GPO Stock Number #033-000-01134-2 $36.00 GPO Order Form. Online version available.

* Hansen, James R. *Spaceflight Revolution: NASA Langley Research Center from Sputnik to Apollo.* NASA SP-4308, 1995. To purchase a paperback copy of this book contact the Government Printing Office. Order GPO Stock Number#033-000-01149-1 $30.00 GPO Order Form. Online version available.

* Wallace, Lane E. *Flights of Discovery: An Illustrated History of the Dryden Flight Research Center.* NASA SP-4309, 1996. To purchase a hardcover copy of this book contact the Government Printing Office. Order GPO Stock Number #033-000-01167-9 $42.00 GPO Order Form.

* Herring, Mack R. *Way Station to Space: A History of the John C. Stennis Space Center.* NASA SP-4310, 1997. To purchase a hardcover copy of this book contact the Government Printing Office. Order GPO Stock Number #033-000-01185-7$37.00 GPO Order Form

* Wallace, Harold D., Jr. *Wallops Station and the Creation of an American Space Program.* NASA SP-4311, 1997. To purchase a copy of this softcover book contact the Government Printing Office. Order GPO Stock Number #033-000-01186-5$11.00 GPO Order Form

* Wallace, Lane E. *Dreams, Hopes, Realities. NASA's Goddard Space Flight Center: The First Forty Years.* NASA SP-4312, 1999. To purchase a copy of this softcover book contact the Government Printing Office. Order GPO Stock Number #033-000-01206-3 $33.00 GPO Order Form

* Dunar, Andrew J. and Waring, Stephen P. *Power to Explore: A History of Marshall Space Flight Center, 1960-1990* NASA SP-4313, 1999. To purchase a copy of this hardcover book contact the Government Printing Office. Order GPO Stock Number #033-000-01221-7 $49.00 GPO Order Form . Online version available.

* Bugos, Glenn E. *Atmosphere of Freedom: Sixty years at the NASA Ames Research Center* NASA SP-2000-4314, 2000.To purchase a copy of this richly illustrated, softcover book contact the Government Printing Office. Order GPO Stock Number #033-000-01225-0 $39.00 GPO Order Form. Online version available.

* Schultz, James. *Crafting Flight: Aircraft Pioneers and the Contributions of the Men and Women of NASA Langley Research Center* NASA SP-2003-4316, 2003. To purchase a copy of this softcover book contact the Government Printing Office. Order GPO Stock Number 033-000-01257-8 $46.00 GPO Order Form.

General Histories, NASA SP-4400:
* Corliss, William R. *NASA Sounding Rockets, 1958-1968: A Historical Summary.* NASA SP-4401, 1971. **Out of print.** Online version available.

* Wells, Helen T., Susan H. Whiteley, and Carrie Karegeannes. *Origins of NASA Names.* NASA SP-4402, 1976. **Out of print.** Online version available.

* Anderson, Frank W., Jr. *Orders of Magnitude: A History of NACA andNASA, 1915-1980.* NASA SP-4403, 1981. **Out of print.**

* Sloop, John L. *Liquid Hydrogen as a Propulsion Fuel, 1945-1959.* NASASP-4404, 1978. **Out of print.** Online version available.

*   Roland, Alex. *A Spacefaring People: Perspectives on Early Spaceflight.* NASA SP-4405, 1985. **Out of print**.

*   Bilstein, Roger E. *Orders of Magnitude: A History of the NACA and NASA, 1915-1990.* NASA SP-4406, 1989. **Out of print**. Online version available.

*   Logsdon, John M., ed., with Linda J. Lear, Jannelle Warren Findley, Ray A. Williamson, and Dwayne A. Day. *Exploring the Unknown: Selected Documents in the History of the U.S. Civil Space Program, Volume I, Organizing for Exploration.* NASA SP-4407, 1995. To purchase a hardcover copy of this book contact the Government Printing Office. Order GPO Stock Number #033-000-01160-1$43.00 GPO Order Form

*   Logsdon, John M., ed, with Dwayne A. Day, and Roger D. Launius. *Exploring the Unknown: Selected Documents in the History of the U.S. Civil Space Program, Volume II, External Relationships.* NASA SP-4407, 1996. To purchase a hardcover copy of this book contact the Government Printing Office. Order GPO Stock Number #033-000-01174-1 $40.00 GPO Order Form

*   Logsdon, John M., ed., with Roger D. Launius, David H. Onkst, and StephenJ. Garber. *Exploring the Unknown: Selected Documents in the History of the U.S. Civil Space Program, Volume III, Using Space.* NASA SP-4407,1998. To purchase a hard cover copy of this book contact the Government Printing Office. Order GPO Stock Number #033-000-01195-4 $41.00 GPO Order Form

*   Logsdon, John M., ed., with Ray A. Williamson, Roger D. Launius, Russell J. Acker, Stephen J. Garber, and Jonathan L. Friedman. *Exploring the Unknown: Selected Documents in the History of the U.S. Civil Space Program,Volume IV, Accessing Space.* NASA SP-4407, 1999. To purchase a hard cover copy of this book contact the Government Printing Office. Order GPO Stock Number #033-000-01219-5 $55.00 GPO Order Form. Online version available.

*   Logsdon, John M., ed., with Amy Paige Snyder, Roger D. Launius, Stephen J. Garber, and Regan Anne Newport. *Exploring the Unknown: Selected Documents in the History of the U.S. Civil Space Program,Volume V, Exploring the Cosmos.* NASA SP-4407, 2001. To purchase a hard cover copy of this book contact the Government Printing Office. Order GPO Stock Number #033-000-01238-1 $70.00 GPO Order Form. Online version available.

*   Logsdon, John M., ed., with Stephen J. Garber, Roger D. Launius, and Ray A. Williamson. *Exploring the Unknown: Selected Documents in the History of the U.S. Civil Space Program, Volume VI: Space and Earth Science* (NASA SP-2004-4407), 2004. Please order by contacting the NASA Center for Aerospace Information at 7121 Standard Drive Hanover, Maryland 21076, (301) 621-0390 or order online. Please mention the title, volume number, and Document ID # 20040095359. The domestic sales price is $43.00 plus shipping. Online version available.

*   Siddiqi, Asif A., *Challenge to Apollo: The Soviet Union and the Space Race, 1945-1974* (NASA SP-2000-4408). To purchase a hard cover copy of this book contact the Government Printing Office. Order GPO Stock Number #033-000-01231-4 $79.00 GPO Order Form This book is also available in a two-part series from the University Press of Florida.

*   Hansen, James R., ed.. *The Wind and Beyond: Journey into the History of Aerodynamics in America, Volume 1, The Ascent of the Airplane.* NASA SP-2003-4409, 2003. To purchase a hard cover copy of this book contact the Government Printing Office. Order GPO Stock Number #033-000-01268-3 $55.00 GPO Order Form.

Monographs in Aerospace History (SP-4500 Series):
Monographs 2 - 32 are available by sending a self-addressed 9x12" envelope for each monograph with appropriate postage for 17 ounces (typically $3.95 within the U.S., $5.70 for Canada, and $12.15 for overseas - international

customers are asked to purchase U.S. postage through an outlet such as www.stampsonline.com) to the NASA Headquarters Information Center, Code CI-4, Washington, DC 20546.

Monographs 25 and 30 are available by sending a self-addressed 8"x11" flat-rate Priority Mail envelope for each monograph to the NASA Dryden Flight Research Center History Office, Mail Stop 1613, P.O. Box 273, Edwards, CA 93523.

* Launius, Roger D. and Aaron K. Gillette, comps. *Toward a History of the Space Shuttle: An Annotated Bibliography.* Monograph in Aerospace History, No. 1, 1992. **Out of print**. This monograph is available online. Online version available.

* Launius, Roger D., and J.D. Hunley, comps. *An Annotated Bibliography of the Apollo Program.* Monograph in Aerospace History No. 2, 1994. This monograph is available online. Online version available.

* Launius, Roger D. *Apollo: A Retrospective Analysis.* Monograph in Aerospace History, No. 3, 1994. This monograph is available online. Online version available.

* Hansen, James R. *Enchanted Rendezvous: John C. Houbolt and the Genesis of the Lunar-Orbit Rendezvous Concept.* Monograph in Aerospace History, No. 4, 1995. This monograph is available online. Online version available.

* Gorn, Michael H. *Hugh L. Dryden's Career in Aviation and Space.* Monograph in Aerospace History, No. 5, 1996. Online version available.

* Powers, Sheryll Goecke. *Women in Flight Research at NASA Dryden Flight Research Center from 1946 to 1995.* Monograph in Aerospace History, No. 6, 1997.

* Portree, David S.F. and Robert C. Trevino. *Walking to Olympus: An EVA Chronology.* Monograph in Aerospace History, No. 7, 1997. Online version available..

* Logsdon, John M., moderator. *Legislative Origins of the National Aeronautics and Space Act of 1958: Proceedings of an Oral History Workshop.* Monograph in Aerospace History, No. 8, 1998. Online version available.

* Rumerman, Judy A., comp. *U.S. Human Spaceflight, A Record of Achievement 1961-1998.* Monograph in Aerospace History, No. 9, 1998. Online version available.

* Portree, David S. F. *NASA's Origins and the Dawn of the Space Age.* Monograph in Aerospace History, No. 10, 1998. Online version available.

* Logsdon, John M. *Together in Orbit: The Origins of International Cooperation in the Space Station.* Monograph in Aerospace History, No. 11, 1998. . Online version available.

* Phillips, W. Hewitt. *Journey in Aeronautical Research: A Career at NASA Langley Research Center.* Monograph in Aerospace History, No. 12, 1998.

* Braslow, Albert L. *A History of Suction-Type Laminar-Flow Control with Emphasis on Flight Research* Monograph in Aerospace History, No. 13, 1999. Online version available.

* Logsdon, John M., moderator. *Managing the Moon Program: Lessons Learned Fom Apollo.* Monograph in Aerospace History, No. 14, 1999. Online version available.

* Perminov, V.G. *The Difficult Road to Mars: A Brief History of Mars Exploration in the Soviet Union* is Monograph in Aerospace History, No. 15, 1999. Online version available.

* Maisel, Martin, Giulanetti, Demo J., and Dugan, Daniel C. *The History of the XV-15 Tilt Rotor Research Aircraft: From Concept to Flight* is Monograph in Aerospace History, No. 17, 2000 (NASA SP-2000-4517). On line version available.

* Jenkins, Dennis R., *Hypersonics Before the Shuttle: A Concise History of the X-15 Research Airplane* is Monograph in Aerospace History, No. 18, 2000 (NASA SP-2000-4518). Online version available.

* Chambers, Joseph R. *Partners in Freedom: Contributions of the Langley Research Center to U.S. Military Aircraft of the 1990s* is Monograph in Aerospace History, No. 19, 2000 (NASA SP-2000-4519). Online version available.

* Waltman, Gene L. *Black Magic and Gremlins: Analog Flight Simulations at NASA's Flight Research Center* is Monograph in Aerospace History, No. 20, 2000 (NASA SP-2000-4520). Online version available.

* Portree, David S.F.. *Humans to Mars: Fifty Years of Mission Planning, 1950-2000* is Monograph in Aerospace History, No. 21, 2001 (NASA SP-2001-4521). Online version available.

* Thompson, Milton O. with J.D. Hunley. *Flight Research: Problems Encountered and What they Should Teach Us* is Monograph in Aerospace History, No. 22, 2001 (NASA SP-2001-4522). Online version available.

* Tucker, Tom. *The Eclipse Project* is Monograph in Aerospace History, No. 23, 2001 (NASA SP-2001-4523). Online version available.

* Siddiqi, Asif A. *Deep Space Chronicle: A Chronology of Deep Space and Planetary Probes 1958-2000* is Monograph in Aerospace History, No. 24, 2002 (NASA SP-2002-4524). Online version available.

* Merlin, Peter W. *Mach 3+: NASA/USAF YF-12 Flight Research, 1969-1979* is Monograph in Aerospace History, No. 25, 2001 (NASA SP-2001-4525) . Online version available.

* Anderson, Seth B. *Memoirs of an Aeronautical Engineer: Flight Tests at Ames Research Center: 1940-1970* is Monograph in Aerospace History, No. 26, 2002 (NASA SP-2002-4526)

* Renstrom, Arthur G. *Wilbur and Orville Wright: A Bibliography Commemorating the One-Hundredth Annversary of the First Powered Flight on December 17, 1903* is Monograph in Aerospace History, No. 27, 2002 (NASA SP-2002-4527). Online version available.

* No monograph 28.

* Chambers, Joseph R. *Concept to Reality: Contributions of the NASA Langley Research Center to U.S. Civil Aircraft of the 1990s* is Monograph in Aerospace History, No. 29, 2003 (SP-2003-4529). Online version available.

* Peebles, Curtis, editor. *The Spoken Word: Recollections of Dryden History, The Early Years* is Monograph in Aerospace History, No. 30, 2003 (SP-2003-4530). Online version available.

* Jenkins, Dennis R., Tony Landis, and Jay Miller. *American X-Vehicles: An Inventory- X-1 to X-50* is Monograph in Aerospace History, No. 31, 2003 (SP-2003-4531). Online version available.

* Renstrom, Arthur G. *Wilbur and Orville Wright: A Chronology Commemorating the One-Hundredth Anniversary of the First Powered Flight on December 17, 1903* is Monograph in Aerospace History, No. 32, 2002 (NASA SP-2003-4532). Online version available.

* Bowles, Mark D. and Arrighi, Robert S. *NASA's Nuclear Frontier: The Plum Brook Research Reactor* is Monograph in Aerospace History, No. 33, 2003 (SP-2004-4533). Online version available.

* McCurdy, Howard E. *Low Cost Innovation in Spaceflight: The History of the Near Earth Asteroid Rendezvous (NEAR) Mission* (NASA SP-2005-4536). Online version available.

* Lambright, W. Henry. *NASA and the Environment: The Case of Ozone Depletion* (NASA SP-2005-4538). Online version available.

* Seamans, Robert C. Jr.*Project Apollo: The Tough Decisions* (NASA SP-2005-4537). Online version available.

Dryden Historical Studies

*   Tomayko, James E., author, and Christian Gelzer, editor. *The Story of Self-Repairing Flight Control Systems* is Dryden Historical Study #1. This study is available from the Dryden Flight Research Center History Ofice by sending a self-addressed 8"x11" flat-rate Priority Mail envelope for each study to the NASA Dryden Flight Research Center History Office, Mail Stop 1613, P.O. Box 273, Edwards, CA 93523.

Electronic Media (SP-4600 Series)

*   *Remembering Apollo 11: The 30th Anniversary Data Archive CD-ROM* (SP-4601, 1999). This CD-ROM is available by sending a self-addressed envelope for each CD-ROM set with appropriate postage (typically $1.90 within the U.S., $2.30 for Canada, and $5.60 for overseas - international customers are asked to purchase U.S. postage through an outlet such as www.stampsonline.com) to the NASA Headquarters Information Center, Mail Code CI-4, 300 E Street SW, Room 1H23, Washington, D.C. 20546-0001

*   *The Mission Transcript Collection: U.S. Human Spaceflight Missions from Mercury Redstone 3 to Apollo 17* (SP-2000-4602, 2001). Now available commerically from CG Publishing. To order send an International Money Order for $8.00 to CG Publishing Inc, Box 62034, Burlington, Ontario, L7R 4K2, Canada or call 905-637-5737.

*   *Shuttle-Mir: the United States and Russia Share History's Highest Stage* (SP-2001-4603, 2002). This CD-ROM is available from NASA CORE for $5 per copy plus shipping and handling (within the U.S., $6 for up to $25 order). To order the CD-ROM, please mail a check, money order or school purchase order to: NASA CORE, Lorain County JVS, 15181 Route 58 South, Oberlin, OH 44074, 440-775-1400, toll free 1-866-776-CORE, FAX 440-775-1460, nasaco@leeca.org, or http://core.nasa.gov on the Web. CORE also accepts orders by credit card (VISA or MasterCard).

*   *U.S. Centennial of Flight Commission presents Born of Dreams - Inspired by Freedom* (SP-2004-4604, 2004). This DVD data disk is available by sending a self-addressed envelope for each DVD with appropriate postage (typically $1.90 within the U.S., $2.30 for Canada, and $5.60 for overseas - international customers are asked to purchase U.S. postage through an outlet such as www.stampsonline.com) to the NASA Headquarters Information Center, Mail Code CI-4, 300 E Street SW, Room 1H23, Washington, D.C. 20546-0001.

*   *Of Ashes and Atoms: A Documentary on the NASA Plum Brook Reactor Facility* (NASA SP-2005-4605). Of Ashes and Atoms was produced and directed by James Polaczynski and written by him with Robert Arrighi. Narrated by Kate Mulgrew (Captain Janeway of the Star Trek Voyager series), this documentary illustrates the history behind Plum Brook Reactor Facility, operating from 1962-1973 as one of the first nuclear test reactors built in the United States and the only one built by NASA. While the reactor never reached its full potential, the personnel who have worked there made great achievements in terms of scientific discovery, as well as building, operating, and safely deconstructing a nuclear reactor. Plum Brook's rich history has significant lessons in terms of management, environmental stewardship, painstaking engineering, and scientific investigation. This DVD is available by sending a self-addressed envelope for each CD with appropriate postage (typically $1.90 within the U.S., $2.30 for Canada, and $5.60 for overseas - international customers are asked to purchase U.S. postage through an outlet such as www.stampsonline.com) to the NASA Headquarters Information Center, 300 E Street SW, Room 1H23, Washington, D.C. 20546-0001, 202-358-0000.

*   Taming Liquid Hydrogen : The Centaur Upper Stage Rocket Interactive CD-ROM. (SP-2004-4606, 2004). This CD-ROM is available by sending a self-addressed envelope for each CD-ROM set with appropriate postage (typically $1.90 within the U.S., $2.30 for Canada, and $5.60 for overseas - international customers are asked to purchase U.S. postage through an outlet such as www.stampsonline.com) to the NASA Headquarters Information Center, Mail Code CI-4, 300 E Street SW, Room 1H23, Washington, D.C. 20546-0001.

* *Fueling Space Exploration: The History of NASA's Rocket Engine Test Facility DVD* (NASA SP-2005-4607). This DVD contains a 25-minute and a condensed 7-minute documentary video on the RETF, which used to be a part of the NASA Glenn Research Center. RETF employees performed pioneering research from 1957 to 1995 on liquid hydrogen propulsion on the Centaur and Saturn rockets, as well as the Space Shuttle. Declared a National Historic Landmark in 1984, the RETF officially closed in 1995 and was torn down in 2003 to make way for the Cleveland airport's expansion. This DVD is available by sending a self-addressed envelope for each CD with appropriate postage (typically $1.90 within the U.S., $2.30 for Canada, and $5.60 for overseas - international customers are asked to purchase U.S. postage through an outlet such as www.stampsonline.com) to the NASA Headquarters Information Center, 300 E Street SW, Room 1H23, Washington, D.C. 20546-0001, 202-358-0000.

Historical Reports (NASA HHR)

* NASA Office of Defense Affairs: The First Five Years (HHR-32, 1970) by W. Fred Boone. Admiral Boone led the Office of Defense Affairs from December 1, 1962 through January 1, 1968, a formative early period in space history when cooperation between NASA, a civilian agency, and the military was especially important. This significant narrative charts these early efforts in coordination. Special thanks to volunteer Chris Gamble for scanning and formatting this book for the Web. . Online version available.

* *Research in NASA History: A Guide to the NASA History Program.* NASA HHR-64, revised June 1997. This monograph-sized publication is available by sending a stamped (for 8 ounces), self-addressed 9x12 inch envelope to the NASA History Division, Code IQ, Washington, DC 20546. . Online version available.

NASA Special Reports (NASA SP-4900)

* *Unmanned Space Project Management: Surveyor and Lunar Orbiter.* Washington,D.C.:NASA SP-4901, 1972. By Erasmus H. Kloman. NASA commissioned the National Academy of Public Administration to undertake this study to look at its innovative management techniques on these complex technological projects. **Out of print**. Online version available.

Other NASA Special Publications
(not in the formal NASA History Series)

* *Results of the Second Manned Suborbital Space Flight, July 21, 1961.* NASA, 1961. **Out of print**. Online version available.

* *The Impact of Science on Society.* NASA SP-482 by James Burke, Jules Bergman, and Isaac Asimov, 1985. Online version available.

* *Space Station Requirements and Transportation Options for Lunar Outpost.* NASA, 1990. Online version available.

* *Space Station Freedom Accommodation of the Human Exploration Initiative.* NASA, 1990. Online version available.

* *Why Man Explores.* NASA EP-125, 1976.

* *Results of the Second Manned Suborbital Space Flight, July 21, 1961.* NASA, 1961. **Out of print**. Online version available.

* *Apollo 13 "Houston, we've got a problem."* NASA EP-76, 1970. **Out of print**. Online version available.

* *Results of the Second U.S. Manned Orbital Space Flight.* NASA SP-6, 1962. **Out of print**. Online version available.

* *Results of the Third U.S. Manned Orbital Space Flight.* NASA SP-12, 1962. **Out of print**. Online version available.

* *MercuryProject Summary including Results of the Fourth Manned Orbital Flight.* NASA SP-45, 1963. **Out of print**. Online version available.

* *X-15 Research Results With a Selected Bibliography.* NASA SP-60, 1965. **Out of print**. Online version available.

* Exploring Space with a Camera. NASA SP-168, 1968. Online version available.

* Aerospace Food Technology. NASA SP-202, 1969. Online version available.

* What Made Apollo a Success? NASA SP-287, 1971. Online version available.

* Evolution of the Solar System NASA SP-345, 1976. Online version available.

* *Pioneer Odyssey* (NASA SP-349/396, revised edition, 1977) by Richard Fimmel, William Swindell, and Eric Burgess. Online version available.

* *Apollo Expeditions to the Moon.* NASA SP-350, 1975. **Out of print**. Online version available.

* *Apollo Over the Moon: A View From Orbit* (NASA SP-362, 1978) edited by Harold Masursky, G.W. Colton, and Farouk El-Baz. Online version available.

* *Introduction to the Aerodynamics of Flight* (NASA SP-367, 1975) by Theodore A. Talay. Online version available.

* Biomedical Results of Apollo (NASA SP-368, 1975) , edited by Richard S. Johnston, Lawrence F. Dietlein, M.D., and Charles A. Berry, M.D. Online version available.

* *Skylab: Our First Space Station* (NASA SP-400, 1977), edited by Leland F. Belew. Online version available.

* *Skylab, Classroom in Space* (NASA SP-401, 1977), edited by Lee Summerlin. Online version available.

* *A New Sun: Solar Results from Skylab* (SP-402, 1979) by John A. Eddy and edited by Rein Ise. Online version available.

* *Skylab's Astronomy and Space Sciences* (NASA SP-404, 1979), edited by Charles A. Lundquist. Online version available.

* *The Space Shuttle* (SP-407, 1976) . Online version available.

* *The Search For Extraterrestrial Intelligence* (NASA SP-419, 1977) , edited by Philip Morrison, John Billingham, and John Wolfe. Online version available.

* Atlas of Mercury (SP-423, 1978) by Merton E. Davies, Stephen E. Dwornik, et. al. Online version available.

* *The Voyage of Mariner 10: Mission to Venus and Mercury* (NASA SP-424, 1978)by James A. Dunne and Eric Burgess. Online version available.

* *The Martian Landscape* (NASA SP-425, 1978)

* *The Space Shuttle at Work* (NASA SP-432/EP-156 1979) by Howard Allaway. Online version available.

* *Project Orion: A Design Study of a System for Detecting Extrasolar Planets* (NASA SP-436, 1980), edited by David C. Black. Online version available.

* *Wind Tunnels of NASA.* NASA SP-440, 1981. **Out of print**. Online version available.

* Viking Orbiter Views of Mars (NASA SP-441, 1980) . Online version available.

* *The High Speed Frontier: Case Histories of Four NACA Programs, 1920-1950.* (NASA SP-445, 1980.) . Online version available.

* The Star Splitters: The High Energy Astronomy Observatories (SP-466, 1984) by Wallace H. Tucker. Online version available.

* Planetary Geology in the 1980s (SP-467, 1985) by Joseph Veverka.

* *Quest for Performance: The Evolution of Modern Aircraft.* (NASA SP-468,1985.) **Out of print**. Online version available.

* *The Long Duration Exposure Facility (LDEF): Mission 1 Experiments* (SP-473, 1984) ed. by Lenwood G. Clark, William H. Kinar, et. al. . Online version available.

* *Voyager 1 and 2, Atlas of Saturnian Satellites* (NASA SP-474, (NASA SP-474, 1984) edited by Raymond Batson. Online version available.

* *Far Travelers: The Exploring Machines* (NASA SP-480, 1985) by Oran W. Nicks. Online version available.

* *Living Aloft: Human Requirements for Extended Spaceflight.* (NASA SP-483, 1985.) **Out of print**. Online version available.

* Space Shuttle Avionics System (SP-504, 1989) by John F. Hanaway and Robert W. Moorehead. Online version available.

* Life Into Space: Space Life Sciences Research, Volumes I and II, 1965-1998 (SP-534, 1995, 2000). Online version available.

* Flight Research at Ames, 1940-1997 (SP-3300, 1998). Online version available.

* The Planetary Quarantine Program (SP-4902, 1974). Online version available.

* *Spaceborne Digital Computer Systems* (NASA SP-8070, 1971). Online version available.

* *Magellan: The Unveiling of Venus* (JPL-400-345, 1989) . Online version available.

* *Guide to Magellan Image Interpretation* (JPL-93-24) by John Ford, Jeffrey Plaut, et. al. Online version available.

* *The Apollo Program Summary Report* (Document # JSC-09423, April 1975) . Online version available.

* *Saturn Illustrated Chronology* (MHR-5, Marshall Space Flight Center, fifth edition, 1971) prepared by David S. Akens. Online version available.

* *Celebrating a Century of Flight* (NASA SP-2002-09-511-HQ). Edited by Tony Springer. Online version available.

* *Present and Future State of the Art in Guidance Computer Memories* (NASA TN D-4224, 1967) by Robert C. Ricci. Online version available.

NASA Educational Publications

Skylab: A Guidebook (NASA EP-107, 1973), by Leland F. Belew and Ernst Stuhlinger. Special thanks to Chris Gamble for formatting this book for the Web.

Spacelab: An International Short-Stay Orbiting Laboratory (NASA EP-165) by Walter Froehlich. The full text and rich images from this informative book about Europe's first major undertaking in human spaceflight are now available on-line thanks to volunteer Chris Gamble's expert help.

A Meeting with the Universe: Science Discoveries from the Space Program (NASA EP-177,1981). Written by a group of NASA scientists for a popular audience, this attractive photo book is not a formal NASA history, but a "history of space exploration--by NASA, by universities, by other government agencies, and by industries--all of whom have played major roles." Warm thanks to Hans-Peter Engel, who scanned and formatted this special book for the Web.

NASA Publications (NPs)

Science in Orbit: The Shuttle & Spacelab Experience: 1981-1986 (NASA NP-119, Marshall Space Flight Center, 1988). Provided by the European Space Agency, the Spacelab entails both an enclosed laboratory and an exposed platform for scientific experiments in space. Thanks to volunteer Chris Gamble for scanning and formatting this informative guide to this unique facility.

NASA Conference Proceedings

Life in the Universe : Proceedings of a conference held at NASA Ames Research Center Moffet Field, California, June 19-20, 1979 (NASA CP-2156, 1981), edited by John Billingham. Special thanks to Chris Gamble for formatting this volume for the Web.

Proceedings of the X-15 First Flight 30th Anniversary Celebration of June 8, 1989 These proceedings include comments by historians, pilots, and others with keen insights on the truly historic X-15 program that bridged aeronautics with astronautics during NASA's first decade.

NASA Technical Memoranda

*   *Destination Moon: A History of the Lunar Orbiter Program* . Washington, D.C.: NASA TM-3487,1977. Written by Bruce Byers, this technical memorandum is a book-length scholarly work detailing the history of the robotic Lunar Orbiter Program, which provided very useful mission planning data for the Apollo program. Without the Lunar Orbiters' mapping of the lunar surface, it would have been extremely difficult, if not impossible, for Apollo planners to decide where to land the Apollo spacecraft on the Moon. A special thanks to Chris Gamble for formatting this document's complete text and illustrative diagrams for the Web. **Out of print**.

Contractor Reports

*   *Computers in Spaceflight: The NASA Experience.* James E. Tomayko wrote this contractor report in 1988. A relatively unique document, this report covers computers in the Gemini, Apollo, Skylab, and Shuttle programs, as well as for robotic spacecraft and ground systems. Chris Gamble deserves kudos for his excellent work formatting the text of this prime reference document for the Web. This document should be available in hard copy, with photographs, in late February 1998 from NASA's Center for Aerospace Information (CASI). Contact CASI at 800 Elkridge Landing Road, Linthicum Heights, MD 21090, 301-621-0100 or email at help@sti.nasa.gov

Other Government Publications Related to Aerospace History

History of Research in Space Biology and Biodynamics at the Air Force Missile Development Center, Holloman Air Force Base, New Mexico, 1946-1958. This early Air Force report contains information that NASA built upon in developing Project Mercury. It may be of special interest to some historians and buffs because of John Glenn's flight on STS-95 and because of the fortieth anniversary of the Mercury Seven selection in 1999. A very special thanks to

Chris Gamble for formatting the complete text of this report for the Web.

Report of the Apollo 13 Review Board (a.k.a. the Cortright Commission): This is the report issued after the Apollo 13 accident which prevented the mission from landing on the moon and nearly cost the lives of the astronauts involved. Special thanks to Colin Fries and Sivram Prasad of the History Division for scanning and formatting this report for the Web.

Report of the Presidential Commission on the Space Shuttle Challenger Accident (commonly called the Rogers Commission Report), June 1986 and Implementations of the Recommendations, June 1987. Online version available.

Transiting from Air to Space: The North American X-15 This case study by Robert S. Houston, Richard P. Hallion, and Ronald G. Boston is a long chapter in *The Hypersonic Revolution: Case Studies in the History of Hypersonic Technology* (AirForce History and Museums Program: 1998). A key contribution to the literature on the X-15, one of NASA's most successful research aircraft programs, this case study was previously published as a stand-alone volume. Special thanks to Hans-Peter Engel, who formatted this work for the Web.

Space Handbook: Astronautics and its Applications. This 1959 publication was a staff report of the Congressional Select Committe on Astronauticsand Space Exploration. An interesting historical document, this Handbook includes much information about astronomy and astronautics that we now know to be incorrect. Nevertheless, this document provides a snapshot of the beginning of the space era. Special thanks to John Henry, who scanned and formatted this document.

The First Century of Flight: NACA/NASA Contributions to Aeronautics. This is an informative and attractive Web exhibit set up in a timeline format. Special thanks to Tony Springer, who supplied the content; Ray Brown, who created the hard copy version; and Douglas Ortiz, who created the Web version.

New Series in NASA History Published by the Johns Hopkins University Press:
These books are available by calling 410-516-6956 or see http://www.press.jhu.edu/books/

*   Cooper, Henry S. F., Jr. *Before Lift-off: The Making of a Space Shuttle Crew.* Baltimore: Johns Hopkins University Press, 1987.
*   McCurdy, Howard E. *The Space Station Decision: Incremental Politics andTechnological Choice.* Baltimore: Johns Hopkins University Press, 1990.
*   Hufbauer, Karl. *Exploring the Sun: Solar Science Since Galileo.* Baltimore: JohnsHopkins University Press, 1991.
*   McCurdy, Howard E. *Inside NASA: High Technology and Organizational Change in the U.S. Space Program.* Baltimore: Johns Hopkins UniversityPress,1993.
*   Lambright, W. Henry. *Powering Apollo: James E. Webb of NASA.* Baltimore: Johns Hopkins University Press, 1995.
*   Bromberg, Joan Lisa. *NASA and the Space Industry.* Baltimore: Johns Hopkins University Press, 1999.
*   Beattie, Donald A. *Taking Science to the Moon: Lunar Experiments and the Apollo Program* . Baltimore: Johns Hopkins University Press, 2001.

* McCurdy, Howard E. *Faster, Better, Cheaper: Low-Cost Innovation in the U.S. Space Program*. Baltimore: Johns Hopkins University Press, 2001.

* Johnson, Stephen B. *The Secret of Apollo: Systems Management in American and European Space Programs*. Baltimore: Johns Hopkins University Press, 2002.

* Lambright, W. Henry, editor. *Space Policy in the 21st Century*. Baltimore: Johns Hopkins University Press, 2002.

* Bilstein, Roger E. *Testing Aircraft, Exploring Space: An Illustrated History of NACA and NASA*. Baltimore: Johns Hopkins University Press, 2003.

* Butrica, Anderw J. *Single Stage to Orbit: Politics, Space Technology, and the Quest for Reusable Rocketry*. Baltimore: Johns Hopkins University Press, 2005.

* Conway, Erik M. *High-Speed Dreams: NASA and the Technopolitics of Supersonic Transportation, 1945-1999*. Baltimore: Johns Hopkins University Press, 2005.

New Series in NASA History Published by Texas A&M University Press

* Schorn, Ronald A. *Planetary Astronomy: From Ancient Times to the Third Millennium*. College Station: Texas A&M University Press, 1998. To order, see http://www.tamu.edu/upress/BOOKS/1998/schorn.htm

New Series in NASA History Published by The University Press of Kentucky

* Gorn, Michael H. *Expanding the Envelope: Flight Research at NACA and NASA*. Lexington: The University Press of Kentucky, 2001. To order see http://www.kentuckypress.com/index.cfm.

* Reed, R. Dale. *Wingless Flight: The Lifting Body Story*. Lexington: The University Press of Kentucky, 2002. To order see http://www.kentuckypress.com/index.cfm

* Ed. by Launius, Roger D. and Dennis R. Jenkins. *To Reach the High Frontier: A History of U.S. Launch Vehicles*. Lexington: The University Press of Kentucky, 2002. To order see http://www.kentuckypress.com/index.cfm.

New Series in NASA History Published by the University Press of Florida

* Ed. by Swanson, Glen W. *"Before This Decade is Out...": Personal Relections on the Apollo Program*. Gainesville: The University Press of Florida, 2002. To order see http://www.upf.com/index.shtml

* Benson, Charles D. and William B. Faherty. *Moon Launch!: A History of the Saturn-Apollo Launch Operations*. Gainesville: The University Press of Florida, 2001. To order see http://www.upf.com/index.shtml

* Benson, Charles D. and William B. Faherty. *Gateway to the Moon: Building the Kennedy Space Center Launch Complex*. Gainesville: The University Press of Florida, 2001. To order see http://www.upf.com/index.shtml.

* Bilstein, Roger E. *Stages to Saturn: A Technological History of the Apollo/Saturn Launch Vehicles*. NASA SP-4206, 1980, 1996, and 2003. Gainesville: The University Press of Florida, 2003. To order see http://www.upf.com/index.shtml.

* Siddiqi, Asif A. *The Soviet Space Race with Apollo*. Gainesville: The University Press of Florida, 2003. To order see http://www.upf.com/index.shtml.

* Siddiqi, Asif A. *Sputnik and the Soviet Space Challenge*. Gainesville: The University Press of Florida, 2003. To order, see http://www.upf.com/index.shtml.

New Series in NASA History Published by Harwood Academic Press

* Ed. by Roger D. Lanius, John M. Logsdon and Robert W. Smith. *Reconsidering Sputnik:*

Forty Years Since the Soviet Satellite. London: Harwood Academic Press, 2000. To order, see http://www.taylorandfrancisgroup.com/

New Series in NASA History Published by the University of Illinois Press

* Ed. by Roger D. Launius and Howard McCurdy. *Spaceflight and the Myth of Presidential Leadership.* Urbana, IL: University of Illinois Press, 1997. To order, see http://www.press.uillinois.edu/f97/launius.html

New Series in NASA History Published by Greenwood Press

* Launius, Roger D. *Frontiers of Space Exploration.* Westport, CT: Greenwood Press, 1998. To order, see http://www.greenwood.com/default.asp

New Series in NASA History Published by the Smithsonian Institution Press

* Heppenheimer, T.A. *Development of the Shuttle, 1972-1981.* Washington, DC: Smithsonian Institution Press, 2002. To order, see http://www.si.edu/.

* Dethloff, Henry C. and Ronald A. Schorn. *Voyager's Grand Tour: To the Outer Planets and Beyond.* Washington, DC: Smithsonian Institution Press, 2003. To order, see http://www.si.edu/.

* Hallion, Richard P. and Michael H. Gorn. *On the Frontier: Experimental Flight at NASA Dryden.* Washington, DC: Smithsonian Institution Press, 2003. To order, see http://www.si.edu/.

New Series in NASA History Published by CG Publishing, Inc.

* *The Mission Transcript Collection: U.S. Human Spaceflight Missions From Mercury Redstone 3 to Apollo 17* (NASA SP-2000-4602). To order send an International Money Order for $8.00 to CG Publishing Inc, Box 62034, Burlington, Ontario, L7R 4K2, Canada or call 905-637-5737.

Miscellaneous Publications of NASA History

* Dawson, Virginia. *Ideas Into Hardware: A History of the Rocket Engine Test Facility at the NASA Glenn Research Center* Cleveland, 2004. Online version available.

About the Author

Joseph R. Chambers is an aviation consultant who lives in Yorktown, Virginia. He retired from the NASA Langley Research Center in 1998 after a 36-year career as a researcher and manager of military and civil aeronautics research activities. He began his career as a specialist in flight dynamics as a member of the staff of the Langley 30- by 60-Foot (Full-Scale) Tunnel, where he conducted research on a variety of aerospace vehicles including V/STOL configurations, re-entry vehicles, and fighter aircraft configurations. He later became a manager of research projects in the Full-Scale Tunnel, the 20-Foot Vertical Spin Tunnel, flight research at Langley, and piloted simulators. When he retired from NASA, he was manager of a group responsible for conducting systems analysis of the potential payoffs of advanced aircraft concepts and NASA research investments.

Mr. Chambers is the author of over 60 technical reports and publications, including NASA Special Publications: SP-514 Patterns in the Sky on the subject of airflow condensation patterns over aircraft; SP-2000-4519 Partners in Freedom on contributions of the Langley Research Center to U.S. military aircraft of the 1990s; and SP-2003-4529 Concept to Reality on contributions of the Langley Research Center to U.S. civil aircraft of the 1990s. He has made presentations on research and development programs to audiences as diverse as the Von Karman Institute in Belgium and the annual Experimental Aircraft Association (EAA) AirVenture Convention at Oshkosh, WI. He has served as a representative of the United States on international committees in aeronautics and has given lectures in Japan, China, Australia, the United Kingdom, Canada, Italy, France, Germany, and Sweden.

Mr. Chambers received several of NASA's highest awards, including the Exceptional Service Medal and the Outstanding Leadership Medal. He also received the Arthur Flemming Award in 1975 as one of the 10 Most Outstanding Civil Servants for his management of NASA stall/spin research for military and civil aircraft. He has a bachelor of science degree from the Georgia Institute of Technology and a master of science degree from the Virginia Polytechnic Institute and State University (Virginia Tech).

REPORT DOCUMENTATION PAGE

1. REPORT DATE (DD-MM-YYYY)	2. REPORT TYPE	3. DATES COVERED (From - To)
01- 12 - 2005	Special Publication	

4. TITLE AND SUBTITLE

Innovation in Flight: Research of the Langley Research Center on Revolutionary Advanced Concepts for Aeronautics

5a. CONTRACT NUMBER

5b. GRANT NUMBER

5c. PROGRAM ELEMENT NUMBER

6. AUTHOR(S)

Chambers, Joseph R.

5d. PROJECT NUMBER

5e. TASK NUMBER

5f. WORK UNIT NUMBER

7. PERFORMING ORGANIZATION NAME(S) AND ADDRESS(ES)

NASA Langley Research Center
Hampton, VA 23681-2199

8. PERFORMING ORGANIZATION REPORT NUMBER

L-19100

9. SPONSORING/MONITORING AGENCY NAME(S) AND ADDRESS(ES)

National Aeronautics and Space Administration
Washington, DC 20546-0001

10. SPONSOR/MONITOR'S ACRONYM(S)

NASA

11. SPONSOR/MONITOR'S REPORT NUMBER(S)

NASA/SP-2005-4539

12. DISTRIBUTION/AVAILABILITY STATEMENT

Unclassified - Unlimited
Subject Category 01
Availability: NASA CASI (301) 621-0390

13. SUPPLEMENTARY NOTES

An electronic version can be found at http://ntrs.nasa.gov

14. ABSTRACT

The goal of this publication is to provide an overview of the topic of revolutionary research in aeronautics at Langley, including many examples of research efforts that offer significant potential benefits, but have not yet been applied. The discussion also includes an overview of how innovation and creativity is stimulated within the Center, and a perspective on the future of innovation. The documentation of this topic, especially the scope and experiences of the example research activities covered, is intended to provide background information for future researchers.

15. SUBJECT TERMS

Aeronautics; Technical innovations; Research aircraft

16. SECURITY CLASSIFICATION OF:			17. LIMITATION OF ABSTRACT	18. NUMBER OF PAGES	19a. NAME OF RESPONSIBLE PERSON
a. REPORT	b. ABSTRACT	c. THIS PAGE			STI Help Desk (email: help@sti.nasa.gov)
U	U	U	UU	398	19b. TELEPHONE NUMBER (Include area code) (301) 621-0390